UMTS Networks

UMTS Networks

Architecture, Mobility and Services

Second Edition

Heikki Kaaranen
Oy Aqua Records Ltd, Finland

Ari Ahtiainen
Nokia Research Center, Finland

Lauri Laitinen
Nokia Research Center, Finland

Siamäk Naghian
Nokia Networks, Finland

Valtteri Niemi
Nokia Research Center, Finland

JOHN WILEY & SONS, LTD

Other Wiley Editorial Offices

John Wiley & Sons, Inc., 111 River Street, Hoboken, NJ 07030, USA

Jossey-Bass, 989 Market Street, San Francisco, CA 94103-1741, USA

Wiley-VCH Verlag GmbH, Boschstr. 12, D-69469 Weinheim, Germany

John Wiley & Sons Australia Ltd, 33 Park Road, Milton, Queensland 4064, Australia

John Wiley & Sons (Asia) Pte Ltd, 2 Clementi Loop #02-01, Jin Xing Distripark, Singapore 129809

John Wiley & Sons Canada Ltd, 22 Worcester Road, Etobicoke, Ontario, Canada M9W 1L1

Wiley also publishes its books in a variety of electronic formats. Some content that appears
in print may not be available in electronic books.

British Library Cataloguing in Publication Data

A catalogue record for this book is available from the British Library

ISBN 0-470-01103-3

Project management by Originator, Gt Yarmouth, Norfolk (typeset in 10/12pt Times).
Printed and bound in Great Britain by Antony Rowe Ltd, Chippenham, Wiltshire.
This book is printed on acid-free paper responsibly manufactured from sustainable forestry
in which at least two trees are planted for each one used for paper production.

Contents

Preface

The world's first public GSM call was made on 1 July 1991 in a city park in Helsinki, Finland. This event is now hailed as the birthday of second-generation mobile telephony. GSM has been an overwhelming success, which was difficult to predict at that early stage. In the past 10 years GSM has become a truly global system for mobile communications. We now have cellular phone penetration rates exceeding 70% in many countries and approaching 90% in the Nordic countries, while, globally, the number of mobile phones has already passed the number of fixed phones, exceeding an expected figure of 1.5 billion in the near future.

A decade later GSM has brought us to the early stages of the third-generation mobile communications system—the Universal Mobile Telecommunications System (UMTS). The first networks have begun operations and a new generation of fancy mobile phones has appeared. By the end of October 2004 some 50 UMTS commercial networks were open for business around the world.

UMTS networks are introducing a completely new, high bit-rate radio technology— Wideband Code Division Multiple Access (WCDMA)— for wide area use. Nevertheless, the core network part of the UMTS system is firmly founded on the successful GSM network, which has evolved from the circuit-switched voice network into a global platform for mobile packet data services like short messaging, mobile Web browsing and mobile email access.

The latest estimates show that packet-switching traffic in mobile core networks will exceed circuit-switching traffic in the near future. This transition is enabled by the UMTS system, which makes it possible for network operators to provide equally robust circuit-switched and packet-switched domains to meet data speed and capacity demands. Most voice and time-critical data services may still use circuit-switching, while less time-sensitive data pass through the UMTS mobile packet core network.

One of the key advantages of UMTS mobile computing and communications devices is the ability to deliver information to users at almost anytime and anywhere. In the UMTS the mobile phone is becoming regarded as a personal trusted device, a life management tool for work and leisure. Among the new possibilities for communication, entertainment and business are new kinds of rich call and multimedia data services, fuelled by the mobility and personalisation of users and their terminals.

This is a book about the way in which UMTS networks can be used as a third-generation platform for mobility and services. It aims to provide a comprehensive overview of the system architecture and its evolution and to serve as a guidebook to those who need to study specifications from the Third Generation Partnership Project (3GPP). The content of the book is divided into three parts.

The first part consists of Chapters 1 and 2, which serve as an executive summary of the UMTS system. Chapter 1 introduces the UMTS technical and service architecture and key system concepts. Chapter 2 is an illustrated history of mobile network evolution from second-generation GSM to the first UMTS multi-access release and beyond to full IP mobility networks.

The second part consists of Chapters 3–9, which examine the radio technology aspects, radio access and core network as well as, to a certain extent, the terminal in more detail. It also explains the functions and services provided to end users. Chapter 3 on the key architecture design challenges of cellular networks provides an overview of the fundamental challenges facing cellular networks and the way they have been resolved, particularly in the UMTS network.

Chapter 4 presents an overview of UMTS access technologies, including the latest enhancements in WCDMA technology within the scope of 3GPP Release 5. In addition, it addresses the other access technologies, like GSM/EDGE and WLAN, as complementary components of the UMTS multi-access network.

Chapters 5 and 6 describe the functional split between controlling functions distributed among the UMTS network elements in the radio access and core network parts. Chapter 7 provides an overview of UMTS user equipment, focusing on those aspects that are most visible to the rest of the UMTS network. In Chapter 8 the UMTS network is examined as a network for services. It addresses service realisation by describing Quality of Service (QoS) and giving some examples of services that can be brought about by UMTS. The advanced security solutions of the UMTS network are then discussed in Chapter 9.

The remaining chapters (Chapters 10 and 11) form the third part of the book. In these chapters we take a protocol-oriented view to describe the system-wide interworking between the different architectural elements. Chapter 10 first elaborates on the basic UMTS protocol architecture and then introduces the individual system protocols one by one. Chapter 11 returns to the network-wide view of earlier chapters by showing selected examples of the system procedures that describe how transactions are carried out across UMTS network interfaces under the coordination of system protocols.

At such an early stage of third-generation mobile communications the success of UMTS will be further enhanced by the thousands of leading system and software engineers, content providers, application developers, system integrators and network operators. We hope this book will help all of them reach their targets and let them enjoy and benefit from the UMTS networking environment.

This book represents the views and opinions of the authors and, therefore, does not necessarily represent the views of their employers.

What's New in the Second Edition?

Since the first edition of this book in 2001, much has happened in wireless communications, in general, and in UMTS network development, in particular. The move towards a data-centric service has been gaining momentum; the UMTS network has become a reality in several countries; short-range radios, such as WLAN and Bluetooth, have become integral components of mobile phones; Internet usage has rapidly spread; and the marriage between mobile networks and IP has become ever more evident. These have all been realised in one way or another in the latest development of 3GPP Release 5. In this new edition we try to reflect these changes while taking care of the main objective of this book: to stay as a comprehensive text of UMTS system architecture. We have also received much invaluable feedback from the readers of the first edition of the book which has come from all four corners of the world. We are very grateful for these insightful comments and have taken them into account in the writing process of the second edition. The level of this feedback has made us confident that the original purpose of the book (i.e., to serve as a UMTS system architecture book) was well received by the readership. The first edition has also been used as a course book for many training sessions, institutes and universities. We have also tried to keep this aspect in mind while taking on board the feedback from readers. In addition, more effort has been made to assure the overall quality of the second edition. To achieve this, more attention has been paid to editing and proof-reading of the text by both the authors and the publisher.

In this edition, every chapter has been revised to reflect the development in 3GPP standards up to Release 5. Some chapters have been radically reorganised and enhanced. We can summarise the changes to the second edition as:

- The first edition considered the UMTS network as a single access that only recognised the WCDMA UTRAN access network, and all topics were written on this basis. This edition recognises the other access technologies as well. The role of basic GSM is made more prominent since it forms the basic coverage anyway. UTRAN has been addressed at the same level as before. Complementary accesses are briefly described because their interworking has become an integral part of 3GPP evolution.
- Chapters 1 and 2 have undergone minor editing changes and some figures have been modified to make them compatible with 3GPP R5.
- Due to these various accesses, Chapter 3 in the first edition has now been split into two new chapters (Chapters 3 and 4). Chapter 3 gives an overview of the radio network challenges that arise from radio communication constraints, device mobility, transport, network management and scarcity of the radio spectrum. The new Chapter 4 provides an overview of selected UMTS access technologies—WCDMA and its enhancements HSPDA, GSM/EDGE and WLAN.
- Chapter 5, UTRAN, has been revised and fine-tuned and HSPDA has been included. Chapter 6, Core Network, has undergone heavy editing and IMS architecture and functions have been described. The additions reflect the main outcomes of R5.

- Chapter 7, Terminal, has not involved any marked changes and only IMS-related aspects have been added. Chapter 8, Services, has been completely rewritten.
- Chapters 9, 10 and 11 have been updated to 3GPP R5.

Throughout the book, these changes have led to about 100 additional pages compared with the first edition, resulting in this edition being fully compatible with 3GPP R5.

In addition, a PDF slideset is available from Heikki Kaaranen—for further information and ordering details please email *heikki.kaaranen@aquarecords.fi* or visit the website *www.aquarecords.fi*

Acknowledgements

While writing the first edition of *UMTS Networks* the team of authors and contributors had the pleasure of following the exciting finalisation of UMTS system specifications. During production of the second edition we're witnessing yet another exciting break-through, the rolling out of UMTS networks around the world. Many colleagues, both from Nokia and outside, provided valuable input and comments on various aspects of the book. We would in particular like to thank Seppo Alanara, Mika Forssell, Harri Holma, Kaisu Iisakkila, Tatjana Issayeva, Sami Kekki, Pekka Korja, Jan Kåll, Juho Laatu, John Loughney, Atte Länsisalmi, Anna Markkanen, Tomi Mikkonen, Juha Mikola, Ahti Muhonen, Aki Niemi, Mikko Puuskari, Mikko J. Rinne, Ville Ruutu, Juha Sipilä, Janne Tervonen, Mikko Tirronen, Ari Tourunen, Jukka Vialén and Andrei Zimenkov.

The inspiring working environment and close contacts with the R&D and standard-isation programmes within Nokia were made possible by the following managers of those programmes: Kari Aaltonen, Heikki Ahava, Tapio Harila, Reijo Juvonen, Jari Lehmusvuori, Juhani Kuusi, Yrjö Neuvo, Tero Ojanperä, Lauri Oksanen, Pertti Paski, Tuula-Mari Rautala, Tuomo Sipilä, Jukka Soikkeli, Jari Vainikka and Asko Vilavaara. The publishing team led by Mark Hammond and Sarah Hinton at John Wiley & Sons, Ltd gave us excellent support in the production of the second edition of the book. Their hard-working spirit made it possible to keep the demanding schedule in the publication process. The invaluable editing effort by Bruce Shuttlewood and the team from Originator Publishing Services helped us to improve the readability and language format of the text.

We must not forget that this is a book about UMTS networks and that these networks are based on the joint design and engineering effort of many colleagues of ours; it was their joint expertise that made it happen. Without being able to list all the experts from the early 1990s, those in the 3GPP organisation and those otherwise involved in UMTS development, we would like to thank all of them for their dedicated work in creating a new era in mobile communications.

Finally, we want to express loving thanks to all the members of our families for the patience and support shown during the long days and late nights of the book-writing effort. Among them, Mrs Satu Kangasjärvelä-Kaaranen deserves special thanks; her help in word-processing and the graphical design of many figures was invaluable in putting the manuscript together.

As we are committed to the continuous improvement of the book, the authors once again welcome any comments and suggestions for improvements or changes that could be implemented in future editions of this book. The email address for gathering such input is *umtsnetworks@pcuf.fi*

The authors of *UMTS Networks*
Helsinki, Finland

Part One

1

Introduction

Ari Ahtianen, Heikki Kaaranen and Siamäk Naghian

Nowadays, it is widely recognised that there are three different, implemented generations as far as mobile communication is concerned (Figure 1.1). The first generation,
1G, is the name for the analogue or semi-analogue (analogue radio path, but digital
switching) mobile networks established in the mid-1980s, such as the Nordic Mobile
Telephone (NMT) system and the American Mobile Phone System (AMPS). These
networks offered basic services for users and the emphasis was on speech and speech-
related services. 1G networks were developed with national scope only and very often
the main technical requirements were agreed between the governmental telecom operator and the domestic industry without wider publication of the specifications. Due to
national specifications, 1G networks were incompatible with each other and mobile
communication was considered at that time to be some kind of curiosity and added
value service on top of the fixed networks.

Because the need for mobile communication increased, also the need for a more
global mobile communication system arose. International specification bodies started
to specify what the second generation, 2G, mobile communication system should look
like. The emphasis for 2G was on compatibility and international transparency; the
system should be regional (e.g., European-wide) or semi-global and the users of the
system should be able to access it basically anywhere within the region. From the end-
user's point of view, 2G networks offered a more attractive "package" to buy; besides
the traditional speech service these networks were able to provide some data services
and more sophisticated supplementary services. Due to the regional nature of standardisation, the concept of globalisation did not succeed completely and there are some 2G
systems available on the market. Of these, the commercial success story is the Global
System for Mobile Communications (GSM) and its adaptations: it has clearly exceeded
all the expectations set, both technically and commercially.

The third generation, 3G, is expected to complete the globalisation process of mobile
communication. Again, there are national and regional interests involved and difficulties can be foreseen. Anyway, the trend is that 3G will mostly be based on GSM
technical solutions for two reasons: GSM technology dominates the market and the
great investments made in GSM should be utilised as much as possible. Based on this,
the specification bodies created a vision about how mobile telecommunication will

UMTS Networks Second Edition   H. Kaaranen, A. Ahtiainen, L. Laitinen, S. Naghian and V. Niemi
© 2005 John Wiley & Sons, Ltd   ISBN: 0-470-01103-3

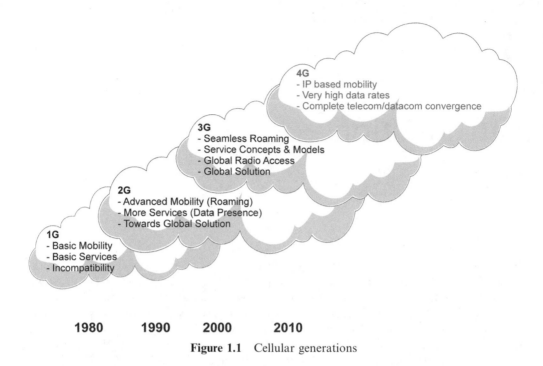

1980   1990   2000   2010

Figure 1.1 Cellular generations

develop within the next decade. Through this vision, some requirements for 3G were shortlisted as follows:

1. The system must be fully specified (like GSM) and major interfaces should be standardised and open. The specifications generated should be valid worldwide.
2. The system must bring clear added value to GSM in all aspects. However, at the start the system must be backward-compatible at least with GSM and ISDN (Integrated Services Digital Network).
3. Multimedia and all of its components must be supported throughout the system.
4. The radio access of 3G must provide wideband capacity that is generic enough to become available worldwide. The term "wideband" was adopted to reflect the capacity requirements between 2G narrowband capacity and the broadband capacity of fixed communications media.
5. The services for end-users must be independent of radio access technology details and the network infrastructure must not limit the services to be generated. That is, the technology platform is one issue and the services using the platform are totally another issue.

While 3G specification work was still going on, the major telecommunication trends changed too. The traditional telecommunication world and up to now the separate data communications (or the Internet) have started to converge rapidly. This has started a development chain, where traditional telecommunication and Internet Protocol (IP) technologies are combined in the same package. This common trend has many names depending on the speaker's point of view; some people call the target of this development the "Mobile Information Society" or "Mobile IP", others say it is "3G All IP" and in some commercial contexts the name "E2E IP" (End-to-End IP) is used as well. From a 3G point of view, a full-scale IP implementation is defined as a single targeted phase of the 3G development path.

The 3G system experiences evolution through new phases and, actually, the work aiming to establish 4G specifications has already started. Right now it may be too early to predict where the 3G evolution ends and 4G really starts. Rather, this future development can be thought of as an ongoing development chain where 3G will continue to introduce new ways of handling and combining all kinds of data and mobility. 4G will then emerge as a more sophisticated system concept bringing still more capacity and added value to end-users.

1.1 Specification Process for 3G

The uniform GSM standard in European countries has enabled globalisation of mobile communications. This became evident when the Japanese 2G Pacific Digital Communications (PDC) failed to spread to the Far East and the open GSM standard was adopted by major parts of the Asian markets and when its variant became one of the nationally standardised alternatives for the US Personal Communication System (PCS) market too.

A common, global mobile communication system naturally creates a lot of political desires. In the case of 3G this can be seen even in the naming policy of the system. The most neutral term is "third generation", 3G. In different parts of the world different issues are emphasised and, thus, the global term 3G has regional synonyms. In Europe 3G has become UMTS (Universal Mobile Telecommunication System), following the European Telecommunications Institute (ETSI) perspective. In Japan and the US the 3G system often carries the name IMT-2000 (International Mobile Telephony 2000). This name comes from the International Telecommunication Union (ITU) development project. In the US the CDMA2000 (Code Division Multiple Access) is also an aspect of 3G cellular systems and represents the evolution from the IS-95 system. In this book, we will describe the UMTS system as it has been specified by the worldwide 3G Partnership Project (3GPP). To bring some order to the somewhat confusing naming policy, 3GPP launched a decision where it stated that the official name of 3G is the "3GPP System". This name should be followed by a release number describing the specification collection. With this logic, the very first version of the European-style UMTS network takes the official name "3GPP System Release 99". Despite this definition, the above-mentioned names UMTS and IMT-2000 are still widely used.

At the outset UMTS inherited plenty of elements and functional principles from GSM and the most considerable new development is related to the radio access part of the network. UMTS brings into the system an advanced access technology (namely, the wideband type of radio access). Wideband radio access is implemented using Wideband Code Division Multiple Access (WCDMA) technology. WCDMA evolved from CDMA, which, as a proven technology, has been used for military purposes and for narrowband cellular networks, especially in the US.

UMTS standardisation was preceded by several pre-standardisation research projects founded and financed by the EU. Between 1992 and 1995 a Research in Advanced Communications in Europe (RACE) MoNet project developed the modelling technique describing the function allocation between the radio access and core parts of the network. This kind of modelling technique was needed, for example, to compare Intelligent Network (IN) and GSM Mobile Application Part (MAP) protocols as mobility management solutions. This was, besides the discussion on the broadband versus narrowband ISDN, one of the main dissents in MoNet. In addition, discussions about the use of ATM (Asynchronous Transfer Mode) and B-ISDN as fixed transmission techniques arose at the end of the MoNet project.

Between 1995 and 1998 3G research activities continued within the Advanced Communications Technology and Services (ACTS) Future Radio Wideband Multiple Access System (FRAMES) project. The first years were used for selecting and developing a suitable multiple access technology, considering mainly the TDMA (Time Division Multiple Access) versus CDMA. The big European manufacturers preferred TDMA because it was used also in GSM. CDMA-based technology was promoted mainly by US industry, which had experience with this technology mainly due to its early utilisation in defence applications.

ITU dreamed of specifying at least one common global radio interface technology. This kind of harmonisation work was done under the name "Future Public Land Mobile Telephony System" (FPLMTS) and later IMT-2000. Due to many parallel activities in regional standardisation bodies this effort turned into a promotion of common architectural principles among the family of IMT-2000 systems.

Europe and Japan also had different short-term targets for 3G system development. In Europe a need for commercial mobile data services with guaranteed quality (e.g., mobile video services) was widely recognised after the early experiences from narrowband GSM data applications. Meanwhile, in the densely populated Far East there was an urgent demand for additional radio frequencies for speech services. The frequency bands identified by ITU in 1992 for the future 3G system called "IMT-2000" became the most obvious solution to this issue. In early 1998 a major push forward was achieved when ETSI TC-SMG decided to select WCDMA as its UMTS radio technology. This was also supported by the largest Japanese operator NTT DoCoMo. The core network technology was at the same time agreed to be developed on the basis of GSM core network technology. During 1998 the European ETSI and the Japanese standardisation bodies (TTC and ARIB) agreed to make a common UMTS standard. After this agreement, the 3GPP organisation was established and the determined UMTS standardisation was started worldwide.

From the UMTS point of view, the 3GPP organisation is a kind of "umbrella" aiming to form compromised standards by taking into account political, industrial and commercial pressures coming from the local specification bodies:

- ETSI (European Telecommunication Standard Institute)/Europe.
- ARIB (Association of Radio Industries and Business)/Japan.
- CWTS (China Wireless Telecommunication Standard group)/China.
- T1 (Standardisation Committee T1—Telecommunications)/US.

- TTA (Telecommunication Technology Association)/Korea.
- TTC (Telecommunications Technology Committee)/Japan.

As this is a very difficult task an independent organisation called the "OHG" (Operator Harmonisation Group) was established immediately after the 3GPP was formed. The main task for 3GPP is to define and maintain UMTS specifications, while the role of OHG is to look for compromise solutions for those items the 3GPP cannot handle internally. This arrangement guarantees that 3GPP's work will proceed on schedule.

To ensure that the American viewpoint will be taken into account a separate 3GPP Number 2 (3GPP2) was founded and this organisation performs specification work from the IS-95 radio technology basis. The common goal for 3GPP, OHG and 3GPP2 is to create specifications according to which a global cellular system having wideband radio access could be implemented. To summarise, there were three different approaches towards the global cellular system, 3G. These approaches and their building blocks are, on a rough level, presented in Table 1.1.

When globality becomes a reality, the 3G specification makes it possible to take any of the switching systems mentioned in the table and combine them with any of the specified radio access parts and the result is a functioning 3G cellular network. The second row represents the European approach known as "UMTS" and this book gives an overview of its first release.

The 3GPP originally decided to prepare specifications on a yearly basis, the first specification release being Release 99. This first specification set has a relatively strong "GSM presence". From the UMTS point of view the GSM presence is very important; first, the UMTS network must be backward-compatible with existing GSM networks and, second, GSM and UMTS networks must be able to interoperate together. The next release was originally known as "3GPP R00", but, because of the multiplicity of changes proposed, specification activities were scheduled into two specification releases 3GPP R4 and 3GPP R5. 3GPP R4 defines optional changes in the UMTS core network circuit-switched side; these are related to the separation of user data flows and their control mechanisms. 3GPP R5 aims to introduce a UMTS network providing mechanisms and arrangements for multimedia. This entity is known as the "IP Multimedia Subsystem" (IMS) and its architecture is presented in Chapter 6. IP and the overlying protocols will be used in network control too and user data

Table 1.1 3G variants and their building blocks

Variant	Radio access	Switching	2G basis
3G (US)	WCDMA, EDGE, CDMA2000	IS-41	IS-95, GSM1900, TDMA
3G (Europe)	WCDMA, GSM, EDGE	Advanced GSM NSS and packet core	GSM900/1800
3G (Japan)	WCDMA	Advanced GSM NSS and packet core	PDC

flows are expected to be mainly IP-based as well. In other words, the mobile network implemented according to the 3GPP R5 specification will be an end-to-end packet-switched cellular network using IP as the transport protocol instead of SS7 (Signalling System #7), which holds the major position in existing circuit-switched networks. Naturally, the IP-based network should still support circuit-switched services too. 3GPP R4/R5 will also start to utilise the possibility of new radio access techniques. In 3GPP R99 the basis for the UMTS Terrestrial Access Network (UTRAN) is WCDMA radio access. In 3GPP R4/5 another radio access technology derived from GSM with Enhanced Data for GSM Evolution (EDGE) is integrated to the system in order to create the GSM/EDGE Radio Access Network (GERAN) as an alternative to building a UMTS mobile network.

1.2 Introduction to the 3G Network Architecture

The main idea behind 3G is to prepare a universal infrastructure able to carry existing and also future services. The infrastructure should be designed so that technology changes and evolution can be adapted to the network without causing uncertainties in the existing services using the current network structure. Separation of access technology, transport technology, service technology (connection control) and user applications from each other can handle this very demanding requirement. The structure of a 3G network can be modelled in many ways, and here we introduce some ways to outline the basic structure of the network. The architectural approaches to be discussed in this section are:

- Conceptual network model.
- Structural network architecture.
- Resource management architecture.
- UMTS bearer architecture.

1.2.1 Conceptual Network Model

From the above-mentioned network conceptual model point of view, the entire network architecture can be divided into subsystems based on the nature of traffic, protocol structures and physical elements. As far as the nature of traffic is concerned, the 3G network consists of two main domains, packet-switched (PS) and circuit-switched (CS) domains. According to 3GPP specification TR 21.905 a *domain* refers to the highest level group of physical entities and the defined interfaces (reference points) between such domains. The interfaces and their definitions describe exactly how the domains communicate with each other.

From the protocol structure and their responsibility point of view, the 3G network can be divided into two strata: the access stratum and the non-access stratum. A *stratum* refers to the way of grouping protocols related to one aspect of the services provided by one or several domains (see 3GPP specification TR 21.905). Thus, the access stratum contains the protocols that handle activities between the User Equipment (UE) and the access network. The non-access stratum contains the protocols that

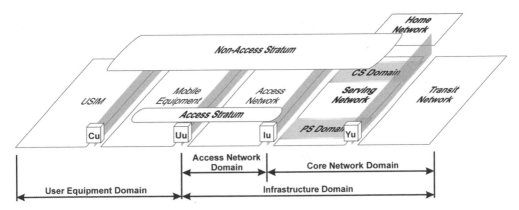

Figure 1.2 UMTS architecture—conceptual model

handle activities between the UE and the core network (CS/PS domain), respectively. For further information about strata and protocols see Chapter 10.

The part of Figure 1.2 called "Home Network" maintains static subscription and security information. The serving network is the part of the core network + domain which provides the core network functions locally to the user. The transit network is the core network part located on the communication path between the serving network and the remote party. If, for a given call, the remote party is located inside the same network as the originating UE, then no particular instance of the transit network is needed.

1.2.2 Structural Network Architecture

In this book we mainly present the issues from the network structural architecture perspective. This perspective is presented in Figure 1.3. In UMTS the GSM technology plays the remarkable role of the background and, actually, UMTS aims to reuse everything, which is reasonable. For example, some procedures used within the non-access stratum are, in principle, reused from GSM but naturally with required modifications.

The 3G system terminal is called "UE" and it contains two separate parts, Mobile Equipment (ME) and the UMTS Service Identity Module (USIM).

The new subsystem controlling wideband radio access has different names, depending on the type of radio technology used. The general term is "Radio Access Network" (RAN). When we talk in particular about UMTS with WCDMA radio access, the name "UTRAN" or "UTRA" is used. The other type of RAN included in UMTS is GERAN. GERAN and its definitions are not part of 3GPP R99, though they are referred to as possible radio access alternatives, which may be utilised in the future. The specification of GERAN and its harmonisation with UTRAN is done in 3GPP R4 and 3GPP R5.

UTRAN is divided into Radio Network Subsystems (RNSs). One RNS consists of a set of radio elements and their corresponding controlling element. In UTRAN the radio element is Node B, referred to as Base Station (BS) in the rest of this book, and the controlling element is the Radio Network Controller (RNC). The RNSs are connected

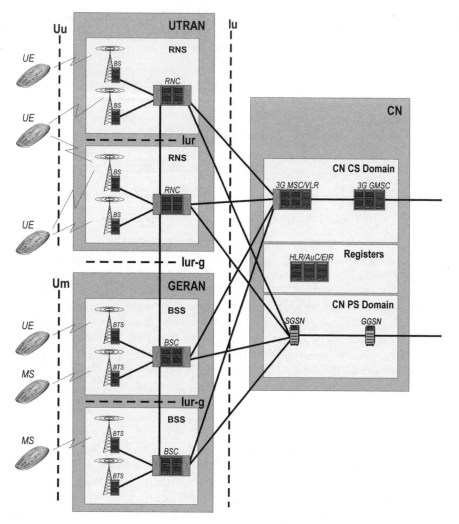

Figure 1.3 UMTS network architecture—network elements and their connections for user data transfer

to each other over the access network internal interface Iur. This structure and its advantages are explained in more detail in Chapter 5.

The other access network shown in Figure 1.3, GERAN, is not handled in detail in this book. Readers interested in GERAN should consult, e.g., Halonen et al. (2002).

The term "Core Network" (CN) covers all the network elements needed for switching and subscriber control. In early phases of UMTS, part of these elements were directly inherited from GSM and modified for UMTS purposes. Later on, when transport technology changes, the core network internal structure will also change in a remarkable way. CN covers the CS and PS domains defined in Figure 1.3. Configuration alternatives and elements of the UMTS core network are discussed in Chapter 6.

The part of Figure 1.3 called "Registers" is the same as the Home Network in the preceding 3G network conceptual model. This part of the network maintains static subscription and security information. Registers are discussed in more detail in Chapter 6.

The major open interfaces of UMTS are also presented in Figure 1.3. Between the UE and UTRAN the open interface is Uu, which in UMTS is physically realised with WCDMA technology. Some additional information about WCDMA on a general level is provided in Chapters 3 and 4. On the GERAN side the equivalent open interface is Um. The other major open interface is Iu located between UTRAN/GERAN and CN.

The RNSs are separated from each other by an open interface Iur. Iur is a remarkable difference when compared with GSM; it brings completely new abilities for the system to utilise: so-called macro diversity as well as efficient radio resource management and mobility mechanisms. When the Iur interface is implemented in the network, the UE may attach to the network through several RNCs, each of which maintains a certain logical role during radio connection. These roles are Serving RNC (SRNC), Drifting RNC (DRNC) and Controlling RNC (CRNC). CRNC has overall control of the logical resources of its UTRAN access points, being mainly BSs. An SRNC is a role an RNC can play with respect to a specific connection between the UE and UTRAN. There is one SRNC for each UE that has a radio connection to UTRAN. The SRNC is in charge of the radio connection between the UE and the UTRAN. It also maintains the Iu interface to the CN, which is the main characteristic of the SRNC. A DRNC plays the logical role used when radio resources of the connection between the UTRAN and the UE need to use cell(s) controlled by another RNC rather than the SRNC itself. UTRAN-related issues in general are discussed in Chapter 5.

Access networks also have connections between themselves through an interface Iur-g. Iur-g is used for radio-resource-management-related information transfer. The difference between Iur and Iur-g is that Iur transfers both signalling and user data, while Iur-g only transfers signalling.

In addition to the CS and PS domains presented in Figure 1.3, the network may contain other domains. One example of these is the broadcast messaging domain, which is responsible for multicast messaging control. However, in this book we concentrate on the UMTS network as presented in Figure 1.4. As far as the various RANs are concerned, we concentrate on UTRAN and highlight some specific items related on UTRAN–GERAN co-existence and co-operation.

1.2.3 Resource Management Architecture

The network element-centric architecture described above results from functional decomposition and the split of responsibilities between major domains and, ultimately, between network elements. Figure 1.4 illustrates this split of major functionalities, which are:

- Communication Management (CM).
- Mobility Management (MM).
- Radio Resource Management (RRM).

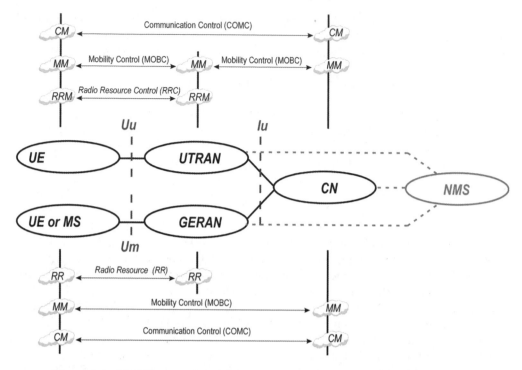

Figure 1.4   UMTS network architecture—management tasks and control duties

CM covers all of the functions and procedures related to the management of user connections. CM is divided into several sub-areas, such as call handling for CS connections, session management for PS connections, as well as handling of supplementary services and short-message services. MM covers all of the functions and procedures needed for mobility and security (e.g., connection security procedures and location update procedures). Most of the MM procedures occur within the CN and its elements, but in the 3G part of the MM functions are also performed in UTRAN for PS connections. The principles underlying CM and MM are discussed in Chapter 6.

RRM is a collection of algorithms UTRAN uses for management of radio resources. These algorithms handle, for instance, the power control for radio connections, different types of handovers, system load and admission control. RRM is an integral part of UTRAN and basic RRM is discussed more closely in Chapter 5. Some system-wide procedure examples about CM, MM and RRM functioning are given in Chapter 11.

Although these management tasks can be located within specific domains and network elements, they need to be supported by communication among the related domains and network elements. This communication is about gathering information and reporting about the status of remote entities as well as about giving commands to them in order to execute management decisions. Therefore, each of the management tasks is associated with a set of control duties such as:

- Communication Control (COMC).
- Mobility Control (MOBC).
- Radio Resource Control (RRC).

COMC maintains mechanisms like call control and packet session control. MOBC maintains mechanisms which cover, for example, execution control for location updates and security. Radio resources are completely handled between UTRAN and the UE. The control duty called "RRC" takes care of, for example, radio link establishment and maintenance between UTRAN and the UE. These collections of control duties are then further refined into a set of well-specified control protocols. For more detailed information about protocols see Chapter 10.

When compared with the traditional GSM system, it is apparent that this functional architecture has undergone some rethinking. The most visible change has to do with mobility management, where responsibility has been split between UTRAN and the CN. In addition, with regards to the RRM, the UMTS architecture follows more strictly the principle of making UTRAN alone responsible for all radio resource management. This is underlined by the introduction of a generic and uniform control protocol for the Iu interface.

1.2.4 Bearer Architecture

As stated earlier in this chapter, the 3G system mainly acts as an infrastructure providing facilities, adequate bandwidth and quality for end-users and their applications. This facility provision, bandwidth allocation and connection quality together is commonly called Quality of Service (QoS). If we think of an end-to-end service between users, the used service sets its requirements concerning QoS and this requirement must be met everywhere in the network. The various parts of the UMTS network contribute to fulfilling the QoS requirements of the services in different ways.

To model this, the end-to-end service requirements have been divided into three entities: the local bearer service, the UMTS bearer service and the external bearer service. The local bearer service contains mechanisms on how the end-user service is mapped between the terminal equipment and Mobile Termination (MT). MT is the part of the UE that terminates radio transmission to and from the network and adapts the terminal equipment capabilities to those of radio transmission. The UMTS bearer service in turn contains mechanisms to allocate QoS over the UMTS/3G network consisting of UTRAN and CN. Since the UMTS network attaches itself to external network(s), end-user QoS requirements must be handled towards the other networks too. This is taken care of by the external bearer service.

Within the UMTS network, QoS handling is different in UTRAN and CN. From the CN point of view, UTRAN creates an "illusion" of a fixed bearer providing adequate QoS for the end-user service. This "illusion" is called the radio access bearer service. Within the CN, its own type of bearer service called the "CN bearer service" is used. This division between the radio access bearer and the CN bearer service is required since the QoS must be guaranteed in very different environments and both of these environments require their own mechanisms and protocols. For instance, the CN

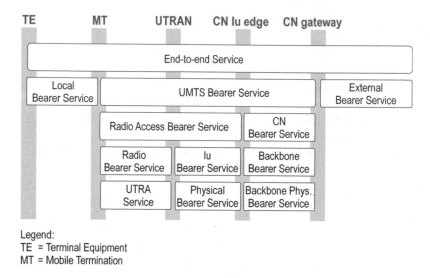

Figure 1.5   Bearer architecture in UMTS

bearer service is quite constant in nature since the backbone bearer service providing the physical connections is also stable. Within UTRAN the radio access bearer experiences more changes as a function of time and movement of the UE, and this sets different challenges for QoS. This division also pursues the main architectural principle of the UMTS network (i.e., independence of the entire network infrastructure from radio access technology).

The structure presented in Figure 1.5 is a network architecture model from the bearer and QoS point of view. Since QoS is one of the most important issues in UMTS, QoS and bearer concepts are handled throughout this book.

The rest of the book uses these architectural approaches as cornerstones when exploring UMTS networks and their implementations.

2

Evolution from GSM to UMTS Multi-access

Heikki Kaaranen

Evolution is one of the most common terms used in the context of Univeral Mobile Telecommunication System (UMTS). Generally, it is understood to mean technical evolution (i.e., how it has evolved and what kind of equipment and in which order it is brought to the existing network, if any). This is partly true, but, in order to understand the impact of evolution, a broader context needs to be examined. Evolution as a high-level context covers not only the technical evolution of network elements but also expansions to network architecture and services. When these three evolution types go hand in hand the smooth migration from 2G to 3G will be successful and generate revenue.

Technical evolution means the development path of how network elements will be implemented and with which technology. This is a very straightforward development and strictly follows general, common technology development trends. Because network elements together form a network, in theory the network will evolve accordingly. In this phase one should bear in mind that a network is only as strong as its weakest element and due to the open interfaces defined in the specifications many networks are combinations having equipment provided by many vendors. Technical evolution may proceed, however, at different rates in association with the different vendors' equipment, and when adapting evolution-type changes between several vendors' equipment the result may not be as good as expected.

Service evolution is not such a straightforward issue. It is based on demands generated by end-users and these demands could be real or imagined; sometimes network operators and equipment manufacturers offer services way beyond subscriber expectations. If end-users' needs and operators' service palettes do not match each other, difficulties with cellular business can be expected. These three dimensions of evolution are shown in Figure 2.1.

UMTS Networks Second Edition   H. Kaaranen, A. Ahtiainen, L. Laitinen, S. Naghian and V. Niemi
© 2005 John Wiley & Sons, Ltd   ISBN: 0-470-01103-3

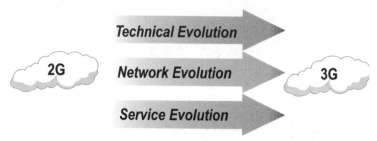

Figure 2.1   Technical, network and service evolution

2.1 From Analogue to Digital

The main idea behind the Global System for Mobile Communication (GSM) specification was to define several open interfaces, which determine the standardised
components of the GSM system. Because of this interface openness, the operator
maintaining the network may obtain different components of the network from different GSM network suppliers. Also, when an interface is open it defines strictly how
system functions are proceeding at the interface and this in turn determines which
functions are left to be implemented internally by the network elements on both
sides of the interface.

As was experienced when operating analogue mobile networks, centralised intelligence generated a lot of load in the system, thus decreasing overall system performance.
This is why the GSM specification in principle provided the means to distribute intelligence throughout the network. The above-mentioned open interfaces are defined in
places where their implementation is both natural and technically reasonable.

From the GSM network point of view, this decentralised intelligence is implemented
by dividing the whole network into four separate subsystems:

- Network Subsystem (NSS).
- Base Station Subsystem (BSS).
- Network Management Subsystem (NMS).
- Mobile Station (MS).

The actual network needed for call establishing is composed of the NSS, the BSS and
the MS. The BSS is the part of the network responsible for radio path control. Every
call is connected through the BSS. The NSS is the network part that takes care of call
control functions. Every call is always connected by and through the NSS. The NMS is
the operation and maintenance-related part of the network. It is also needed for the
whole network control. The network operator observes and maintains the quality of the
network and services through the NMS. The open interfaces in this concept are located
between the MS and the BSS (Um interface) and between the BSS and the NSS (A
interface). Um is actually very much like the Integrated Services Digital Network
(ISDN) terminal interface, U; it implements very similar facilities to this system and
also lower level signalling is adapted from narrowband ISDN. The small "m" after U in
the name stands for "modified". The interface between the NMS and the NSS/BSS was

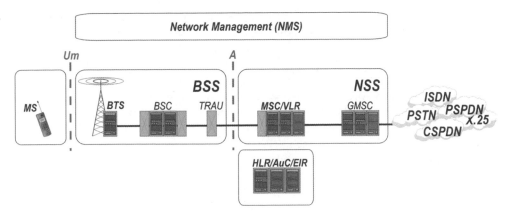

Figure 2.2 The basic GSM network and its subsystems

expected to be open, but its specifications were not ready in time and this is why every manufacturer implements NMS interfaces with their own proprietary methods.

The MS is a combination of a terminal's equipment and a subscriber's service identity module. The terminal equipment as such is called "Mobile Equipment" (ME), and the subscriber's data is stored in a separate module called the Service Identity Module (SIM). Hence, ME + SIM = MS. Please note that SIM officially stands for Subscriber Identity Module. We prefer the name "Service Identity Module", since it better describes the SIM functionality.

The Base Station Controller (BSC) is the central network element of the BSS and it controls the radio network. This means that the following functions are the BSC's main responsibility areas: maintaining radio connections towards the MS and terrestrial connections towards the NSS. The Base Transceiver Station (BTS) is a network element maintaining the air interface (Um interface). It takes care of air interface signalling, ciphering and speech processing. In this context, speech processing means all the methods BTS performs in order to guarantee error-free connection between the MS and the BTS. The Transcoding and Rate Adaptation Unit (TRAU) is a BSS element that takes care of speech transcoding (i.e., capable of converting speech from one digital coding format to another and vice versa).

The Mobile Services Switching Centre (MSC) is the main element of the NSS from the call control point of view. MSC is responsible for call control, BSS control functions, interworking functions, charging, statistics and interface signalling towards BSS and interfacing with the external networks (PSTN/ISDN/packet data networks). Functionally, the MSC is split into two parts, though these parts could be in the same hardware. The serving MSC/VLR is the element maintaining BSS connections, mobility management and interworking. The Gateway MSC (GMSC) is the element participating in mobility management, communication management and connections to the other networks. The Home Location Register (HLR) is the place where all the subscriber information is stored permanently. The HLR also provides a known, fixed location for subscriber-specific routing information. The main functions of the HLR are subscriber data and service handling, statistics and mobility management. The Visitor Location Register (VLR) provides a local store for all the variables and functions needed to handle calls to and from mobile subscribers in the area related to the VLR. Subscriber-related information remains in the VLR as long as the mobile subscriber visits the area. The main functions of the VLR are subscriber data and service handling and mobility management. The Authentication Centre (AuC) and Equipment Identity Register (EIR) are NSS network elements that take care of security-related issues. The AuC maintains subscriber-identity-related security information together with the VLR. The EIR maintains mobile-equipment-identity-(hardware)-related security information together with the VLR.

When thinking of the services, the most remarkable difference between 1G and 2G is the presence of a data transfer possibility; basic GSM offers 9.6 kb/s symmetric data connection between the network and the terminal. The service palette of the basic GSM is directly adopted from Narrowband ISDN (N-ISDN) and then modified to be suitable for mobile network purposes. This idea is visible throughout the GSM implementation; for example, many message flows and interface-handling procedures are adapted copies of corresponding N-ISDN procedures.

2.2 From Digital to Reachability

The very natural step to develop the basic GSM was to add service nodes and service centres on top of the existing network infrastructure. The GSM specification defines some interfaces for this purpose, but the internal implementation of service centres and nodes is not the subject of this specification. The common name for these service centres and nodes is Value Added Service (VAS) platforms and this term adequately describes the main point of adding this equipment to the network.

A minimum VAS platform typically contains two pieces of equipment:

- The Short Message Service Centre (SMSC).
- The Voice Mail System (VMS).

Technically speaking, VAS platform equipment is relatively simple and meant to provide a certain type of service. It uses standard interfaces towards the GSM network and may or may not have external interfaces towards other network(s).

From the service evolution point of view, VAS is the very first step toward generating revenue with services and partially tailoring them. The great success story in this sense has been the SMS, which was originally planned to be a small add-in to the GSM system. Nowadays, it has become extremely popular among GSM subscribers.

Basic GSM and VAS were originally intended to produce "mass services for mass people", but, due to the requirements of end-users, a more individual type of service was required. To make this possible, the Intelligent Network (IN) concept was integrated together with the GSM network. Technically, this means major changes in switching network elements in order to add the IN functionality; moreover, the IN platform itself is a relatively complex entity. IN enables service evolution to take major steps towards individuality ("mass service for individual people"); furthermore, with IN the operator is able to carry out business in a more secure way (e.g., prepaid subscriptions are mostly implemented with IN technology).

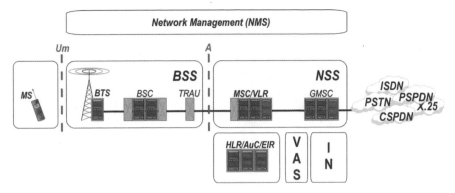

Figure 2.3   Value-added service platforms

IN as a technology has its roots in Public Switched Telephone Networks (PSTNs) and as such does not meet all mobile network requirements. Due to this, the original IN concept has been enhanced and introduced as CAMEL (Customised Applications for Mobile network Enhanced Logic). CAMEL eliminates the failings of IN, such as its lack of support for service mobility.

2.3 Jump to Packet World and Higher Speeds

At the outset, GSM subscribers have used the 9.6-kb/s circuit-switched (CS) symmetric "pipe" for data transfer. Due to the Internet and electronic messaging the pressures for mobile data transfer have increased a lot and this development was maybe under-estimated at the time when the GSM system was first specified. To ease this situation, a couple of enhancements have been introduced. First, channel coding is optimised. By doing this the effective bit rate has increased from 9.6 kb/s up to ≈14 kb/s. Second, to put more data through the air interface, several traffic channels can be used instead of one. This arrangement is called "High Speed Circuit Switched Data" (HSCSD). In an optimal environment an HSCSD user may reach data transfer using 40–50-kb/s data rates. Technically, this solution is quite straightforward, but, unfortunately, it wastes resources and some end-users may not be happy with the pricing policy of this facility; the use of HSCSD very much depends on the price the operators set for its use. Another issue is the fact that most of the data traffic is asymmetric in nature; that is, typically a very low data rate is used from the terminal to the network direction (uplink) and higher data rates are used in the opposite direction (downlink).

The CS symmetric Um interface is not the best possible access media for data connections. Furthermore, when we consider that the great majority of data traffic is packet-switched (PS) in nature, something more had to be done to "upgrade" the GSM network to make it more suitable for more effective data transfer. The way to do this is by using General Packet Radio Service (GPRS). GPRS requires two additional mobile-network-specific service nodes: Serving GPRS Support Node (SGSN) and Gateway GPRS Support Node (GGSN). By using these nodes the MS is able to form a PS

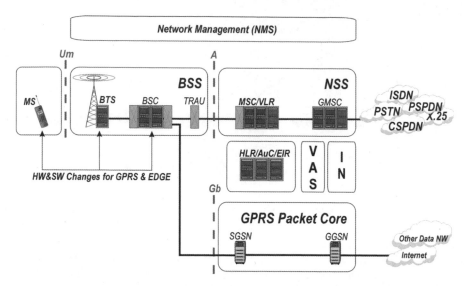

Figure 2.4   General Packet Radio Service (simplified illustration)

connection through the GSM network to an external packet data network (the Internet).

Figure 2.4 shows a simplified diagram of a GPRS network when it is implemented using basic GSM. Please note that a fully functioning GPRS network requires additional equipment, like firewalls for security reasons, DNS (Domain Name Server) for routing enquiries with the GPRS network, DHCP (Dynamic Host Config- uration Protocol) server for address allocation and so on. These pieces of equipment are not mobile-specific and they function exactly the same way as their "cousins" on the traditional Internet side.

GPRS has the potential to use asymmetric connections when required and in this way network resources are better utilised. GPRS is a step that brings Internet Protocol (IP) mobility and the Internet *closer* to the cellular subscriber, but is not a complete IP mobility solution. From the service point of view, GPRS starts a development path where increasingly traditional CS services are converted to be used over GPRS, because these services were originally more suitable for PS connections. One example of this is the Wireless Application Protocol (WAP), the potential of which is amply discovered when using GPRS. In addition, the greatest killer service in GSM (namely, SMS) behaves more optimally when transferring over a GPRS connection.

When PS connections are used, Quality of Service (QoS) becomes a very essential issue. In principle, GPRS supports the QoS concept, but in practice it does not. The reason here is that GPRS traffic is always second-priority traffic in the GSM network: it uses otherwise unused resources in the Um interface. Because the amount of unused resources is not exactly known in advance, no one can guarantee a certain bandwidth for GPRS in a continuous way and, thus, QoS cannot be guaranteed either. There are some ways to avoid this problem. The most cost-effective is to dedicate, for instance, one radio channel per cell for GPRS use only. By doing this, the operator is able to guarantee at least some kind of GPRS capacity for the mobiles camped on this particular cell. This method, however, does not provide any solid solution for QoS problems; it only eases the situation and improves the *probability* of gaining GPRS service in crowded, populated cells.

So far in this evolution chain, the GSM air interface has used traditional GSM modulation; the only other way to transfer data would be by means of either CS (HSCSD) or PS (GPRS) services. When using GPRS, the packet data transfer rate starts to be an issue, especially in the downlink direction. By applying a completely new air interface modulation technique, Octagonal Phase Shift Keying (8-PSK), where one air interface symbol carries a combination of three information bits, the bit rate in the air interface can be remarkably increased. When this is combined with very sophisticated channel-coding technique(s), one is able to achieve a data rate of 48 kb/s compared with conventional GSM which can carry 9.6 kb/s per channel and one information bit represents one symbol in the air interface. These technical enhancements are called "Enhanced Data Rates for Global/GSM Evolution" (EDGE).

The primary target with EDGE is to use it to enhance packet transfer data rates. This is why EDGE is often commercially introduced as E-GPRS (Enhanced GPRS). The implementation of EDGE as a technology requires some other changes in the network, especially on the transport mechanisms and transmission topology; the bit rates available with the BSS for basic GSM purposes are not enough. This problem will especially come to the fore when the operator increases site density and introduces EDGE technology simultaneously. These two changes together may increase the average bit rate per end-user to amounts the transmission is unable to handle without any changes.

When EDGE is implemented within the BSS its name changes to GERAN (GSM/EDGE Radio Access Network). With the channel-coding methods that have been introduced and 8-PSK modulation, the GPRS terminal could in theory achieve a 384-kb/s data transfer rate. This requires that the GPRS terminal gets eight air interface time slots with the best channel-coding method available for its use. Thus, the data rate could be 8×48 kb/s = 384 kb/s. It should be noted that the EDGE-capable terminals in the market are not able to do this; commercial terminals are able to utilise four channels simultaneously at a maximum.

From a network evolution point of view, EDGE in general has its pros and cons. A good point is the data rate(s) achieved; these are getting close to UMTS urban coverage requirements. The disadvantage with EDGE is that the data rates offered are not necessarily available throughout the cell. If EDGE is to be offered with complete coverage, the amount of cells will increase dramatically. In other words, EDGE may be an expensive solution in some cases. The future of EDGE is nowadays seen as a complementary technology enabling better interworking with the Wideband Code Division Multiple Access (WCDMA)-based UTRAN and the GSM-based GERAN. These two access networks constitute the basic accesses defined for UMTS networks.

2.4 3GPP Release 99

3G introduced the new radio access method, WCDMA. WCDMA and its variants are global; hence, all 3G networks should be able to accept access by any 3G network

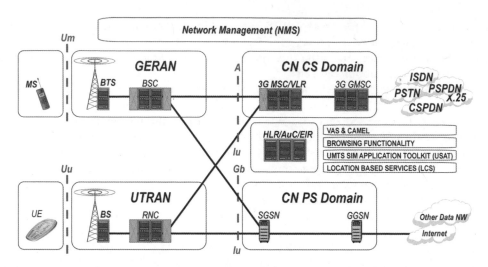

Figure 2.5 3G network implementation on the 3GPP R99

subscriber. In addition to its global nature, WCDMA has been thoroughly studied in the laboratory and it has been realised that it has better spectral efficiency than Time Division Multiple Access (TDMA) (under certain conditions) and it is more suitable for packet transfer than TDMA-based radio access. WCDMA and radio access equipment as such are not compatible with GSM equipment, and this is why, when adding the WCDMA to the network, one must add two new elements: the Radio Network Controller (RNC) and the Base Station (BS). The network part that contains these elements and maintains the WCDMA radio technology is called the "UMTS Terrestrial Radio Access Network" (UTRAN).

On the other hand, one of the key requirements for UMTS is GSM/UMTS interoperability. One example of interoperability is inter-system handover, where the radio access changes from GERAN to UTRAN and vice versa during the transaction. This interoperability is taken care of by two arrangements. First, the GSM air interface is modified in such a way that it is able to broadcast system information about the WCDMA radio network in the downlink direction. Naturally, the WCDMA radio access network is able to broadcast system information about the surrounding GSM network in the downlink direction, too. Second, to minimise implementation costs, 3GPP specifications introduce possibilities to arrange interworking functionality with which the evolved 2G MSC/VLR becomes able to handle wideband radio access, UTRAN.

So far, the abilities provided by the IN platform have been enough from the service point of view. The concept of IN is directly adopted from PSTN/ISDN networks and, thus, it has some deficiencies as far as mobile use is concerned. The major problem with standard IN is that IN as such is not able to transfer service information between networks. In other words, if a subscriber uses IN-based services they work well but only within his or her home network. This situation can be handled by using CAMEL, as explained earlier. CAMEL is able to transfer service information between networks.

Later on, the role of CAMEL will increase a lot in 3G implementation; actually, almost every transaction performed through the 3G network will experience CAMEL involvement, at least to some extent.

Transmission connections within the WCDMA radio access network are implemented by using ATM (Asynchronous Transfer Mode) on top of a physical transmission medium in a 3GPP R99 implementation. A pre-standardisation project FRAMES (1996–1998) discussed at great length whether to use ATM in the network or not. The final conclusion favoured using ATM for two reasons:

- ATM cell size and its payload are relatively small. The advantage here is that the need for information buffering decreases. When buffering is excessive, expected delays will easily increase and the static load in the buffering equipment will likewise increase. One should bear in mind that buffering and, thus, generated delays have a negative impact on the QoS requirements of real-time traffic.
- The other alternative, IP (especially its version IPv4), was also considered, but IPv4 has some serious drawbacks, being limited in its addressing space and missing QoS. On the other hand, ATM and its bit-rate classes match the QoS requirements very well. This leads to the conclusion that where ATM and IP are combined (for packet traffic), IP is used on top of ATM. This solution combines the good points of both protocols: IP qualifies the connections with the other networks and ATM takes care of connection quality and also routing. As a result of IPv4 drawbacks a compromise has been made. Certain elements of the network use fixed IPv4 types of addresses, while the real end-user traffic uses dynamically allocated IPv6 addresses, which are valid within the 3G network. To adapt the 3G network to other networks in this case, the 3G IP backbone network must contain an IPv4 $\leftrightarrow$ IPv6 address conversion facility, because external networks may not necessarily support IPv6.

Core Network (CN) nodes have also evolved technically. CS domain elements are able to handle both 2G and 3G subscribers. This requires changes in MSC/VLR and HLR/AC/EIR. For example, security mechanisms during connection set-up are different in 2G and 3G and now these CS domain elements must be able to handle both of them. The PS domain is actually an evolved GPRS system. Though the names of the elements here are the same as those in 2G, their functionality is not. The most remarkable changes concern the SGSN, whose functionality is very different from that in 2G. In 2G, the SGSN is mainly responsible for Mobility Management (MM) activities for a packet connection. In 3G, the MM entity is divided between the RNC and SGSN. This means that every cell change the subscriber does in UTRAN is not necessarily visible to the PS domain, but RNC handles these situations.

The 3G network implemented according to 3GPP R99 offers the same services as those of GSMPhase2+. That is, all the same supplementary services are available, teleservices and bearer services have different implementation, but this is not visible to the subscriber; a speech call is still a speech call, no matter whether it is done through a traffic channel (GSM) or by using 3G bandwidth. In addition to GSM, the 3G network in this phase may offer some other services not available in GSM (e.g., video calls); various streaming-type services and multimedia messaging that utilises location services (LCSs) could be good examples of these. In this phase the majority of services are moved/transferred/converted to the PS domain whenever reasonable and applicable.

The new services require new platforms for their implementation. Let us start with WAP: it had already been introduced in the context of GSM and GPRS. WAP had its own limitations and the content presented for end-users was not acceptable in all cases. Development on the terminal side has made it possible to use more sophisticated methods within the network. Instead of pure WAP, we could say that terminals utilise a browser functionality supported by the network. This browser functionality in turn implements XML definitions. End-users see this as a complete browsing experience in colour display with formatted documents looking very similar to those gathered from the Internet via a normal, traditional desktop computer.

This browser functionality is a kind of cornerstone for the other services the 3G network offers to end-users. Another very attractive platform for services in this phase is the UMTS SIM Application Toolkit (USAT) which arranges the possibility of handling SIM cards over the air. In general, service personalisation will be a very interesting issue. One branch of services in this respect is that the delivered content depends on the location of the end-user. For this arrangement the 3G network contains a platform enabling the use of LCSs.

2.5 3GPP Release 4

In order to simplify matters, we could say that the 3GPP R99 implementation is a GSM-based, GSM-evolved mobile network containing two different access networks and delivering both CS and PS traffic with variable speeds.

According to published work on 3GPP evolution, 3GPP R4 contains some major items to be implemented. The most important from the network architecture point of view are the UTRA Frequency Division Duplex (FDD) repeater function, IP transport for CN protocols and bearer-independent CS CN.

WCDMA radio access provides excellent possibilities to extend coverage and "transfer" capacity within the radio coverage area, unlike GSM radio access. This is not, however, a simple issue as such, and the use of repeaters has its effect on, for instance, LCSs. 3GPP R4 offers an option to convert protocol stacks in such a way that the transport protocols become IP-based. The third mentioned item, bearer-independent CS CN brings scalability to the system. The traditional MSC contains both connection capacity and connection control capacity, but these two capacity types do not necessarily go hand in hand. 3GPP R4 defines the way to split these two capacity types into two different nodes.

The node that maintains CS connection capacity is called the "Circuit Switched Media Gateway" (CS-MGW) and it takes care of all physical connection set-up matters. The node that maintains connection control capacity is called the "MSC server". The MSC Server and CS-MGW have a one-to-many relationship (i.e., one MSC server could control numerous CS-MGWs). With this arrangement the operator is able to optimise the physical length of the user plane within its network. This in turn helps us to migrate to the IP-based transport network.

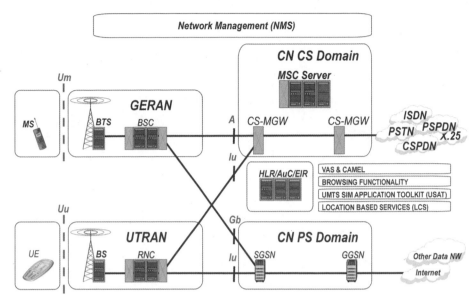

Figure 2.6   3GPP R4 implementation scenario

2.6 3GPP Release 5

After 3GPP R4 the aim was to implement the following major items:

- IP transport over the whole system from the BS up to the network border gateway.
- To introduce an IP Multimedia Subsystem (IMS) in order to start wide use of various multimedia services.
- To unify the open interface between the various access and core networks.
- To gain more capacity in the UTRAN air interface in the downlink direction.

The major items defined to be implemented in 3GPP R5 aim to simplify the network structure; making the transport protocol environment uniform enables more straight-forward solutions to be used than those used in R3 implementation. The first main item mentioned, IP transport throughout the whole network starting from the BS, has the aim of simplifying this transport network structure.

From the service point of view, the IMS will play a major role in R5 and further implementations. IMS is a separate system solution that is able to utilise various networks itself; one of these is UMTS. With IMS, end-users will be able to use sophisticated multimedia and messaging services. IMS architecture is described in Chapter 6 and some related service aspects are discussed in Chapter 8.

From the end-user point of view, the UMTS aims to harmonise the network structure (i.e., the end-user is not necessarily aware through which access he or she is using the network and retrieving services). To make this experience as smooth as possible, the various access networks must be harmonised as much as possible. Within the network the "harmonisation point" is the Iu interface between the CN and the access networks.

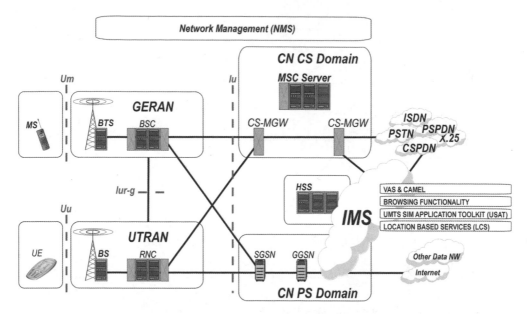

Figure 2.7   3GPP R5 implementation scenario

UTRAN has had the Iu interface from the very beginning, while GERAN has not. One of the key topics in 3GPP R5 is to make changes in GERAN in such a way that it functions similarly to UTRAN. When these changes have been made, GERAN works in *Iu mode*. In order to transfer end-user-related radio functionality within the network as smoothly as possible, UTRAN and GERAN have a defined interface between themselves named "Iur-g". Iur-g does not transfer a user plane—this is provided for signalling purposes only.

In this phase it is expected that the majority of services used are asymmetric in nature (i.e., the downlink direction carries a heavier load than the uplink direction). To handle this situation better in UTRAN, a collection of changes related to the UTRAN radio path are introduced. This collection is known as "High Speed Downlink Packet Access" (HSDPA). This affects the radio path, BS and transport channel arrangement in UTRAN. HSDPA is discussed in more detail in Chapter 4.

2.7 Trends beyond 3GPP Release 5

What happens after 3GPP R5? There is no exact answer to this question, but some trends are visible.

IP is expected to replace other technologies throughout the whole transport network and is also expected to be the key method when implementing services, accessing them and controlling the network. In other words, all IP solutions, implemented in one way or another, are likely to be among the major trends after 3GPP R5.

On the service side the expectation is that there will be so-called symmetric services utilising IP. These kinds of services could be, for instance, interactive video meetings with real-time pictures. When these services become available to the public, air interface capacity will once more rear its head. In 3GPP R5 the HSDPA introduces capacity enhancement in the downlink direction. In order to use the services mentioned above, the uplink on the radio path will have to be able to carry more capacity. For this purpose, a feature called "High Speed Uplink Packet Access" (HSUPA) is introduced.

The third trend will be the ongoing development of harmonisation. For this purpose. 3GPP is continuously working with study items related to multi-access and interworking arrangements between various access technologies.

Part Two

3

The Key Challenges Facing the Mobile Network Architecture

Siamäk Naghian and Heikki Kaaranen

The purpose of this chapter is to introduce, albeit briefly, the fundamental challenges facing the mobile network architecture so that we can provide the reader, first, with the reasons behind both restrictions and opportunities when considering the basic architecture of any radio communication system, and, second, to provide the reader with an overview of how these issues are addressed in general (when relevant, particularly in the Universal Mobile Telecommunication System, or UMTS, network).

Figure 3.1 shows the key components of a mobile network that could become the architectural bottlenecks of the network. The basic constraints of radio communication, the issues that arise from backhaul transport, the role and necessity of network management and finally limitations down to the scarcity of the radio spectrum are the key factors that set the boundary conditions for mobile network architecture development. Optimal network architecture requires that all these components evolve hand in hand and solutions are developed taking into account the consistency of the overall architecture.

3.1 Radio Communication Constraints

The usage of radio communication has been ongoing since Hertz experimentally showed the relationship between light and electricity in 1887 after Maxwell had illustrated the principal equations of electromagnetic fields in 1864. Then, Marconi used the "Hertzian" wave to communicate, inventing wireless telegraphy in 1896.

The fundamental principles of radio communication rely on utilising radio waves as a transmission medium. As a natural phenomenon, radio waves originate from electromagnetic fields. Under certain circumstances, time-dependent electromagnetic fields produce waves that radiate from the source to the environment. This source can be,

UMTS Networks Second Edition   H. Kaaranen, A. Ahtiainen, L. Laitinen, S. Naghian and V. Niemi
© 2005 John Wiley & Sons, Ltd   ISBN: 0-470-01103-3

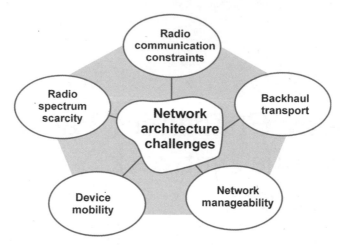

Figure 3.1   Key challenges in designing mobile network architecture

for example, a transmitter, like a base station or a mobile handset. Like their originators' radio waves and their characteristics are strictly dependent on the environment where the waves propagate. Correspondingly, a radio-wave-based communication system is vulnerable to environmental factors (e.g., mountains, hills, huge reflectors like buildings, the atmosphere, and so on).

Every radio communication system consists of at least two elements: the transmitter and the receiver. As in mobile systems, these two elements can also be integrated in one device (transceiver), enabling it to operate both as a transmitter and a receiver. An example of such a device is the base station and mobile handset in any advanced public mobile system. Figure 3.2 illustrates the simplest radio communication system, consisting of one base station and one mobile handset. Suppose that the base station acts as a transmitter source for a specific time in a certain environmental condition. The radio signal propagates from the base station to the mobile station at the speed of light and the received signal strength at the handset depends basically on the distance from the base station, the wavelength and the communication environment.

Environmental factors impeding radio wave propagation include any man-made or natural obstacle, like high buildings, terrain and weather conditions, etc. They affect the path, phase and time of signal propagation when they traverse between the transmitter and receiver. Also, system parameters (e.g., antenna height and beam direction) have their own effect on propagation distance, signal strength and attenuation. Therefore, the nature of radio communications inherently brings about some thorny limitations. The main problems that every radio communication system faces are as follows:

- The multipath propagation phenomenon.
- The fading phenomenon.
- Radio resource scarcity.

Amongst them, multipath propagation is also considered by many as an advantage to

Figure 3.2 Basic building blocks of a radio communication system

radio communication because it enables the radio receiver to hear the base station even without signal Line Of Sight (LOS). Despite this, it brings more complexity to the system by setting specific requirements and constraints on receiver and transmitter architecture. In order to understand the nature of a radio communication system the above-stated characteristics need to be well understood.

The factors that affect radio propagation are extremely dynamic, unpredictable and diverse. In spite of that and in order to model the propagation phenomena, it is necessary to sort and model them so that they can be treated appropriately. As shown in Figure 3.3, on its way from the transmitter to the receiver the radio wave experiences reflection, diffraction and scattering phenomena in addition to those components that travel directly (LOS) to the receiver.

These propagation mechanisms lead to multipath propagation, which causes fluctuations in the received signal's amplitude, phase and angle of arrival, referred to as multipath fading. Reflection is the consequence of a collision between the electromagnetic wave and an obstruction whose dimensions are very large in comparison with the wavelength of the radio wave. The result of this phenomenon is reflected radio waves, which can be captured constructively at the receiver (e.g., the mobile or the base station). Diffraction, also called "shadowing", in turn, is the consequence of a collision

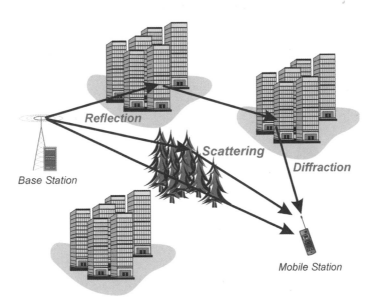

Figure 3.3 Propagation mechanisms: reflection, scattering and diffraction (LOS in the figure)

between the radio wave and an obstruction, which is impossible to penetrate. Scattering, on the other hand, is the consequence of a collision between the radio wave and obstructions whose dimensions are somehow equal to or less than the wavelength of the radio wave. These together explain how radio waves can travel in a radio network environment even without an LOS path.

From the receiver's perspective and depending on the existing preconditions of any of the above-mentioned propagation events, the received signal power is affected randomly by each or a combination of them. In addition, device mobility, indoor and outdoor coverage, and hierarchical network structure raise some specific aspects about the propagation environment, which makes the situation more complicated to cope with.

But, how can the ultimate impact of these on the radio signal be predicted? There are ways to treat the effect of propagation characteristics on the signal strength of the radio channel: link budget and time dispersion. The basic idea behind link budget is to determine the expected signal level at a particular distance or location from a transmitter, like a base station or a mobile station. By modelling the link budget, the essential parameters of the radio network, such as transmitter power requirements, coverage area and battery life, can be defined. Link budget calculation can be carried out by estimating signal path loss. Furthermore, path loss estimation can be done based on the free space model, defining that in an idealised free space model the attenuation of signal strength between the base station and mobile station basically follows an inverse square law.

Due to differences in the mobile network environment, it is almost impossible to predefine all the parameters affecting radio channel modelling and system designing. Therefore, some general models, such as the Okumura–Hata model, have been developed that, to some degree, could map the most typical cases. In order to consider the entire effect of radio channel fading, however, link budget is not adequate on its own. Instead, the effect of multipath propagation in terms of time dispersion should also be considered. This can be done by estimating the variation in propagation delay related to replicas of the transmitted signal that reach the receiver.

A typical fading process is illustrated in Figure 3.4. As shown, any fading process has curves of two different simple shapes; that is, downward slopes fade, referring to the deterioration of the signal strength, and upward slopes represent the undesired interference. Therefore, a combination of these simple curves can be employed to approximate the envelope of any fading process and the control actions for the fading process need to be accordingly determined to compensate for signal fluttering around the desirable average.

Figure 3.5 illustrates the main fading types that may appear in any radio network in one way or another, depending on the environmental condition. The main classes of radio channel fading are large-scale and small-scale fading. The former represents the average signal power attenuation or path loss due to device movement over areas between the base station and the mobile station. Small-case fading, on the other hand, is the result of rapid variations in signal amplitude and phase that can be experienced between the base station and mobile station. Small-scale fading is also called "Rayleigh fading" or "Rician fading" depending on the Non Line Of Sight (NLOS) or LOS characteristic of the reflective paths. Small-scale fading can be

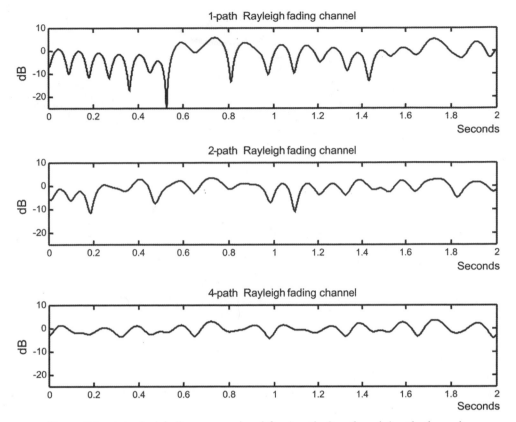

Figure 3.4   A typical fading power signal for 1-path, 2-path and 4-path channels

further divided into frequency selective fading, flat fading, fast fading and slow fading. The radio channel needs to be modelled on the type of fading it experiences.

In general, the mobile radio channel experiences signal fading that most likely consists of the Rayleigh-faded component without LOS, making the channel more difficult to cope with. As a result, a signal arrives at a mobile station from many directions with different delays, causing significant phase differences in signals travelling on different branches. Moreover, the behaviour of radio channel fading is strictly dependent on the physical position and motion of the mobile station. Even a small change in position or motion of the mobile station may result in a different phase for each signal branch. Thus, the motion of the mobile station through the radio network may result in fast fading, which can be extremely disruptive to radio signals, setting strict requirements for radio network planning and optimisation.

If we assume that a radio signal is transmitted from a base station to a mobile station with constant transmitted power, then the effect of fading on the strength of signal transmitted by the base station can be perceived as an inverse function of distance at the mobile station receiver; as the distance between the mobile station and base station increases, the received signal strength decreases at the mobile station. Assuming also that the bandwidth for the uplink and downlink remains the same, the average path

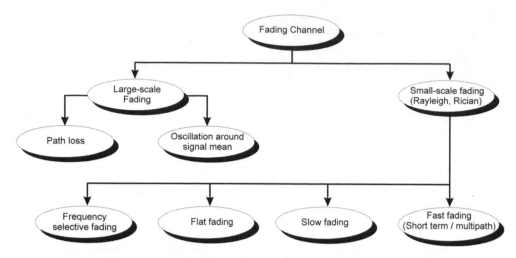

Figure 3.5   Radio channel fading classification

loss—which basically follows the log-normal distribution of average path loss—is the same for both directions. If different bandwidths are used for uplink and downlink, Rayleigh fading is typically independent of each link, meaning that when the uplink channel is fading the downlink channel is not necessarily fading at the same time, and vice versa.

In addition to fading and multipath phenomena, interference is a thorny problem for every radio communication system. The basic reason behind interference is the simultaneous connections to the base station and when they use a common shared bandwidth the problem gets worse. This forms perhaps the most dominant interference source in multi-access radio systems. Minimising the undesirable impact of fading and interference as well as optimising scarce radio resource usage are very dependent on the radio network planning employed, utilised radio access techniques, and algorithms used for controlling radio resources, cellular concepts, modulation techniques, advanced antenna techniques, etc. Therefore, before describing further details of the interference aspects of a radio system, we proceed next by introducing the concept of a cellular system as a fundamental solution for alleviating the declining capacity of radio systems, as a result of path loss, fading and interference.

3.2 Cellular Radio Communication Principles

The simple radio communication system illustrated earlier in this chapter is unable to provide access service to a large number of end-users and soon after network take-off it would face capacity limitation. Let us outline the basic problems inherited from the fundamental characteristics of radio communications to such a simple system.

First, public radio communications should offer duplex communication in order to provide simultaneous two-way communication. Therefore, it is not practical to offer such services by, for instance, broadcasting information. Moreover, the signal strength

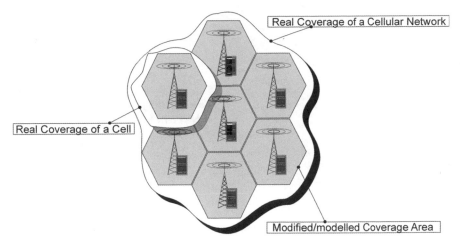

Figure 3.6 A cluster of cells in a cellular network

at the receiver deteriorates as the distance between the transmitter and receiver increases, resulting in an unacceptable Qualify of Service (QoS) in areas distant from the transmitter. Second, every transmitter is capable of offering a limited number of radio links/channels to end-users who intend to have simultaneous calls. There is no other way than to adapt the system architecture so that these limitations are removed.

A very fundamental architectural solution is the cellular concept. The main idea is simple. Suppose we are planning a radio network for a large city where there are millions of mobile users. Based on the cellular concept, the large area is divided into a number of sub-areas called *cells*. Each cell has its own base station that can provide capacity for a specific number of simultaneous users by emitting a low-level transmitted signal. Figure 3.6 illustrates an example of the 7-cell cluster of a cellular network.

The nature of radio communication and the cellular concept are the primary starting points that have led to the current architecture of mobile networks. Figure 3.7 illustrates the basic architecture of a cellular concept-based radio network. As shown, the basic architecture of any advanced cellular system consists of base stations, a switching network and fixed network functionality for backbone transmission.

The cellular solution resolves the basic problems of radio systems in terms of radio system capacity constraints, but it encounters new problems, such as:

- Interference due to the cellular structure, including both inter- and intra-cell interference.
- Problems due to mobility.
- Per-cell radio resource scarcity.

Assume a cellular system with asynchronous users sharing the same radio bandwidth and served by the same radio base station in each coverage area or cell, each base station not only receives interference from mobiles in the home cell but also from terminals and base stations located in neighbouring cells. Depending on the source

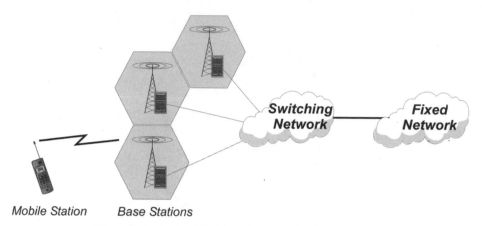

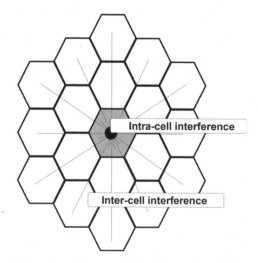

Figure 3.7   A simplified architecture of cellular network

of interference, they can be classified as intra-cell/co-channel, inter-cell/adjacent cell and interference due to thermal noise. Therefore, as illustrated in Figure 3.8, in order to cope with the total interference received from terminals, interference within the home cell and neighbouring cells should be investigated.

In order to alleviate channel interference, both multipath and path losses in the cellular network *frequency reuse technique*, in which in each cell of the cluster pattern a different frequency is used, can be applied. Therefore, the frequency reuse factor is an essential parameter for cellular networks and can partly explain the spectrum efficiency of the network. By optimising the frequency reuse factor the problems of adjacent channel and co-channel hazards can be greatly reduced, resulting in capacity improvement, a crucial performance metric for the public radio network. Figure 3.9 shows an example of one–seven and one–one Frequency Reuse (FR) patterns, respectively.

Figure 3.8   Inter-cell and intra-cell interference in cellular network

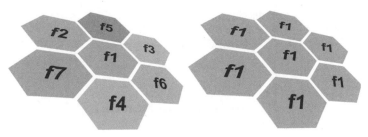

Figure 3.9 An example of frequency re-use pattern, FR = 7 and FR = 1

The cellular concept increases the radio system capacity, especially when utilising higher frequency reuse. The smaller the cells, the more efficiently the radio spectrum is used, but the cost of the system increases accordingly because more base stations are needed. Multilayer network designing with a macro-, micro- and pico-cell structure brings further enhancements to the cellular and frequency reuse concept to improve system capacity.

It is the subject of network planning to optimise the combination of network structure and frequency reuse to increase the system capacity cost-effectively while avoiding an undesirable increase in the number of required base stations. This, however, implies that the primary architecture of the network allows such combination and optimisation. These solutions set new requirements for the system to handle; for example, the device mobility and handover control functions in the network should be in place.

3.3 Multi-access Techniques

The primary objective of the cellular concept was to tackle capacity limitation, but it does not help per-cell capacity limitation on its own as far as simultaneous users are concerned. From the radio spectrum's standpoint, it is extremely important to know how radio resources are allocated to simultaneous users. Controlling radio resources has become one of the most critical features of any mobile network that serves a large number of subscribers.

Multi-access techniques have been developed to combat the problem of simultaneous radio access allocation to access requesters. The primary principle of any multi-access scheme is to provide a response to how the available bandwidth is shared. Again, the primary problems encountered by multi-access solutions are related to the intrinsic characteristics of radio systems: bandwidth limitation, multipath fading, and interference from other users in the cellular network in which the cellular concept and frequency reusing have been utilised. Effective use of frequency means that there should be as many simultaneous users as possible utilising the fixed bandwidth within regulated parameters.

To address this, different multi-accesses have been developed. The multi-access method widely used in analogue cellular networks is called Frequency Division Multiple Access (FDMA) where every user employs their own frequency or channel or a set of them, as shown in Figure 3.10. When considering the number of radio connections this means one frequency = one user = one channel. This multiple access

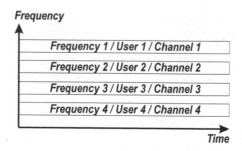

Figure 3.10 Frequency Division Multiple Access (FDMA)

technique, however, is not an efficient way of sharing the limited radio resources in a public mobile system.

Therefore, a more effective way to utilise frequency resources, which enables an increase in system capacity, called "Time Division Multiple Access" (TDMA) was developed at an earlier phase of radio communication systems. TDMA is the most common multiple access method used in 2G cellular systems, such as the Global System for Mobile Communication (GSM). Also, TDMA-based transmission systems are widely used. Maybe the best known system is the G.703 specification-based Pulse Code Modulation (PCM) transmission.

In the GSM system, the users of a given frequency divide it in time: every user has a little slice of time (timeslot) for different operations. As shown in Figure 3.11, this timeslot is repeated frequently and this generates the impression of continuous connection. This system makes it possible to have several users—as many as the number of timeslots, which is eight for GSM—simultaneously using the same frequency.

Code Division Multiple Access (CDMA) is another multi-access technique which has been used for similar purposes to FDMA and TDMA. However, it approaches the same problem by utilising a totally different spectrum-sharing strategy. As a spread-spectrum-based radio access scheme, CDMA is perhaps one of the most sophisticated schemes that has been used in different mobile systems (particularly in UMTS).

Figure 3.12 illustrates the basic radio resource strategy of the CDMA scheme. Unlike in the TDMA and FDMA schemes, in CDMA the radio resource is allocated on codes. Thus, all simultaneous users can occupy the same bandwidth at the same time. Every user is assigned a code/codes varying per transaction and these codes are used for cell, channel and user separation. Every user uses the same frequency band simultaneously

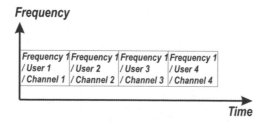

Figure 3.11 Time Division Multiple Access (TDMA)

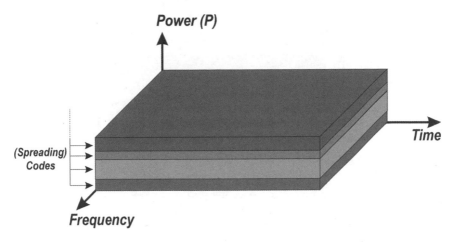

Figure 3.12 Code Division Multiple Access (CDMA)

and, hence, there are no timeslots or frequency allocation in the same sense as in TDMA- and FDMA-based systems, respectively:

- If the originating bit rate is low, it can be spread well and, thus, the power required for transmission will be small. This kind of case can be seen as the thin layer in the figure.
- If the originating bit rate is high it cannot be spread as well and, thus, the power required for transmission will be higher. This kind of case can be seen as the thick layer in the figure.

Unlike bandwidth-limited multi-access, such as FDMA and TDMA, that suffers mainly from co-channel interference due to the high range of frequency reusing, in the CDMA systems inter-user uplink interference is the most crucial type of experienced interference. One basic reason is that inter-user uplink interference increases on an accumulated power basis, and the performance of each user becomes poorer as the number of simultaneous users increases in a single cell.

Depending on the spreading signal used for modulation, the CDMA scheme can be categorised into the following groups:

- Direct Sequence CDMA (DS-CDMA).
- Frequency Hopping CDMA (FH-CDMA).
- Time Hopping CDMA (TH-CDMA).
- Hybrid Modulation CDMA (HM-CDMA).
- MultiCarrier CDMA (MC-CDMA).

In a **DS-CDMA** scheme, the data signal is scrambled by the user-specific pseudo noise (PN) code at the transmitter side (e.g., the mobile or base station) when spreading the signal with a desirable chip rate and process gain. At the receiver (the mobile or base station), the original signal is extracted by exactly the same spreading code sequence.

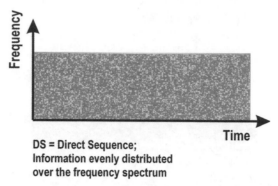

Figure 3.13   Direct Sequence Code Division Multiple Access (DS-CDMA)

As a result, every signal is assumed to spread over the entire bandwidth of the radio connection. As illustrated in Figure 3.13, interference may therefore be generated from all directions in contrast to narrowband cellular systems.

Frequency reuse for DS-CDMA is one, meaning that all users have the same frequency band to transfer their information. The inherent advantage of the DS-CDMA is its multipath-fading tolerability. Indeed, in a TDMA-based system like GSM a similar advantage is achieved by utilising a frequency-hopping technique, resembling the spread spectrum scheme. On the other hand, in the case of a common shared frequency the system is more vulnerable to multi-user interference than TDMA and FDMA.

As shown in Figure 3.14, in FH-CDMA, changing the carrier frequency over the transmitted time of the signal produces the spread bandwidth. That is, during a specific period, the carrier frequency is kept the same, but afterwards the carrier is hopped to another frequency based on the spreading code, resulting in a spread signal.

Depending on the hopping rate of the carrier signal, FH-CDMA is divided into two groups: fast-frequency-hopping and slow-frequency-hopping methods. For the fast-frequency-hopping modulation scheme the hopping rate is typically greater than the symbol rate, while, for the slow-frequency scheme, the hopping rate is typically smaller than the symbol rate. Some essential procedures of the CDMA scheme, like power control, are much easier to implement and handle in FH-CDMA than in DS-CDMA. This is partly due to the frequency division aspect of this technique when it is applied to cellular systems. On the other hand, producing high-rate frequency hopping is, however, a more complex issue in FH-CDMA.

In the TH-CDMA scheme, the transmitted signal is divided into frames, which are further divided into time intervals. During the transmission time, the data burst is hopped between frames basically by utilising code sequences. Principally, the aim is to select the code sequence so that simultaneous signal transmission in the same frequency may be minimised. The most important idea behind this approach is to create synergy between the strengths of the previously described schemes to provide convenient schemes for a set of specific applications. The combinations of these CDMA schemes leads to different hybrid schemes, such as DS/FH, DS/TH, FH/TH and DS/FH/TH.

Unlike DS-CDMA, in MC-CDMA (shown in Figure 3.15) the entire frequency band is used with several carriers instead of one. Therefore, the MC-CDMA transmitter

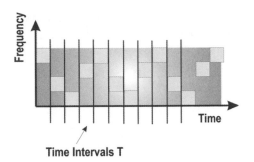

Figure 3.14 Frequency Hopping Code Division Multiple Access (FH-CDMA)

basically spreads the original data over different subcarriers using different frequency bands by denoting a spreading code in the frequency demand. The MC-CDMA schemes are mainly categorised into two groups. One spreads the original data stream using a given spreading code, and then modulates it over different subcarriers while the other one is basically similar to DS-CDMA.

In addition to the widely used multi-access techniques discussed previously, "Orthogonal Frequency Division Multiple Access" (OFDMA) has recently engaged the interest of communication industry. As shown in Figure 3.16, in an OFDMA method an individual tone, or groups of tones, is assigned to different simultaneous users. As the shared bandwidth is based on the orthogonal division of the tones, efficient use of bandwidth can be achieved. The number of allocated tones can be adjusted dynamically

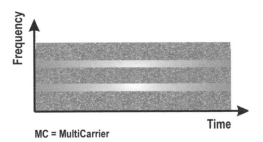

Figure 3.15 MultiCarrier Code Division Multiple Access (MC-CDMA)

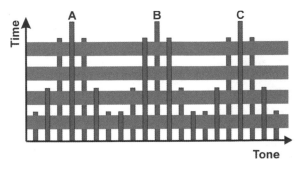

Figure 3.16 Orthogonal Frequency Division Multiple Access (OFDMA)

according to the amount of data to be transferred. OFDMA can be equipped with different hopping techniques and also benefits from time division channel sharing. When combined with frequency hopping it would easily benefit from the main advantages of the spread spectrum method.

3.4 Device Mobility

Although mobility allows end-users to be reachable anywhere and anytime, nevertheless, it brings strict requirements for the cellular network architecture. In addition to the serious consequences of device motion on the multipath behaviour of the channel, such as fading and Doppler spread, keeping the moving device connected to the network and managing its mobility is one of the most challenging, though essential, functions of any cellular network.

Generally, the network architecture embraces this challenge by supporting such functions as paging, location updating and connection handover. The handover mechanism guarantees that whenever the mobile is moving from one base station area/cell to another, radio connection is handed over to the target base station without interruption. The location update procedure, on the other hand, enables the network to keep track of the subscriber camping within the coverage of the network, while paging is used to reach the handset to which a call is destined. Location update and paging mechanisms guarantee that the mobile station can be reached even though there is no continuous active radio link between the mobile and network. In association with all these functions, special attention should be paid to minimise the power consumption of the mobile device.

In order to keep track of the mobile device, the whole geographical area is divided into Location Areas (LAs) that consist of a logical group of cells (Figure 3.17). Location updating takes place when a handset is turned on or it changes LA. The handset triggers the location update procedure by requesting a location update from the network. Once the subscriber information has been checked out the network responds to the device location update request, resulting in providing new location information and cancelling the old data. This may be carried out either periodically or on a predefined timing.

In paging, when a mobile device is the destination, its current position is determined by interrogating the Home Location Register (HLR) and the Visitor Location Register (VLR). If the entire LA is paged for this purpose the technique is known as *simultaneous paging*. Also, a portion of the LA, known as a "Paging Area" (PA), can be paged to reach the mobile device. If the called party is not located by initial attempts, paging is carried out sequentially over the paging areas until the location process is completed. This method is known as *sequential paging*.

These simple mechanisms, though, cause serious challenges when the number of handsets increases in the cellular system network, the cost of control signalling grows and radio resource is wasted. The reason is mainly due to the increase in the number of incoming calls and the boundary-crossing rate. This time, splitting the cells of the networks to gain more capacity leads to an increase in the signalling overhead as a result of mobility handling. In order to minimise overhead signalling and optimise

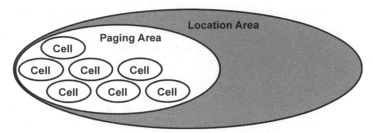

Figure 3.17   Relationship between Cell, Paging Area (PA) and Location Area (LA)

network capacity, the cell size, LA and PA need to be fine-tuned. In the case of simultaneous paging, and assuming that LA = PA, as long as the LA is changed, the generated paging and the overhead access traffic behave in a counteractive fashion. In other words, while small-size LAs cause less paging traffic they generate more access traffic compared with large-size LAs. In contrast, the selection of a larger LA leads to an increase in the traffic intensity of broadcast paging messages, while the traffic intensity of location-updating messages is reduced.

To what extent the cost of mobility handling could be minimised without causing drawbacks to the overall performance of the network depends on the mobility management architecture employed in the network. In a centralised architecture with hierarchical Mobility Management (MM) overhead signalling is accumulated in the hubs at the higher levels of the hierarchy, while in a decentralised network architecture overhead signalling is more distributed, though this may not necessarily lead to more efficient MM.

3.5 Network Transport

In addition to limitations that arisen from the basic characteristics of radio communication and device mobility, network transport and backhaul adaptation is another vital factor that sets strict requirements for the network architecture and the way it could be designed and deployed. Though the terms "transport" and "transmission" are occasionally used interchangeably, they convey different meanings and purposes. Transport is the functional process and the protocols used by transmission. Therefore, it is not necessarily tied to the physical connections used in the transmission process. Although it may have consequences for transmission means, to a great extent it can be independently evolved. On the other hand, "transmission means" refers to the action of conveying signals from one point or multipoints. In this respect, the process of transmission can be affected directly or indirectly, with or without intermediate storage and forwarding. Therefore, transmission techniques can be distinguished according to switching methods (e.g., packet-switched transmission, circuit-switched transmission or the techniques used for transmission, such as fibre transmission).

The paradox is that, though transmission infrastructure is one of the most costly parts of the network, both in terms of Capital Expenditure (CAPEX) and Operation and Maintenance Expenditure (OPEX), and causes most of the operation bottlenecks, unfortunately the transmission part of the network architecture often remains invisible to network architects at the early state of network development. As a result, difficulties arise at the time of network deployment, when invariably it is too late to treat. Lastly, at this step the real (or practical) problems will appear. One key reason is that network transmission evolution follows its own evolutionary path with a different life cycle when compared with other domains of network architecture (e.g., the air interface). Emerging networks should be adapted to the limitations and possibilities that the available transmission infrastructure sets. Renewing the transmission infrastructure requires a lot of investment and requires a commitment from many different parties other than just the network operators, making it a complex reality.

The overall network architecture, types of services offered by the network and the transport backbone should all fit well together in order to accomplish an optimised network solution. The network hierarchy, the way of conveying user and control information, network scalability, desired services and QoS requirements should be considered while at the same time keeping an eye on the transport backbone and its evolution potential.

Consider Figure 3.18 which adequately reflects the starting point of the transmission backbone for a UMTS network that has a fully centralised network architecture, like UMTS Release 4 or 5. As shown, the transmission backbone media and its topology differ markedly depending on the requisite capacity, current backbone facilities, radio access coverage, etc. In the access network the physical topology can be a tree, star, chain, loop or a combination of these with a logical star, while at the higher level close to the core network a loop topology with a logical mesh is often used.

In the UMTS network, user plane traffic is processed at different hierarchical levels, from Node B up to the Core Network (CN). Also, control traffic is strictly tied to the user plane. Irrespective of the reasons behind the current hierarchical architecture, for the transmission backbone it means the higher the level of network hierarchy the greater the amount of accumulated traffic generated. Therefore, higher level network elements will readily become the bottleneck of the network. Therefore, transmission capacity should be fitted to the network hierarchy; at higher levels high-capacity transmission means, such as fibre, are needed, but when it comes to the edge of the network microwave transmission becomes a more flexible and cost-efficient substitution, particularly in terms of capacity extending.

3.6 Transport Alternatives for UMTS

At the present time, the main transport technology used for backbone transmission in cellular networks relies on Time Division Multiplexing (TDM) technology. This is regardless of the recent network architecture evolution trends shifting from the circuit-centric to the date-centric paradigm. Even emerging packet-switched (PS) technology elements in the network architecture, such as the General Packet Radio Service (GPRS), have to adopt legacy transport facilities. In addition, diverse services are conveyed over conventional transmission technologies, such as Plesiochronous Digital Hierarchy (PDH) and Synchronous Digital Hierarchy (SDH). As network evolution movement towards a multimedia and data-centric solution is getting faster,

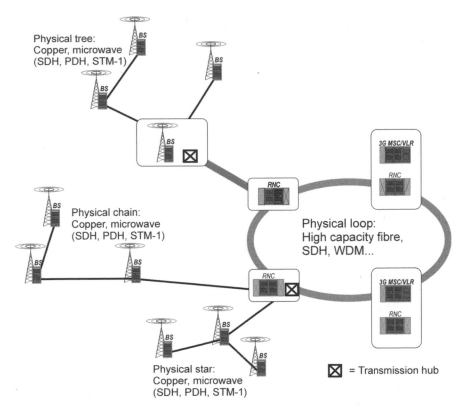

Figure 3.18 An example of backhaul transport network for UMTS

the current transport backhaul cannot meet data service requirements; TDM-based transport technology was originally developed for circuit-switched (CS) services and, therefore, is too limited to fit the flexibility and efficiency needed for data services. TDM technology allocates transport resources based on the time division approach and treats voice and data services equally, in terms of bandwidth sharing, and does so without considering the inherent nature of bursty traffic in data services. Increasing data traffic will require the transport network to be capable of handling data packet flows more flexibly and efficiently. TDM's equal treatment of data and voice services and its stiff bandwidth allocation scheme cannot support good granularity and, therefore, wastes the leftover bandwidth that could otherwise be used.

On the other hand, the starting point for developing a Wideband CDMA (WCDMA) air interface was based on the assumption that the majority of services will be bursty, yet voice services should still be handled concurrently. This led to a flexible radio access bearer architecture for UMTS that in turn also requires a flexible transport technology rather than, for example, E1 on the network side in order to benefit from the design advantages of the system.

The above-mentioned limitations drove the development of alternative transport technologies for UMTS networks. Two alternative solutions were considered: at the start of network development the favoured choice was Asynchronous Transfer Mode

(ATM) technology, but gradually Internet Protocol (IP) technology received more attention.

3.6.1 Asynchronous Transfer Mode in UMTS

Unlike TDM transport technology, ATM technology benefits from the principle of PS technology by carrying user information in a flow of "cells". This removes the limitations of legacy transport networks and adapts it to the requirements of UMTS air interface flexibility.

Third Generation Partnership Project (3GPP) and Number 2 (3GPP2) standards specify that all the elements of the Radio Access Network should be interconnected by utilising an ATM transport network. As a result, currently available UMTS Radio Access Network (RAN) elements from different manufacturers are equipped with ATM technology. Practically, this means that RAN elements include an ATM access port and employ ATM signalling protocols and switching technology. It is also possible to use IMA (Inverse Multiplexing for ATM) in RAN. This in turn brings about the possibility of carrying Iub interface traffic over GSM transmission routes, allowing network operators to reuse currently available transmission networks.

In addition to flexible resource sharing, the most important benefit of using ATM in RAN comes from the way data packets are formed, classified and carried across the network. Because each ATM connection is uniquely identified with specific classes, the network is capable of supporting an efficient QoS mechanism. This fits well with the bearer architecture thinking of UMTS defined for different classes of QoS, such as Real Time (RT) and Non-Real Time (NRT) services, to be supported simultaneously. These also lead to more efficient resource management, as the Radio Network Controller (RNC) in the RAN will be able to schedule and optimise traffic load across the network.

Although ATM will remove the main constraints behind TDM technology, new limitations arise: ATM signalling is too complicated and causes undesirable delays and loads to the network. In addition, ATM may lag behind the reach of economies of scale, an important advantage of such emerging technologies as IP. Therefore, transition from ATM towards IP became the focus of attention in 3GPP and in association with Releases 4 and 5.

In the current architecture of UMTS, the operator's transport network may include base stations (BSs) or Node Bs, RNCs, Base Station Controllers (BSCs), Mobile Switching Centres (MSCs), Serving GPRS Support Nodes (SGSNs), and Operation and Management Centres (OMCs). Therefore, ATM-based traffic in the radio access network can travel over E1 User Network Interface (UNI) ATM or E1 IMA, or over one or multiple E1 lines and Synchronous Transport Module 1 (STM-1) links. In order to support different types of traffic ATM Adaptation Layer Type 2 (AAL2) and ATM Adaptation Layer Type 5 (AAL5) for voice and data are used. Transmission media could include fibre optics, E1/T1 copper leased lines, microwave or Digital Subscriber Line (DSL), depending on the economies of scale and the availability of media types. It is also possible to combine different kinds of transmission media. While fibre is the best transport option due to its speed and capacity, E1/T1, microwave, and DSL bring their own advantages to the transport network and time after time they are found to be the only options available due to the cost or lack of backbone infrastructure and the co-existence of 2G and 3G networks.

In spite of the dominant presence of ATM in the early phase of UMTS networks, 3GPP has prepared the way for a transition to IP transport. The original design philosophy of ATM was to provide good bandwidth granularity while guaranteeing the delay and jitter requirements for voice services. There is inefficiency in the handling of data packets due to the small size of the packets, which cannot be optimised for different traffic flows. When there is no need for good granularity of ATM (e.g., when the links have a high bit rate and latency is not so critical), the use of IP can become more efficient and lucrative. For the purpose of continuity, however, IP can form an overlay on the existing ATM backbone and, therefore, ATM will keep its role in the cellular network in the near future as well.

3.6.2 IP Transport

With the rapid growth of the global use of IP and the smooth convergence of telecommunication and IP services, the use of IP transport in the cellular network has become ever-current. IP-based transport has already become commonplace in fixed networks. In this way, it has proven its robustness and ability to support a more distributed functional architecture coupled with a more flexible network topology than what is practical in TDM or ATM networks. IP transport benefits from economies of scale, openness, simplicity and easy-to-deploy characteristics. While it satisfactorily meets the requirements of the UMTS network it also alleviates the limitations of an ATM-based transport technology, at least for data-centric traffic.

IP refers to a set of protocol stacks, specified by the Internet Engineering Task Force (IETF), that form the pillars of the Internet. All services (such as email, web browsing, file downloads, gaming, etc.) use the IP to carry data through the global network. As each data packet is identified by a unique global address, user data can be dealt with by simply using addressing and routing mechanisms when transferring data from the source to the destination nodes (Figure 3.19).

Coupled with other higher level protocol stacks, such as Transmission Control Protocols (TCPs), IP constitutes a simple and flexible data delivery mechanism that can convey any type of digital information between devices connected in a world full of heterogeneous networks and subnetworks. When an IP host, such as a PC, mobile or any IP devices equipped with a global IP address, desires to send a message to a destination IP device the message body is encapsulated in an IP packet. Once the packet is provided with the globally unique IP address of the destination device and the IP routers in the Internet are in place, delivering the packet to the target destination can be easily achieved. In principle, the same procedure can be applied for all services provided by the Internet.

The simplicity, openness and flexibility of IP transport, coupled with its global acceptance, were found to be valuable assets for it to be considered for UMTS networks as well. Indeed, the entry of IP into cellular networks had already begun through the emergence of GPRS products. Based on the current protocols employed in GPRS,

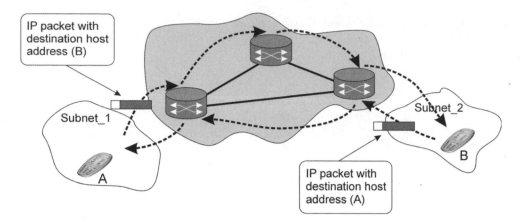

Figure 3.19 A simplified illustration of an IP packet being transferred from the source to the destination host, and vice versa

packet data services are supplied to mobile terminals, for example, by allocating an IP address to them. The UMTS packet CN uses routers when connecting to the Internet. This allows, to some extent, IP mobility and mobile Internet services to be offered to mobile users.

In its current form, however, converging mobile networks with the IP world encounters new barriers; the current version of IPv4 dates back to the early 1980s when it was difficult to foresee large acceptance of the Internet. Ipv4 was not originally designed to support multimedia services, while an essential requirement for the UMTS was to support both CS and PS services simultaneously. Therefore, QoS support coupled with the conventional shortage of Ipv4 (i.e., limited address space) were the basic obstacles to exploration of IP in mobile networks. The emergence of a new version of IP (i.e., IPv6) removes these two obstacles. Therefore, an upgraded, IPv6 packet CN could be well suited to meet the crucial requirements of UMTS networks, allowing the carriage of RT and NRT services. This allows the operators to benefit of an "All IP" CN by utilising a single technology that would simplify their network architecture and alleviate network complexity. Also, it will make the service offering more cost-efficient while supporting multi-services whose QoS is assured.

It is worth keeping in mind that transition from ATM to IP transport will remain challenging due to the massive investment needed for transport network replacement. Furthermore, transmission will most likely happen in a different network hierarchy with different evolution cycles, most likely starting from the transport network, CN and spreading to the Radio Access Network (RAN). Therefore, in practice it will be a long process to realise a fully IP-based mobile network. Regardless of the transition time cycle 3GPP Release 5 allows operators to explore IP in all CN and transmission functions.

Moving towards IP will obviously radically change how the network architecture evolves. The main foreseeable changes will reflect the hierarchical architecture of the network; IP will push network architecture towards a flat structure. Put simply, the transport network will markedly drive network architecture development.

3.7 Network Management

An essential question when designing the network architecture is how to manage and monitor the network's overall operation, and remove flaws from the network when they occur. In general, network management can be perceived as a service that employs a variety of methods and tools, applications, and devices to enable the network operator to monitor and maintain the entire network. For a typically centralized mobile network, network management means the ability to control and monitor the entire network from a specific location and, possibly, remotely. The rapid universal growth of mobile networks has made the role of the network management a key feature that should be carefully taken into account early in the design process of network architecture. In particular, the basic functionality of networks should provide operators with the ability to control and maintain their networks and services. Recent developments indicate that, in practice, OPEX constitutes the major capital investment for network operators.

In this context, network management requires a dedicated management function for each network element, transmission means, and switching and storage subsystems, all of which together form the entire network management system. Yet, it is essential that management entities function consistently to monitor network operation and to remove the failures efficiently. In addition, network management plays a key role when upgrading network software and when managing capacity and services to meet end-user demands. It is also key to adapting and optimising network operation based on traffic behaviour which can vary quite randomly.

Due to the importance of network management in the real life of the network and the wide area it concerns, the International Standards Organisation (ISO) has also made attempts to specify a framework and common paradigm for designing the network management system as part of the network architecture. Despite this, network management systems provided by different vendors may vary depending on their network element management system architecture.

3.7.1 High-level Architecture of a Network Management System

Currently, the basis for communication network management is built on the Telecommunications Management Network (TMN) paradigm. It describes a generic model of how a telecommunication network could be managed. The TMN divides management into several layers located on top of each other:

- *Network Element Layer (NEL)*: A telecommunications network consists of different physical entities, like switches, controllers, terminals, etc. In TMN, all of these are referred to by a common name Network Element (NE) and these NEs together form the Network Element Layer (NEL).
- *Element Management Layer (EML)*: This layer contains all the functions and resources required for a TMN-type interconnection between the NE and the Network Management Layer (NML).

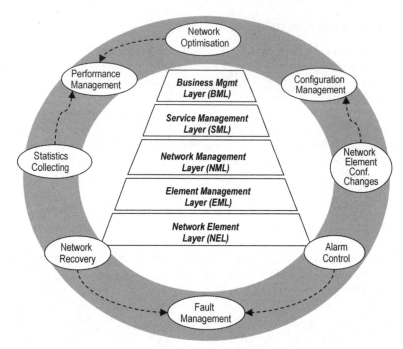

Figure 3.20   TMN pyramid model and some basic functions of a network management system

- *Network Management Layer (NML)*: The NML contains all the functions and resources required for NE management. In this context, three separate management layers/levels are in common use: Fault Management (FM), Performance Management (PM) and Configuration Management (CM).
- *Service Management Layer (SML)*: The SML contains functionality to handle interaction between services and service providers, interfaces with other enterprises, customers and vendors as well as service-level agreements, statistics gathering and analysing, billing and network usage.
- *Business Management Layer (BML)*: This high-level layer consists of the business goal, business planning, product planning, financial issues like budgeting, and legal arrangements and security.

Therefore, network management systems should cover both the management needs at the element level, such as BTS, RNC and CN, as well as at the network and service management levels. The network-element-level management function provides the necessary functions for commissioning and installation, and fault-remedying of individual elements, while the network-level functionality provides means to handle overall operation and maintenance as well as scaling the entire network.

The main functions of the network management system consists of FM, PM and CM:

- FM is a process (or processes) triggering faults/alarms in NEs, collecting them together and sending them to the Network Management Centre (NMC). The NMC is a physical point performing all the activities related to the NML.

- On the other hand, PM is a process (or processes) for collecting statistics from NEs and sending them to the NMC.
- CM is a process (or processes) containing tools for NE configuration changes: addition, deletion and modification. For instance, a change in the cell parameters of a certain cell, as done from the NMC, is one task done via CM.

Keeping in mind the basic architecture of the Network Management System (NMS) it is obvious that its development is tightly linked to the overall architecture design of the network; a network with a hierarchical architecture may not fully fit with a flat NMS and vice versa. It is also crucial that network operation and scalability is suitably aligned with NMS features as the challenges associated with network expansion relate both to network operation management and strategic network planning for foreseeable growth. This should also cover the smooth transition and adaptation from current networks like GSM to the emerging UMTS network.

The massive increase in the number of mobile devices and services is continuing to grow, resulting in very large-scale mobile networks. Efficient management of large mobile networks to which a diverse range of wireless devices are connected is not only essential but also complicated and laborious. Following the current trend in network architecture, the traditional TMN management approach relies heavily on a centralised network management architecture. Alhough new software technologies, like Object Oriented SW, may bring new opportunities for network management to get rid of its dependencc on the network architecture, it may not help avoid it totally; network manageability will remain as an important part of network architecture design well into the future.

3.8 Spectrum and Regulatory

Advances in technology, regulation and spectrum allocation along with economical policies constitute a combination that should be in balance so that the emerging radio system can take off successfully. The primary intent of spectrum regulators, such as ITU, is to harmonise the regional and global use of the radio spectrum and provide rules and recommendations for optimised use of the radio spectrum. Currently, regulation affects the communication market's development and structure more sensitively than ever (as has been recently experienced with 3G spectrum licensing). Unlike the original intents of regulatory bodies, the frequency policy of regulatory bodies could also be used as an instrument to create an unbalanced value chain. This may ease or stall the growth in the communication sector under different circumstances.

From a system architecture standpoint, the air interface architecture, transmitter and receiver structure are those components of the network that are most affected by spectrum rules and plans. The air interface architecture should be capable of efficient use of the spectrum within the constraints defined by the regulatory bodies. Therefore, spectrum efficiency has become a substantial criterion when comparing different radio systems. Furthermore, the developing life cycle of the technology should be harmonised with the spectrum allocation process practised by the World Radiocommunication

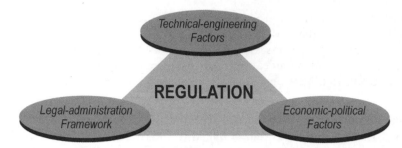

Figure 3.21   Interaction of different factors in the spectrum regulation process

Conference (WRC), especially when there is no other way to deploy the technology than allocating new frequencies.

Regulation standardisation goes hand in hand with the radio spectrum allocation and overall technical standardisation of communication systems. In both areas, the main drivers focus on both technology–engineering and economic–political aspects. The legal elite play an important role in supporting the entire regulation process.

So far, we have described the different techniques applied in radio systems to achieve an efficient radio spectrum. By having all these optimised, radio systems ultimately face fundamental physical limitations, also known as Shannon's limits, in terms of bandwidth capacity that they could provide. Under certain circumstances, one way to increase system radio channel capacity is to increase the bandwidth. However, despite the public misconception, the radio spectrum is not an unlimited natural resource but, rather, is extremely scarce. Although advanced systems, equipped with sophisticated coding and modulation methods, smart antennas and radio resource management, exploit the available radio spectrum more efficiently, nevertheless, the rapid growth in demand for wireless services diminishes the already-scarce available spectrum. Therefore, it is essential to regulate the use of the radio spectrum to avoid non-harmonised and inefficient use of such a scarce natural resource. Figure 3.22 illustrates the electromagnetic spectrum, including the radio wave spectrum from 3 kHz up to 300 GHz. Furthermore, the figure shows an interesting point; currently, the most occupied portion of the radio spectrum allocated for wireless applications ranges from 100 to over 2,000 MHz. The reason is partly because of the characteristics of the radio wave; that is, the higher the radio frequency then the higher the radio wave attenuation, the lower the coverage capability and the higher the power consumption.

Therefore, technically, the higher the radio frequency the more complex and non-cost-effective the system implementation becomes, assuming that other technical characteristics of the system remain unchanged. For cellular systems, a higher frequency application increases the number of BSs needed in the subsystem to provide the requisite coverage, increasing the upfront and operational costs of the system.

The economic–political aspects of the radio spectrum are still more complicated. The general national radio administrations are in charge of the careful regulation of the whole radio spectrum. In addition, many international organisations are involved in radio spectrum regulation. The basic goal is to make the scarce radio resource available in a manner that, on the one hand, the system suppliers and users are content with and, on the other hand, ensures the spectrum can be used efficiently.

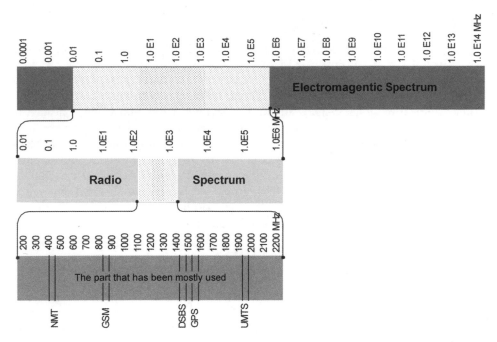

Figure 3.22 The electromagnetic spectrum, the radio spectrum portion and the mostly occupied portion

In addition to these factors, the communication market environment is extremely competition-oriented and ever-changing. Thus, spectrum regulation has also become a key competitive factor both nationally and globally and, hence, the process of deregulation is attracting greenfield operators (i.e., newcomers to the mobile communications business) to the field. Moreover, national interests are not always coherent, resulting in divergent spectrum policies.

Due to the dramatic growth in the wireless communication market, it seems that the scarcity of the radio spectrum will be more crucial in the future. Therefore, it is essential to make a trade-off between technical solutions, wireless applications and regulation to reach a fair compromise. As far as 1G was concerned, the radio spectrum was something to be given away free-of-charge (except in the US, where radio spectrum slots have always been subject to regulation fees). The free-of-charge radio spectrum principle was intended to be continued in 2G and this development was partially successful.

Spectrum regulation for the 3G system brought more competition and expectation to the field of communication systems and business. It also disproved the common misconception that the radio frequency was an endless natural resource. The number of mobile users has now exceeded the billion mark and high-bit-rate services are becoming the main offering of mobile networks. As a result, radio spectrum regulation fees have increased unprecedentedly, particularly in Europe. 3G spectrum regulation aspects are a story of their own and estimating their consequences remains beyond the scope of this book.

3.8.1 UMTS Spectrum Allocation

3G spectrum regulation has been a hot subject worldwide for debate during the last few years. It includes many sophisticated issues, which vary greatly between countries and operators. First, the requisite spectrum room for 3G is not available similarly in different countries as shown in Figure 3.23. Basically, this is because of the spectrum portion occupied by other systems already in use and of the guard bands needed for systems that operate on adjacent frequency bands. Second, the political approaches to spectrum licensing also vary from country to country. Naturally, operators involved in 2G systems want the most advanced technology by participating in 3G development work. In addition, many greenfield operators intend to reinforce their role in the market by getting involved in the 3G system. Therefore, the 3G spectrum regulation environment demands close international and national co-operation in order to harmonise spectrum allocation worldwide and manage it efficiently within countries.

From a 3G point of view, it was decided in mid-1999 by the Operator Harmonisation Group (VOHG) that there will be three CDMA variants in use. These are:

- DS-WCDMA-FDD: Direct Sequence-Wideband Code Division Multiple Access-Frequency Division Duplex.
- DS-WCDMA-TDD: Direct Sequence-Wideband Code Division Multiple Access-Time Division Duplex.
- MC-CDMA: MultiCarrier-Code Division Multiple Access.

In these terms, the first part describes the information-spreading method over the frequency spectrum, the second part indicates the multiple access scheme and the last part expresses how different transmission directions (uplink and downlink) are separated. In this context, the word "Wideband" here does not have any specific meaning.

Originally, it was inserted in the name because the Euro-Japanese version of CDMA used a wider bandwidth than the American one. The wider band makes it possible to insert some attractive features into the system, such as multimedia services with adequate bandwidth and macro diversity. Due to the OHG decision, however, all three CDMA variants use similar bandwidth.

The WCDMA-FDD uses frequencies of 2,110–2,170 MHz downlink (from the BS to the UE) and 1,920–1,980 MHz uplink (from the UE to the BS). Air interface transmission directions are separated by different frequencies and the duplex distance is 190 MHz.

The TDD variant of the WCDMA uses a frequency band located at both sides of the WCDMA-FDD uplink. The lower frequency band offered for the TDD variant is 20 MHz and the higher one is 15 MHz.

For the purposes of comparison it should be mentioned that the GSM1800 system uses frequencies of 1,805–1,880 MHz downlink (from the BTS to the MS) and 1,710–1,785 MHz uplink (from the MS to the BTS). Air interface transmission directions are separated from each other by different frequencies and the duplex distance is 95 MHz.

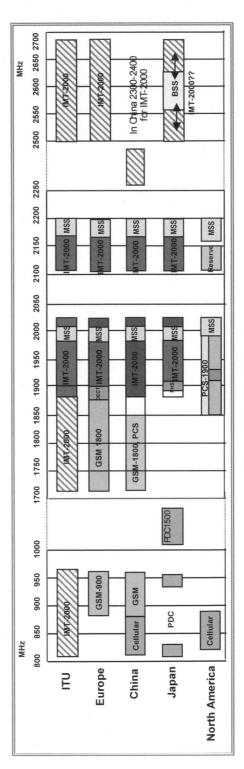

Figure 3.23 The 3G spectrum

4

Overview of UMTS Radio Access Technologies

Siamäk Naghian and Heikki Kaaranen

This chapter provides a short overview of various Universal Mobile Telecommunication System (UMTS) radio access technologies. As pointed out in the 3G Partnership Project (3GPP) specifications, UMTS is more than a single technology or access method. Instead, it is a mixture of selected technologies and the trend is to discover mechanisms whereby these selected access methods can work together so that the end-user is always able to enjoy adequate coverage and services may have the adequate spectrum and platform for their functions.

Currently, the 3GPP specifications recognise three main access technologies:

- Wideband Code Division Multiple Access (WCDMA) and its selected variants discussed in Chapter 3. Of these, the Frequency Division Duplex (FDD) variant is the most meaningful as far as UMTS is concerned. In addition, WCDMA is to be enhanced with High Speed Downlink Packet Access (HSDPA), and access using this combination could offer enhanced data rates in the downlink direction.
- Global System for Mobile Communication/Enhanced Data for GSM Evolution (GSM/EDGE) radio access, which has been the development basis for 3GPP R99.
- Complementary accesses. These are the access candidates for the UMTS system that are being evaluated or will be evaluated in the future. Of these, Wireless Local Area Network (WLAN) and the role it should play are now under study.

4.1 WCDMA Essentials

The main radio technology employed in UMTS is WCDMA whose variants FDD and Time Division Duplex (TDD) were selected by the European Telecommunications Institute (ETSI) in 1998. Although, just like traditional CDMA, the spread spectrum forms the underlying technique for WCDMA but employing a different control channel and signalling, wider bandwidth, and a set of enhanced futures for fulfilling the requirements of 3G systems, it is significantly different from its counterpart. In the following

UMTS Networks Second Edition   H. Kaaranen, A. Ahtiainen, L. Laitinen, S. Naghian and V. Niemi
© 2005 John Wiley & Sons, Ltd   ISBN: 0-470-01103-3

sections we briefly overview the WCDMA radio technology but to an extent that could help the reader to understand the basic structure and operation of the 3G air interface and perceive its requirements of the general network architecture. The basic features of WCDMA, such as soft handovers, power control, code management, traffic-handling mechanisms suitable for the air interface, etc., imply that a set of functions with specific allocations in the overall architecture are in place.

4.1.1 Basic Concepts

The underlying technique utilised in WCDMA is the Direct Sequence Spread Spectrum (DSSS) whose main principles are illustrated in Figures 4.1 and 4.4. Let us assume that the radio signal is transmitted from the Base Station (BS) to the Mobile Station (MS). At the BS, the transmitted signal with rate R is spread by combining it with a wideband-spreading signal, creating a spread signal with bandwidth W. At the mobile, the received signal is multiplied by the same spreading signal. Now, if the spreading signal, locally generated at the mobile, is synchronised with the spread code/signal, the result is the original signal plus possibly some trumped-up higher frequency components which are not part of the original signal and, hence, can easily be filtered. If there is any undesired signal at the mobile, on the other hand, the spreading signal will affect it just as it did the original signal at the BS, spreading it to the bandwidth of the spreading signal.

This basic process makes WCDMA more robust, flexible and resistant to interference, and solid against jamming and adversary interception. However, to realise its efficiency WCDMA occupies a wider bandwidth compared with basic CDMA. By benefiting from the wider bandwidth, WCDMA can use several channels in the radio interface (Uu). Taking into account that the effective bandwidth for a WCDMA air interface is 3.84 MHz and with guard bands the required bandwidth is 5 MHz, as shown in Figure 4.2.

It is also planned that WCDMA radio should be able to operate in different frequency bands as described in Table 4.1.

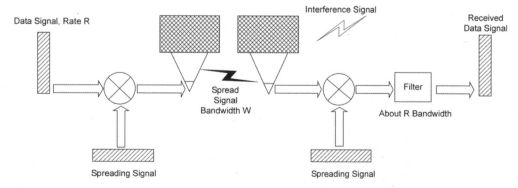

Figure 4.1   The basic technique of the Direct Sequence Spread Spectrum (DSSS)

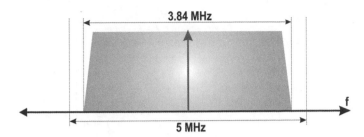

Figure 4.2 One direction of the WCDMA carrier and its dimensions

Table 4.1 3GPP-planned FDD operating bands

Operating band	UL frequencies (UE transmit, BS receive) (MHz)	DL frequencies (UE receive, BS transmit) (MHz)
WCDMA core band	1,920–1,980	2,110–2,170
1,900 MHz	1,850–1,910	1,930–1,990
1,800 MHz	1,710–1,785	1,805–1,880
1.7/2.1 GHz (USA)	1,710–1,770	2,110–2,170
UMTS850	824–849	869–894
UMTS800 (Japan)	830–840	875–885

As in the DS-CDMA scheme, the data signal in WCDMA is scrambled by utilising a user-specific pseudo noise (PN) code at the transmitter side to spread the signal over the entire band. At the receiver the received signal is extracted by using the same code sequence. Thus, following the basic principles of information theory, several simplified conclusions can be reached:

- The information to be transferred represents a certain power (say, P_{inf}).
- The wider the band for information transfer, the smaller the power representing transferred information in a dedicated (small) point within the information transfer band. In other words, the total power P_{inf} is an integral over the information transfer band in this case.
- The more information there is to be transferred, the more power is required. Thus, when power increases momentarily, P_{inf} increases too. In this context, the higher the original bit rate to be transferred, the more power required.

If we take this into account and combine it with the information presented in Figure 3.12, we are able to illustrate how WCDMA treats a single originating piece of user data, called a "bit".

In the air interface, each originating information bit is like a "box" having constant volume but the dimensions of the "box" change depending on the case. By studying Figure 4.3, we can see the depth of the "box" (frequency band) is constant in WCDMA.

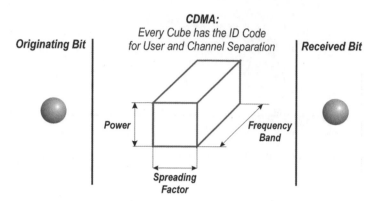

Figure 4.3   WCDMA air interface and bit treatment

The other two dimensions, power and spreading factor are subject to change. Based on this the following conclusions can be made:

- The better the signal can be spread, the smaller the required energy per bit (power). This can be applied if the originating bit rate is low. In other words, the spreading factor increases and power decreases.
- The smaller the spreading factor, the greater the energy required per bit (power). This is applied when the originating bit rate is high. In other words, the spreading factor increases and power decreases.

A confusing issue with WCDMA is that a "bit" is not a "bit" in all cases. The term "bit" refers to the information bit, which is the "bit" that occurs in the original user data flow. The "bit" occurring in the code used for spreading is called a "chip". Based on this definition, we are able to present some basic items needed in WCDMA.

The bit rate of the code used for original signal spreading is, as defined, 3.84 Mb/s. This value is constant for all WCDMA variants used in 3G networks. This is called the "System Chip Rate" (SCR) and is expressed as 3.84 Mchip/s (megachips per second). With this SCR the size of one chip in time is $1/384\,000\,0 = 0.000\,000\,260\,41$ s.

As mentioned previously, the basic idea in WCDMA is that the signal to be transferred over the radio path is formed by multiplying the original, baseband digital signal by another signal that has a much greater bit rate. Because both signals consist of bits, one must make a clear separation of the kinds of bits in question (see Figure 4.4). Hence:

- In the air interface, information is transmitted as *symbols*. Symbol flow is a result of modulation. Before modulation the user data flow consisting of bits underwent channel coding, convolutional coding and rate matching (see Figure 4.3, where the cube in the middle of the picture actually represents 1 symbol). Depending on the modulation method used 1 symbol represents different numbers of bits. In the case of DS-WCDMA-FDD, 1 symbol transmitted in an uplink direction represents 1 bit and

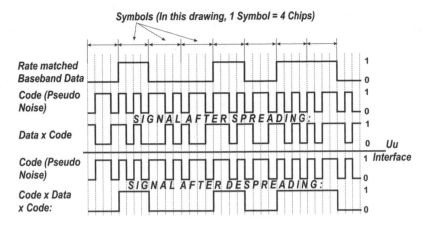

Figure 4.4   Bit, chip and symbol, and the way the signal is spread in WCDMA

1 symbol transmitted in the downlink direction represents 2 bits. This difference is due to the different modulation methods used in uplink and downlink directions.
• One bit of the code signal used for signal multiplying is called a *chip*.

But how can the desired signal be captured at the receiver end? In principle, the process is quite straightforward: every receiver uses its unique code to pick up the desired signal. The received signal is multiplied by a receiver-specific code, resulting in the data to be multiplied. If the right code and desired received signal are multiplied, the result will be post-integrated data with clear peaks on the signal; otherwise, the post-integrated data will not include clear signal peaks for further processing.

How widely the signal is spread depends on the spreading factor used in association with it. The spreading factor is a multiplier describing the number of chips used in the WCDMA radio path per 1 symbol. The spreading factor K can be expressed mathematically as follows:

$$K = 2^k, \qquad \text{where } k = 0, 1, 2, \ldots, 8$$

For instance, if $k = 6$ the spreading factor K gets a value of 64 indicating that 1 symbol uses 64 chips in the WCDMA radio path in the uplink direction (see Tables 4.2 and 4.3). Another name for spreading factor is *processing gain* (G_p), and it can be expressed as a function of the bandwidths used:

$$G_p = \frac{B_{\text{Uu}}}{B_{\text{Bearer}}} = \frac{\text{System chip rate}}{\text{Bearer bit rate}} = \text{Spreading factor}$$

In the formula, B_{Uu} stands for the bandwidth of the Uu interface and B_{Bearer} is the bandwidth of rate-matched baseband data. In other words, B_{Bearer} already contains excessive information, like channel coding and error protection information. Based on the equation above and taking into account the different bit amounts 1 symbol carries in uplink and downlink directions, we are able to calculate the bearer bit rates available in WCDMA.

Table 4.2 Spreading factor–symbol rate–bit rate relationship in the uplink direction

Spreading factor value	Symbol rate (ksymbol/s)	Channel bit rate (kb/s)
256	15	15
128	30	30
64	60	60
32	120	120
16	240	240
8	480	480
4	960	960

Table 4.3 Spreading factor–symbol rate–bit rate relationship in the downlink direction

Spreading factor value	Symbol rate (ksymbol/s)	Channel bit rate (kb/s)
512	7.5	15
256	15	30
128	30	60
64	60	120
32	120	240
16	240	480
8	480	960
4	960	1,920

These figures are indicative because the share of user data (payload) changes according to the radio channel configuration used.

The WCDMA system uses several codes. In theory, one type of code should be enough, but, in practice, radio path physical characteristics require that the WCDMA system use different codes for different purposes, and that these codes have such features as orthogonality and autocorrelation, making them suitable for their specific use. There are basically three kinds of codes available, channelisation codes, scrambling codes and spreading code(s). Their use is shown in Table 4.4.

Another confusing issue with WCDMA is that the same item could have many names. This naming confusion is also present as far as codes are concerned. For

Table 4.4 Code types of WCDMA

	Uplink direction	Downlink direction
Scrambling codes	User separation	Cell separation
Channelisation codes	Data and control channels from the same terminal	Users within one cell
Spreading code	Channelisation code × scrambling code	Channelisation code × scrambling code

example, depending on the type of code used, scrambling code is occasionally named differently: it is also known as "Gold code" (often used in radio-path-related technical publications) and "long code". Of these alternatives, the name "scrambling code" is the preferred name. Scrambling code is used in the downlink direction for cell/sector separation; they are used also in the uplink direction. In this case, the users (i.e., their mobiles) are separated from each other using this code.

Furthermore, as each user's data stream occupies the entire frequency band, the right signal has to be picked up with the minimum distortion. To separate different transmissions spread over the frequency band, spreading codes are used. A spreading code is a unique code assigned to the beginning of the transaction by the network. It can be envisaged as a "key" which is used both by the mobile and the network. Both ends of the connection use this "key" to open the noise-like transmitted wideband signal. Or, to be exact, to extract the correct wideband transmission from the frequency band, since the transmitted wideband signal may contain many mobile network connections.

From the point of view of the spreading code, the capacity of a cell depends on the downlink scrambling code amount assigned for the cell (the minimum is 1). Every downlink scrambling code then has a set of channelisation codes under it and every call/transaction requires one channelisation code to operate. In practice, one spreading code is actually the scrambling code × channelisation code. If channelisation codes are not used, the spreading code is the same as the scrambling code. Also, the spreading code depends on the information type to be delivered. The common information the cell delivers towards the terminals on its area consumes some channelisation codes from the channelisation code set.

4.1.2 WCDMA Radio Channels

WCDMA radio access allocates bandwidth for users and the allocated bandwidth and its controlling functions are handled using the term "Channel". The functionality implemented through WCDMA defines what kinds of channels are required and how they are organised. As shown in Figure 4.5 the channel organisation used by WCDMA comprises three layers: logical channels, transport channels and physical channels. Of these, logical channels describe the types of information to be transmitted, transport channels describe how the logical channels are to be transferred and physical channels are the "transmission media" providing the radio platform through which the information is actually transferred.

Looking back at the architecture issues explained in Chapter 1, the channel structures and their use differs markedly from the GSM. The term "physical channels" means different kinds of bandwidths allocated for different purposes over the Uu interface. In other words, physical channels actually form the physical existence of the Uu interface between the User Equipment (UE) domain and the access domain. In GSM, physical channels and their structure are recognised by the Base Station Controller (BSC), but, in WCDMA, they really exist in the Uu interface, and the Radio Network Controller (RNC) is not necessarily aware of their structure at all.

Instead of physical channels the RNC "sees" transport channels. Transport channels carry different information flows over the Uu interface, and the physical element

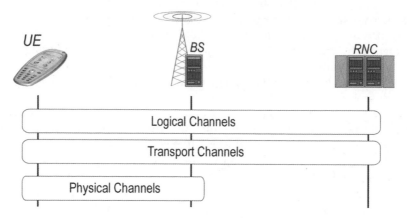

Figure 4.5 Logical, transport and physical channels in WCDMA UTRAN

mapping these information flows to the physical channels is the BS (Base Station). Logical channels are not actually channels as such, rather they can be understood as different tasks the network and the terminal need to carry out at different times. These partially timely structures are then mapped to the transport channels that perform the actual information transfer between the UE domain and access domain.

Concerning the logical channels, the UE and the network have different tasks to do. Thus, the logical, transport and physical channel structures are somewhat different in either direction. Put roughly, the network has the following tasks to perform:

- It must inform the UE about the radio environment. This information consists of, for instance, the code value(s) used in the cell and in the neighbouring cells, allowed power levels, etc. This kind of information the network is provided to the UE through the logical channel called the "Broadcast Control Channel" (BCCH).
- When there is a need to reach a certain UE for communication (e.g., a mobile-terminated call) the UE must be paged in order to discover its exact location. This network request is delivered on the logical channel called the "Paging Control Channel" (PCCH).
- The network may have certain tasks to do, which are or may be common for all the UE residing in the cell. For this purpose the network uses a logical channel called the "Common Control Channel" (CCCH). Since there could be numerous UE using the CCCH simultaneously, the UE must use a UMTS Terrestrial Access Network (UTRAN) Radio Network Temporary Identity (U-RNTI) for identification purposes. By investigating the U-RNTI received, the UTRAN is able to route the received messages to the correct serving RNC. U-RNTI is discussed in Chapter 5.
- When there is a dedicated, active connection, the network sends control information concerning this connection through the logical channel called the "Dedicated Control Channel" (DCCH).
- Dedicated traffic: the dedicated user traffic for one user service in the downlink direction is sent through the logical channel called the "Dedicated Traffic Channel" (DTCH).

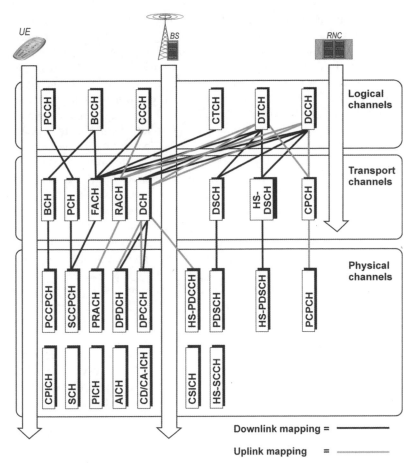

Figure 4.6 Channel types, their location in relation to network elements, and their mapping in downlink and uplink directions in UTRAN

- Common Traffic Channel (CTCH) is a unidirectional channel existing only in the downlink direction; it is used when transmitting information either to all UE or a specific group of UE in the cell.

An overview of WCDMA channelisation arrangements is provided in Figure 4.6. The transport channels shown in the figure, with one exception, are mandatory ones. The mandatory transport channels are the Broadcast Channel (BCH), Paging Channel (PCH), Forward Access Channel (FACH) and Dedicated Channel (DCH). In addition to these, the operator may configure UTRA to use the Downlink Shared Channel (DSCH) and High Speed DSCH (HS-DSCH). In these transport channels, the only dedicated transport channel is the DCH; the others are common ones. In this context the term "dedicated" means that the UTRAN has allocated the channel to be used between itself and certain terminals. The term "common" means that several terminals could use the channel simultaneously.

The BCH carries the content of the BCCH (i.e., the UTRA-specific information to be delivered in the cell). This information consists of, for example, random access codes, access slot information and information about neighbouring cells. The UE must be able to decode the BCH in order to register to the network. The BCH is transmitted with relatively high power because every terminal in the intended cell coverage area has to be able to "hear" it. The PCH carries paging information and is used when the network wishes to initiate a connection with certain UE. The FACH carries control information to the UE known to be in the cell. For example, when the RNC receives a random access message from the terminal, the response is delivered through the FACH. In addition to this, FACH may carry packet traffic in the downlink direction. One cell may contain many FACHs, but one of these is always configured in such a way (i.e., with a low bit rate) that all the terminals residing in the cell area are able to receive it. The DCH carries dedicated traffic and control information (i.e., the DCCH and DTCH). It should be noted that one DCH might carry several DTCHs depending on the situation. For example, a user may have a simultaneous voice call and video call active. The voice call uses one logical DTCH and the video call requires another logical DTCH. Both of these, however, use the same DCH. From the UTRA capacity point of view, the aim is to use common transport channels as much as possible, since dedicated channels will occupy radio resources. The optional DSCH is a target of increasing interest. It carries dedicated user information (i.e., DTCH and DCCH) for packet traffic and several users can share it. In this respect it is better than the DCH, since it saves packet-traffic-related network resources in the downlink direction. Another point is that the maximum bit rate for the DSCH and HS-DSCH can be changed faster than in the DCH. The expected wide use of services producing occasional packet bursts (e.g., web surfing) has increased the interest towards the DSCH and, specifically, the the HS-DSCH channel.

To enhance the data capability of the WCDMA system, 3GPP R5 specifies a number of new transport and physical channels: the HS-DSCH is a transport channel that can be shared by several mobile devices. It is associated with one downlink Dedicated Physical Channel (DPCH), one or several Shared Control Channels (HS-SCCH) which is also a new channel, and the uplink Dedicated Control Channel (HS-DPCCH) for transferring HS-DSCH-related feedback information. The HS-DSCH channel can be transmitted over the entire cell or over only part of the cell using, for example, beam-forming antennas. HS-SCCHs that have a fixed rate (60 kb/s, SF = 128) are used to handle the downlink signalling required for a HS-DSCH. In the physical layer a High Speed Physical Downlink Shared Channel (HS-PDSCH) with a fixed Spreading Factor (SF = 16) is used to carry the HS-DSCH. When a mobile device has the required capability of using multiple channelisation codes in the same HS-PDSCH subframe, the bit rate is considerably improved.

In the uplink direction the logical channel amount required is smaller. There are only three logical channels, CCCH, DTCH and DCCH. These abbreviations have the same meaning as those used in the downlink direction.

There are three mandatory transport channels in the uplink direction: the Random Access Channel (RACH), DCH and the Common Packet Channel (CPCH). The RACH carries control information from the UE to the UTRAN (e.g., connection set-up requests). In addition to this, the RACH may carry small amounts of packet data. The DCH is the same as in the downlink direction (i.e., a dedicated transport channel carrying DCCH and DTCH information). The CPCH is a common transport channel meant for packet data transmission. It is a kind of extension for the RACH, and the CPCH counterpart in the downlink direction is the FACH.

When information is collected from the logical channels and organised for use by the transport channels it is in ready-to-transfer format. Before transmitting, the transport channels are arranged by the physical channels. The other physical channels present are used for physical radio media control, modification and access purposes.

The physical channels are used between the terminal and the BS. Due to the network architecture solution explained at the beginning of this book, the physical access (i.e., the physical channels) arrangement is separated from the other layers. This arrangement makes it possible in theory to swap the physical radio access medium below the other layers. In practice, however, the radio access medium change will impact on higher layers while this modulant minimises these changes.

The Primary Common Control Physical Channel (P-CCPCH) carries the BCH in the downlink direction. The P-CCPCH is available in such a way that all the terminals populated within the cell coverage are able to demodulate its contents. Because of this requirement, the P-CCPCH has some characteristics that actually are limitations when compared with the other physical channels in the system. The P-CCPCH uses a fixed channelisation code and, thus, its spreading code is fixed too. This is a must because, otherwise, the terminals are not able to "see" and demodulate the P-CCPCH. The P-CCPCH bit rate is 30 kb/s with a spreading factor of 256. The bit rate has to be low because this channel is transmitted with relatively high power. If higher bit rates are used, interference starts to increase, thus limiting system capacity. Therefore, in this specific case the relation between the spreading code, the transmitted power and the bit rate can be envisaged as deviating from the basic principle of WCDMA discussed earlier, because, in principle, it would be possible to use lower power with the defined bit rate conveyed via the channel.

The Secondary Common Control Physical Channel (S-CCPCH) carries two transport channels within it: the PCH and the FACH. These transport channels may use the same or a separate S-CCPCH; thus, a cell always contains at least one S-CCPCH. The bit rate of S-CCPCH is fixed and relatively low for the same reasons as the P-CCPCH. At a later phase the S-CCPCH bit rate can be increased by changing system definitions. The configuration of S-CCPCH is variable: depending on the case, the S-CCPCH can be configured differently in order to optimise system performance. For instance, the pilot symbols can be included or not. Referring to the variable configuration alternatives of the S-CCPCH, one alternative to increase system performance is to multiplex the PCH information together with the FACH to the S-CCPCH and PCH-related paging indications to a separate physical channel called a "Paging Indicator Channel" (PICH).

The Dedicated Physical Data Channel (DPDCH) carries dedicated user traffic. The size of the DPDCH is variable and it may carry several calls/connections in it. As the name states, it is a dedicated channel, which means that it is used between the network and one user. Dedicated physical channels are always allocated as pairs for one connection: one channel for control information transfer and the other for actual traffic. The Dedicated Physical Common Channel (DPCCH) transfers control

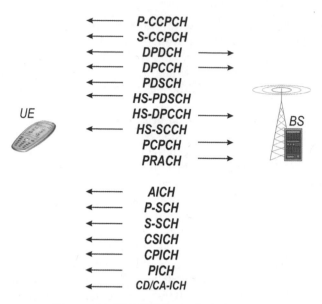

Figure 4.7   WCDMA physical channels

information during the dedicated connection. Figure 4.8 shows how the DPDCH and the DPCCH are handled in both uplink and downlink directions.

In the downlink direction the DPDCH (carrying user data) and the DPCCH (e.g., carrying power control data-rate information) are time-multiplexed. If there is nothing to carry in the DPDCH the transmitted signal is pulse-like causing EMC-type disturbances (this is not a problem in the downlink direction). In the uplink direction the DPDCH and the DPCCH are separated by I/Q modulation. If there are no user data to carry in the DPDCH, neither do pulse-like disturbances exist. The outcome of I/Q modulation in the UE is actually one channel, but carrying two information branches (see Figure 4.8) along with the two information branches consumes one code resource.

Together, the DPCCH and the DPDCH carry the contents of the transport channel DCH. When the dedicated connection uses a high-peak bit rate the system readily starts to suffer from the lack of channelisation codes in the cell. In this case, there are two options: either to add a new scrambling code to the cell or to use common channels for transmission of dedicated data. Adding scrambling codes is not recommended because orthogonality is lost. Instead, using common channel resources for packet data transmission is seen as a better way to increase capacity. The downlink DCH is able to provide information about whether the receiving UE must decode the Physical Downlink Shared Channel (PDSCH) for additional user information. The PDSCH carries the DSCH transport channel and, as was explained earlier, the DSCH is an optional feature the operator may configure for use or not.

If there is a need to send packet data in an uplink direction and the RACH packet transfer capacity is not enough, the UE may use the uplink CPCH. The corresponding physical channel in the uplink direction is the Physical (uplink) Common Packet

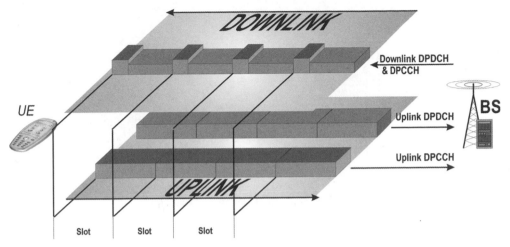

Figure 4.8   DPDCH and DPCCH in uplink and downlink directions

Channel (PCPCH). The counterpart of the CPCH in the downlink direction is the DPCCH. The Packet Random Access Channel (PRACH) carries information about the Random Access Procedure (RAP; see Figure 4.9). With this procedure, terminal accesses to the network and also small data amounts can be transferred. The RAP has the following phases:

1. The UE decodes the BCH information on the P-CCPCH and finds out which RACH slots are available as well as the scrambling code(s).
2. The UE randomly selects one RACH slot for use.
3. The UE sets the initial power level to be used (this is based on the received downlink power level) and sends the so-called "preamble" to the network.
4. The terminal decodes the AICH (Acquisition Indication Channel) to find out whether the network was notified of the sent preamble. If it was not the UE sends the preamble again, but with a higher bit power level.
5. When the AICH indicates that the network was notified of the preamble, the UE sends the RACH information on the PRACH. The length of the RACH information sent is either one or two WCDMA frames, taking either 10 or 20 ms.

The Synchronisation Channel (SCH) provides the cell search information for the UE residing in the cell coverage area. The SCH is actually a combination of two channels, the Primary Synchronisation Channel (P-SCH) and the Secondary Synchronisation Channel (S-SCH). The P-SCH uses a fixed channelisation code, whose length is 256 and this channelisation code is the same in every cell of the system. When the UE has demodulated the P-SCH it has achieved frame and slot synchronisation of the system and also knows to which scrambling code group the cell to be accessed belongs.

The Common Pilot Channel (CPICH) is an unmodulated code channel, which is scrambled using a cell-specific scrambling code. The CPICH is used for dedicated channel estimation (by the terminal) and to provide channel estimation reference(s)

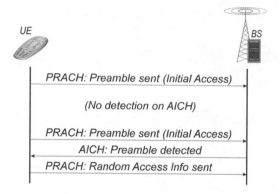

Figure 4.9 Random access basic procedure

when common channels are involved. In this respect, the pilot signal is much the same with the same functionalities as the training sequence included in the middle of a GSM burst. Normally, one cell has only one CPICH but there can be two of them. In this case, these channels are called the "primary CPICH" and the "secondary CPICH". The cell may contain the secondary CPICH, for instance, when the cell contains narrow-beam antenna aiming to offer services to a dedicated "hot spot" area. Thus, a dedicated area uses the secondary CPICH and the primary CPICH offers a pilot for the whole cell coverage area. The terminals listen to the pilot signal continuously and this is why it is used for some of the "vital" purposes in the system (e.g., handover measurements and cell load balancing): from the system point of view, CPICH power-level adjustment balances the load between cells. The UE always searches the most attractive cells and by decreasing the CPICH power level the cell is less attractive.

The other physical channels listed in Figure 4.9 are the CPCH Status Indication Channel (CSICH), CPCH Collision Detection Indicator Channel (CD-ICH) and CPCH Channel Assignment Indicator Channel (CA-ICH). The CSICH uses the free space that occurs in the AICH, and the CSICH is used to inform the UE about CPCH existence and configuration. To avoid collisions (i.e., two lots of UE using same identity patterns), a CD-ICH and a CA-ICH are used. These physical channels transfer the collision detection information towards the UE.

4.1.3 WCDMA Frame Structure

In order for the radio access to handle such control actions as timing, synchronisation arrangements, transmission assurance, etc. between the network and mobile station, burst data should be structured in a well-defined way. Therefore, WCDMA contains a frame structure that is divided into 15 slots, each of length $\frac{2}{3}$ m/s and, thus, the frame length is 10 ms (see Figure 4.10).

Based on this, one WCDMA frame is able to handle:

$$\frac{0.010\ \text{s}}{0.000\ 000\ 260\ 41\ \text{s}} = 38{,}400\ \text{chips}$$

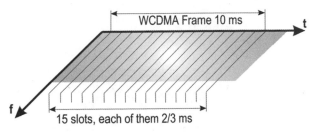

Figure 4.10   WCDMA frame

One slot in the WCDMA frame contains:

$$\frac{38,400 \text{ chips}}{15 \text{ slots}} = 2,560 \text{ chips}$$

Unlike in GSM, WCDMA does not contain any super-, hyper- or multiframe structures. Instead, WCDMA frames are numbered by a System Frame Number (SFN)). An

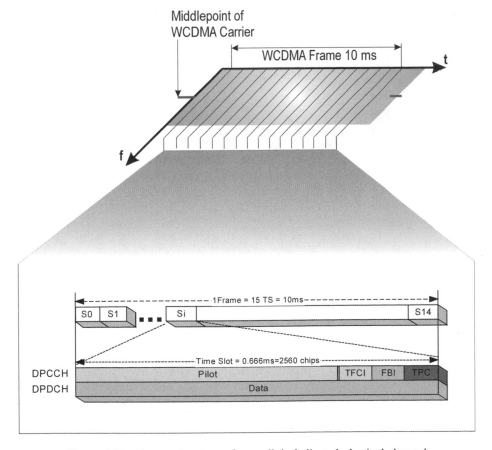

Figure 4.11   Frame structure of an uplink dedicated physical channel

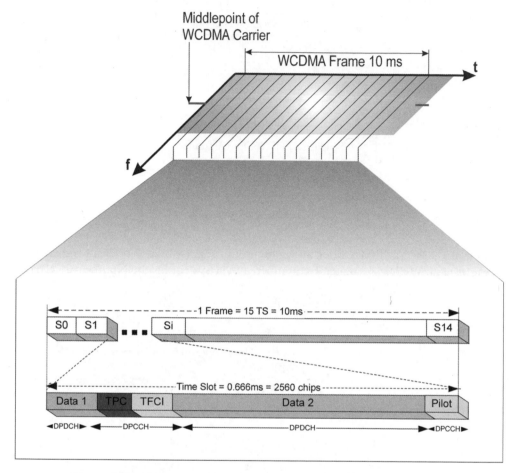

Figure 4.12   Frame structure of a downlink dedicated physical channel

SFN is used for internal synchronisation of the UTRAN and timing of BCCH information transmission.

Figures 4.11 and 4.12 illustrate the uplink and downlink dedicated channel frame
structures, respectively. As shown in the figures, the dedicated physical channels have
different structure for uplink and downlink.

In the uplink case, the basic frame structure of the dedicated physical channel follows
the downlink frame structure, but the main difference is that uplink dedicated channels
cannot be considered as a time multiplex of DPDCH and DPCCH. This leads to the
realisation that multicode operation is possible for uplink dedicated physical channels.
It should be mentioned, however, that when multicode is used several parallel DPDCHs
are transmitted using different channelisation codes with only one DPCCH per
connection.

The uplink DPCCH consists of pilot bits to support channel estimation for coherent
detection: Transmit Power Control (TPC) issues commands for the adjustment of

transmit power, Feedback Information (FBI) and the optional Transport Format Combination Indicator (TFCI) informs the receiver about the instantaneous parameters of the different transport channels multiplexed on the uplink DPDCH and corresponds to the data transmitted in the same frame.

In the case of downlink, every slot includes pilot bits, transmitted power control bits, a transport frame indicator and data.

It needs to be mentioned that within one downlink DPCH, the dedicated data generated at layer 2 and above (i.e., the DCH) is transmitted in a time multiplex with control information generated by layer 1 (known pilot bits, TPC commands and an optional TFCI). Therefore, the downlink DPCH can be seen as a timed multiplex of downlink DPDCH and downlink DPCCH. In addition to the harmonisation compromise in standardisation, the reasons behind this approach are:

- To minimise continuous transmission by mobile handsets.
- To use downlink orthogonal codes more efficiently.
- To minimise the power control delay by utilising the slot offset between uplink and downlink slots.

For the common channels the structure is the same and the main difference between the common and dedicated channels is that in common channels the TPC bits are not used.

Finally, Table 4.5 summarises the main characteristics of the WCDMA-FDD scheme described previously. These technical parameters enable the WCDMA-FDD scheme to meet the requirements of the 3G mobile system.

4.2 WCDMA Enhancement—HSDPA

4.2.1 Introduction

In the early phase of UMTS development it was envisaged that data traffic would follow the trend experienced from fixed networks in which the share of IP traffic was becoming dominant. 3GPP had already initiated the concept of "All IP" that would lead to the emergence of IP traffic handling in the UMTS Core Network (CN), realised by introducing new building blocks, such as IP Multimedia Subsystems (IMSs). In order to promote the overall data capability of the network, it was obvious that the development effort should also have focused on UTRAN evolution and, in particular, its air interface. On the other hand, the data enhancements of UTRAN and its air interface had been initiated in Release 4 too: The DSCH had already been developed and it paved the way for further enhancements toward a higher bit-rate network. The introduction of DSCH had proven that the air interface had the potential of new enhancements. Therefore, advancing the development that was already underway made good sense for both business and technology reasons. The time was ripe for emerging HSDPA.

Consequently, 3GPP Release 5 specified a new HSDPA to serve high-peak data rate users. In order to achieve higher data throughput, reduce delay and high-peak rates, HSDPA employs such techniques as Adaptive Modulation and Coding (AMC), and Hybrid Automatic Repeat Request (HARQ) combined with a fast-scheduling and cell

Table 4.5 WCDMA-FDD: main technical characteristics

Parameter	Specification
Multiple access	FDD: DS-CDMA
Duplex scheme	FDD
Chip rate (Mchip/s)	3.84
Frame length (ms)	10
Channel coding	Convolutional coding ($R = \frac{1}{2}, \frac{1}{3}, \frac{1}{4}, K = 9$); turbo code of $R = \frac{1}{2}, \frac{1}{3}, \frac{1}{4}$ and $K = 4$
Interleaving	Inter/intraframe
Data modulation	FDD—DL: QPSK, 16QAM with HARQ support (for HSDPA); UL: dual-channel QPSK
Spreading modulation	FDD—UL: BPSK; DL: QPSK
Power control	Closed loop (inner loop, and outer loop), open loop. Step size: 1–3 dB (UL); power cycle: 1500/s
Diversity	RAKE in both BS and MS; antenna diversity; transmit diversity
Inter-BS synchronisation	FDD—no accurate synchronisation needed
Detection	MS and BS—pilot-symbol-based coherent detection in UL; CPICH channel estimation in DL
Multiuser detection	Supported (not at the first phases)
Service multiplexing	Variable mixed services per connection is supported
Multirate concept	Is supported by utilising a variable spreading factor and multicode
Handover	Intra-frequency soft and softer handovers are supported, inter-system and inter-frequency handovers are supported

change procedure. Although these techniques could be considered the cornerstones of HSDPA, the overall enhancements in the UTRAN and, particularly, in the air interface far exceed them without sacrificing backward capability with Release 4.

4.2.2 The Benefits and Impacts

The primary benefits of HSDPA, as perceived directly by the end-user, are the approximate five orders of magnitude higher data throughput up to the over 10-Mb/s peak rate with 15 multicodes. These benefits depend on the modulation used for the resource configuration shown in Figure 4.13. Because other factors, such as cell coverage, UE

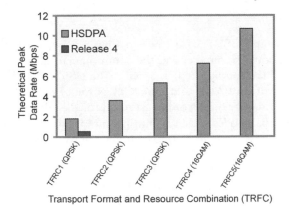

Figure 4.13 Comparison of the achievable data rate in Releases 4 and 5

mobility, UE distance from the BS and the number of simultaneous users, will also affect the achievable peak rate, the practical maximum peak data rate might lag far behind the target, but the improvement is still huge when compared with UTRAN Release 4 data capability. Also, significant lower end-to-end latency and improved cell capacity (up to an order of magnitude when compared with the DSCH of the microcell in Release 4) are other highlights of HSDPA enhancements. These all contribute directly to the spectrum efficiency of the system and radically improve the system capability to fit well with data-centric service offerings. In this regard, the main focus is on the streaming, interactive and background services rather than real-time services.

As a trade-off, HSDPA also has its downside: though kept backward-compatible, upgrades and enhancements to the Release 4 air interface and architecture are required. The implications posed from HSDPA for the UMTS network architecture and, especially, its air interface can be summarized as follows:

- Network architecture: HSDPA requires a significant part of the packet-handling function to be transferred to the edge of network, leading to a more distributed architecture than Release 4.
- Physical layer: new adaptive modulation and coding methods pose significant modifications to the physical layer architecture in terms of channel structure, multiplexing, timing and procedures required for HSDPA operation.
- Fast scheduling means more efficient operation of Medium Access Control (MAC) and its closer interaction with the physical layer. Short-framing may also require more processing capacity from the BS and, to some extent, from the UE.
- Fast retransmission can be realised by employing more control signalling and an advanced retransmission mechanism.

Having these examples as the general implications of HSDPA, let us gain more specific view of this attractive feature of UTRAN by going briefly through its basic concept, characteristics, and principles.

4.2.3 Basic Concept

The primary idea behind HSDPA is shown in Figure 4.14. Originally, in Release 4 of UMTS, data traffic could be handled via the common channels (CCHs), the DCH and the DSCH. In particular, for data-centric services the DSCH is used while low-rate data can also be handled via the FACH and the DCH as well. For the DCH, the channel bit rate upper rate may vary depending on the SF used for the allocated codes, accordingly. What HSDPA principally does is that it simply employs a time-multiplexing approach to transfer data packets on a single shared channel while it uses a multicode with a fixed SF. This seemingly simple operation, however, implies a certain functionality and a set of procedures to make it practical over the air interface: multiplexed data should be efficiently scheduled, modulated, encoded and conveyed over the air interface and the radio link should be adapted, accordingly. As a consequence, enhancements on top of Release 4 UTRAN features became inevitable.

Figure 4.15 illustrates the basic functionalities and functional entities specified in Release 5 needed to bring about the basic HSDPA operation mentioned above. The main functional entities of HSDPA include AMC, Fast Packet Scheduling (FPS), HARQ and a cell change procedure described in the following subsections.

4.2.4 Adaptive Modulation and Coding

The main objective of adaptive modulation and coding is to compensate for radio channel instability by fine-tuning the transmission parameters. There are different ways and instruments, like power control, adaptive antenna, dynamic code and channel allocation, etc., to accomplish radio link adaptation. Although all the techniques pursue ultimately the same goal, they realise it differently and, therefore, they may be employed as complementary methods when beneficial. However, as far as the HSDPA is concerned, applying fast power control was not considered reasonable due

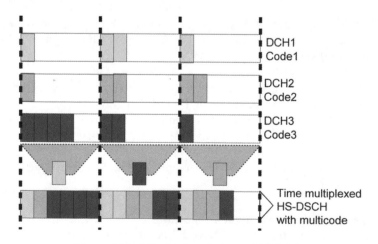

Figure 4.14 The basic principle of HSDPA

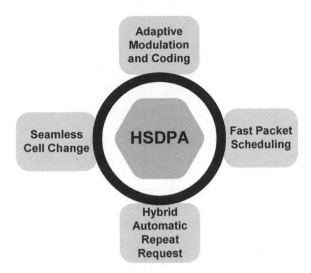

Figure 4.15 The main functional entities of HSDPA

to the unremarkable gain and high complexity when used with AMC and, therefore, was excluded.

As far as the AMC is concerned, its primary characteristic is that it fine-tunes the modulation and coding parameters of the physical layer to compensate for channel variations. This is basically done by making use of the radio channel measurements extracted by the mobile terminal and, for HSDPA in particular, using the Channel Quality Indication (CQI) and retransmission procedure. Armed with these and traffic-related information, such as Quality of Service (QoS) and the state of radio and physical resources, AMC enables the network to select the most suitable modulation and coding methods.

With respect to modulation, Release 5 allows the HSDPA, and more specifically the HS-DSCH, to use either 16-Quadrature Amplitude Modulation (16QAM) or Quadrature Phase Shift Keying (QPSK) modulation. QPSK had already been specified in Release 4, while 16QAM was specifically defined in Release 5 for HSDPA operation. Higher order modulation methods, like 16QAM, provide higher spectral efficiency in terms of data throughput compared with QPSK and, therefore, can be used to improve high-peak data rates. It also allows modulation selection to be combined with the channel-encoding process, occasionally referred to as the "Transport Format and Resource Combination" (TFRC) within the scope of the UMTS specification. As a result, based on the channel measurements, the best combination of multicode, channel rate and modulation can be selected, leading to maximum throughput for a certain channel condition.

Although the benefits of AMC are well known, its performance is vulnerable to the radio channel measurements extracted by the terminal: the measurement cycle may not follow the usual channel variation due to fast-channel fading; in addition, they are not error-free. Unreliable channel-status reports may result in wrong decisions being made in packet scheduling, transmit power setting as well as modulation and coding selecting.

Therefore, HSDPA has been equipped with advanced CQI estimation procedures that use CPICH-received pilot power information, channel timing, adaptive reporting cycle and higher layer interacting to ensure errorless AMC operation. In addition, HARQ helps compensate for the vulnerability of AMC by also bringing link-layer information into the process.

4.2.5 Hybrid Automatic Repeat Request

Due to radio channel instability, explicit radio measurements may not in isolation form a reliable basis for AMC operation and, therefore, complementary mechanisms are needed. HARQ allows the receiving Network Element (NE) to detect errors and when necessary to request retransmissions. Being one of the basic mechanisms used in data transmission, the retransmission technique ensures errorless reception of data packets. When compared with the conventional ARQ, the added value brought by HARQ lies in its ability to combine the initial estimates or explicit information from both the original transmission and the corresponding retransmissions with the link adaptation process. In this way, it helps reduce the number of required retransmissions and also promotes errorless link adaptation regardless of radio channel variations.

Depending on what strategy and protocols are used in the HARQ retransmission process, HARQ can be classified in a number of different variants: Rate Compatible Punctured Turbo Codes, Incremental Redundancy and Chase Combining are just a few examples. While some of them use additional redundant information transmitted in an incremental manner if the decoding fails at an early phase of the process, the others treat each retransmission independently.

As retransmission delay and overhead signalling are the most critical criteria, especially for mobile network applications, one of the most straightforward types of retransmission procedure, called "Stop-and-Wait" (SAW), was selected for HSDPA. In SAW, the transmitter operates on the current block until successful reception of the block by the UE has been assured. It uses an optimised acknowledgement mechanism and message to confirm successful transmission of a data packet while avoiding retransmission. To avoid the additional delay posed by waiting time it employs N channel HARQ accompanied by SAW to make the retransmission process parallel and, thus, save the wasted time and resource.

Therefore, while the HARQ protocol is based on an asynchronous downlink and a synchronous uplink scheme, the combined scheme used in HSDPA relies on the Incremental Redundancy method. When applying Chase Combining, as a specific variant of HARQ, the needed UE soft memory is partitioned across the HARQ processes in a semi-static fashion through an upper layer (e.g., RRC signalling). This is accomplished in association with Transport Format determination and selection.

4.2.6 Fast Scheduling

The efficient operation of HSDPA with regard to AMC and HARQ implies that the packet-scheduling cycle is fast enough to track short-term variations in a UE fading signal. This is very important in the absence of fast-power control and Variable

Spreading Factor (VSF) mechanisms, as they have been replaced by AMC, HARQ and fast-retransmission procedures. This is also the main reason for having the Packet Scheduler (PS) in the Base Transceiver Station (BTS) rather than in the RNC, as was the case for Release 4. In this way the delay in the scheduling process is minimised and the radio measurements also better reflect the radio channel condition, leading to more reliable and fairer scheduling decisions. This, coupled with the fixed (16) code allocation strategy and the reduction in the Transmission Time Interval (TTI) from 10 or 20 ms in Release 4 to a fixed slot of 2 ms in HSDPA, enables the PS to undertake fast scheduling and frame formation. The implemented PS still remains vendor-dependent, as has been the case for the Radio Resource Management (RRM) algorithms used in both 2G and 3G mobile networks.

4.2.7 Seamless Cell Change

Seamless cell change enables the UE to connect to the best cell available to serve it on the downlink, leading to seamless connectivity in HS-PDSCH mode. This will also reduce undesirable interference, particularly in the case of soft handover. Cell change that is actually part of mobility procedures for HS-PDSCH also ensures UE mobility in association with high-speed data connection. To achieve this, the role of serving HS-DSCH cell (i.e., the cell associated with the BS that performs transmission and reception of the serving HS-DSCH radio link for a given UE) is transferred from one of the radio links belonging to the source HS-DSCH cell to a radio link belonging to the target HS-DSCH cell. This needs specific treatment because HS-PDSCH allocation to a given UE only belongs to the serving HS-DSCH radio link assigned to the UE. Like ordinary handovers in UTRAN, the serving cell change may principally be decided by either the UE or the network. However, Release 5 only supports the network-controlled option, which is handled by Radio Resource Control (RRC) signalling (Figure 4.16).

The serving HS-DSCH may be performed based on different criteria and considerations, including physical channel configuration, UE–UTRAN synchronisation and serving BS placement in the network hierarchy. In this context, it is worth emphasising that serving HS-DSCH BS relocation and serving HS-DSCH cell change are two separate procedures, even though serving HS-DSCH BS relocation pre-requests a serving HS-DSCH cell change, but not other way around. Another essential difference is that in association with serving HS-DSCH BS relocation, the HARQ entities located in the source HS-DSCH BS are ceased and new HARQ entities in the target HS-DSCH BS are created. Nevertheless, both procedures pursue UE seamless connectivity in high-speed data mode by providing fast relocation of the serving radio link.

4.2.8 Basic Operation and Architectural Considerations

The advent of HSDPA brought about significant modifications to the Hard Ware (HW) and Soft Ware (SW) modules of both the BS and the UE, though the main functionality of the RNC remains mainly similar to Release 4. Figure 4.17 illustrates the basic procedure and functional allocation of HSDPA entities and functions. As shown, Fast Scheduling is the most remarkable function that has been removed from the

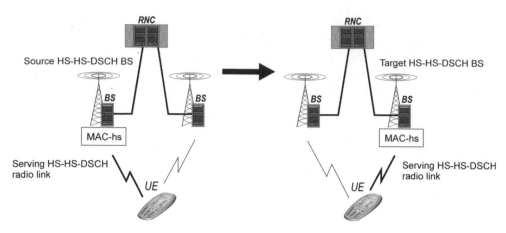

Figure 4.16 Inter-Node B serving HS-DSCH cell change combined with serving HS-DSCH Node B relocation

RNC and given to the BS. In addition, the BS is responsible for AMC handling, like HARQ and link adaptation, which are new functionalities. From the standpoint of the UE, though the main functionality of AMC is handled by the BS, the UE should be capable of providing CQI and AMC signalling, such as ACK/NACK message handling and the HARQ mode of operation. Therefore, in order for the BS and the UE to operate with HSDPA they both need to be upgraded with the new features. From the RNC's standpoint the MAC-d will be retained and the MAC-hs is the only new entity that should be implemented in the MAC layer. In addition, some modification to RRC messages may be needed.

There are also three new channel types involved in HSDPA operation: the HS-DSCH transport channel for downlink is shared by several users, the HS-SCCH logical channel for uplink is for handling control information related to decoding and retransmission (TFRI, HARQ) and HS-DPCCH is an uplink physical channel associated with HS-DSCH for carrying control information related to retransmission (ACK/NACK) and CQI.

One or several HS-PDSCHs together with an associated DPCH are tied to a set of separate HS-SCCH channels, also called an "HS-SCCH" set. The timing between these channels has been specified so that the time offset between the start of HS-SCCH information and the start of the corresponding HS-PDSCH subframe is kept fixed. The channel configuration process is handled by using RRC signalling. In addition, the number of channels in a HS-SCCH set for the UE can vary from 1 to 4.

In a basic HSDPA operation, as shown in Figure 4.17, once the RRC connection is in place the UE provides the serving BS with channel-quality-related and controlling information including UE capability and requested capacity. Based on this information coupled with scheduling-related information, such as determined TTI, radio and physical resources, etc., the BS may choose the HS-DSCH set and parameters, modulation, etc. and start HS-SCCH transmission two timeslots prior to HS-DSCH transmission. Upon reception, the UE decodes the HS-SCCH information. Based on the information (e.g., extracted from TFRI) it will obtain the necessary parameters, such

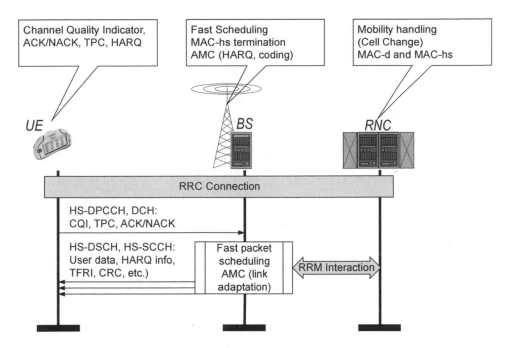

Figure 4.17 The basic procedure and functional allocation of HSDPA

as the dynamic part of the HS-DSCH transport format, including the transport block set size and modulation scheme as well as the channel-mapping scheme in the corresponding HS-DSCH TTI. Once the UE has decoded all the necessary parameters, it can start to engage in the data processing and HARQ process and return ACK/NACK to the BS, accordingly. After completing the process, the timing between HS-SCCH, HS-DSCH and ACK/NACK will play an essential role during the data connection lifetime and, therefore, it should be strictly followed by the UE.

4.3 GSM/EDGE

This section introduces, albeit briefly, GSM radio access technology and its enhancements, General Packet Radio Service (GPRS) and Enhanced Data for GSM Evolution (EDGE).

4.3.1 Basic concepts

GSM radio technology was originally developed for pan-European mobile phone systems during the 1980s. It was a remarkable technology step in many senses. First, it was the first technology that was properly specified before implementation. The earlier technologies were initially designed in laboratories and then a "specification" describing the implementation was written. Second, compared with analogue radio technologies GSM radio was designed in such a way to provide more capacity.

Since there are many books on the market describing GSM radio in detail, we are not going to give any exhaustive description here, but some essential points will be highlighted.

Originally, GSM was specified to be used over the 900-MHz frequency band, but due to commercial success and the need to extend it to other bands, the same technology was quite soon adopted and modified for other frequencies. The first version of GSM with its originally planned frequency band is today called "GSM900". Another band taken into GSM use was around 1,800 MHz (just below the WCDMA-FDD band). This variant was named "GSM1800". In the USA, GSM-based networks use a higher frequency band called "GSM1900". For special purposes (e.g., railway companies), some other bands are considered.

The selected multi-access method is Time Division Multiple Access (TDMA), where users utilise a common frequency resource and are separated timewise from each other. Since GSM traffic is full-duplex, the transmission directions (uplink and downlink) must be separated. This separation is brought about by using separate frequency bands for each transmission direction. The frequency distance (so-called "duplex distance") between the uplink and downlink frequency is 45 MHz in GSM900 and 90 MHz in GSM1800.

The select modulation method for standard GSM is Gaussian Minimum Shift Keying (GMSK). This modulation method allows a 270,833-kb/s bitstream to be carried by the carrier that transports the information. The GSM system was originally designed for circuit-switched (CS) traffic (especially speech); after much consideration and a codec selection process it was decided that one frequency should be able to carry eight channels altogether (i.e., eight separate timeslots, which together form the basic timing structure called a "TDMA frame"). Since the modulator produces an approximate 271-kb/s bitstream, one channel can carry $271 \div 8 * 34$ kb/s bitstream.

User traffic and signalling are carried over the radio interface as bursts. Bursts are smaller in size and time than the timeslot. Certain types of bursts carry only certain types of traffic and also traffic types are separated (i.e., only certain timeslots may contain signalling and others user traffic).

The most basic term in GSM technology is "TRX" (Transmitter–Receiver). Originally it meant the physical transmitter–receiver equipment in the BS, but the same term was adopted for wider use. TRX is very often used as a high-level term describing the GSM radio interface structure as a whole. In this sense, TRX means a combination of two frequencies (uplink and downlink frequency) carrying certain channel definitions (Figure 4.18).

As TRX is the most basic term we can use it to set up a GSM-related terminology structure.

The smallest publicly visible part of a radio network is a cell. Cell coverage is formed by an antenna structure, but the traffic within a cell is maintained by TRX. The minimum number of TRXs in a cell is one and the maximum in real-life implementations is four to six TRXs per cell (Figure 4.19).

The GSM BS is a physical network element, which may consist of a number of cells.

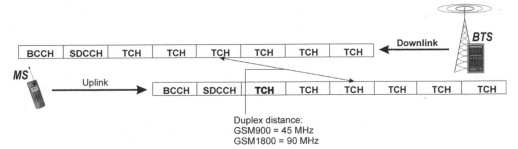

Figure 4.18 GSM TRX example configuration

The minimum is one cell and the maximum in real-life implementations is three cells per BS.

In Chapter 1 we presented a resource management model of a network indicating the various management tasks and control duties involved. Extending this same model to GSM and the terms explained above, the term "TRX" relates to radio resources and the term "cell" relates to Mobility Management (MM) and network addressing. MM and related issues are covered in Chapter 6.

4.3.2 Radio Channels and Frame Structures

Since the multi-access method in GSM is TDMA, network-level synchronisation and various timing structures are extremely vital from the system functionality point of view. The basic timing structure unit is a TDMA frame maintained by TRX.

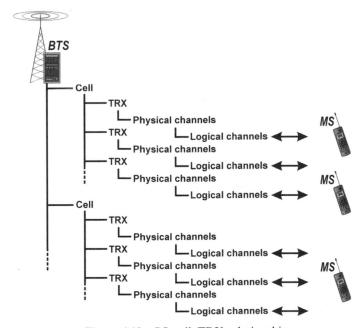

Figure 4.19 BS–cell–TRX relationship

GSM radio has to perform many tasks, and these are tied to further timing structures and channelisation. The channelisation system in GSM is very straightforward and consists of two types of channels to begin with. The radio channels implementing traffic in the radio interface are called *physical channels*.

Physical channels exist from the radio interface up to the BSC and their allocation to traffic or signalling use is controlled by the BSC. There are three types of physical channels:

- BCCH (Broadcast Control Channel): BCCH always has a constant location, which is TDMA frame timeslot 0 (slots are numbered from 0 to 7). This physical channel carries the information that Mobile Stations (MSs) listen to in order to synchronise themselves and access the network.
- SDCCH (Stand-alone Dedicated Control Channel): when a transaction is to be set up, the MS moves to the SDCCH and carries out transaction-related signalling activities. SDCCH could be physically located to the same timeslot as BCCH. If so, the combination is called a "combined BCCH/SDCCH". In larger cells having more than one TRX, the BCCH and SDCCH are normally separated from each other. In this case the SDCCH is in the BCCH TRX timeslot 1.
- TCH (Traffic Channel): when signalling is complete and the actual transaction with user traffic starts, the MS is transited or forced to a TCH for user traffic. Note that all transactions do not require the TCH. For example, sending an SMS only requires SDCCH resources, the same applies to MM-related registration procedures.

Both the MS and the network have logical issues that need to be carried out, and mostly (but not always) these are tied to timing structures. Such logical issues are called *logical channels*.

Logical channels use physical channel resources to undertake their tasks, and the structure used to pack logical channels onto physical channels is predefined with some options. In standard GSM, the following logical channels exist:

- SCH (Synchronisation Channel): the MS uses this to get into synchronisation with the network. Place: physical BCCH.
- FCCH (Frequency Correction Channel): this channel updates MSs about frequency information. Place: physical BCCH.
- BCCH (logical Broadcast Control Channel): this logical channel contains system information. The MS first opens the SCH and the FCH so that it can read the BCCH information and, thus, camp on the cell. Place: physical BCCH.
- RACH (Random Access Channel): the MS uses this channel to perform an initial access procedure or service (e.g., location update, voice call, etc.). Place: physical BCCH.
- AGCH (Access Grant Channel): the network uses this logical channel for commanding the mobile station to a logical SDCCH. Place: physical BCCH.
- PCH (Paging Channel): the network uses this logical channel to send paging information to an MS in the case of a mobile-terminated transaction. Place: physical BCCH.
- SDCCH (logical Stand-alone Dedicated Control Channel): one physical SDCCH may contain either four or eight logical SDCCH blocks depending on the selected channel configuration. Transaction signalling always requires one logical SDCCH

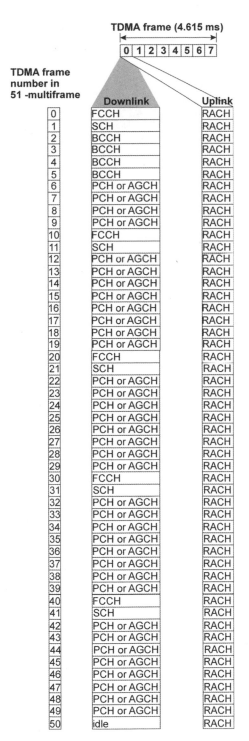

Figure 4.20 Logical channel mapping onto physical channels in the case of stand-alone BCCH

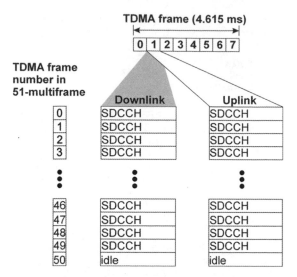

Figure 4.21   Logical channel mapping onto physical channels in the case of stand-alone SDCCH

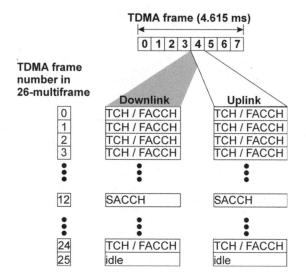

Figure 4.22   Logical channel mapping onto physical channels in the case of a full-rate CS traffic channel

block. Thus, one physical SDCCH may contain either four or up to eight simultaneous transactions during the signalling phase. Place: physical SDCCH.

- TCH (logical Traffic Channel): depending on the traffic type the physical TCH may contain several types of traffic channels. These are TCH/F (full-rate), TCH/H (half-rate) or TCH/D (dual-rate). The decision to use logical TCH is a result of negotiation between the MS and the network.
- SACCH (Slow Associated Control Channel): when a transaction is either on the

SDCCH or on the TCH, the radio connection is continuously measured. The MS sends measurement reports through this logical channel to the network. The SACCH is tied timewise to the TCH frame structure.

- FACCH (Fast Associated Control Channel): during a transaction there may be a need to exchange control information faster than the SACCH makes possible. In these cases the system contains an FACCH, which in fact occasionally steals time from the TCH. An FACCH is mostly used in the context of handovers. In these cases user traffic is carried over TCH to the old cell, and at the same time the MS gains access to the new cell through the FACCH.

As stated earlier, the basic timing structure unit is one TDMA frame. For logical and physical channel organisation, GSM radio has two parallel frame structures. The so-called "51-frame" consists of 51 TDMA frames and this frame structure maintains the control- and signalling-related channels (i.e., physical BCCH and SDCCH follow this structure). Within the 51-frame the logical channels using these physical resources are repeated in a certain, predefined order. For traffic channels the system provides a 26-frame structure. The 26-frame defines how logical channels using physical TCH resources behave (i.e., where the SACCH is located in this frame structure).

The next step in the frame structure is to combine these two frame models into one. The result is called the "superframe" which consists of 26 "51-frame" structures or 51 "26-frame" structures depending on the case and selected channel.

The highest level framing structure is formed by taking 2,048 superframes; this is called the "hyperframe". The hyperframe is also called the "Ciphering Sequence". This is because of the radio interface ciphering method GSM uses. Every TDMA frame is numbered starting from 0. The highest possible TDMA frame number could thus be $26 * 51 * 2,048 - 1 = 2,715,647$. When a transaction is changed to ciphered mode, the network and the MS agree on the TDMA frame number where ciphering will start.

4.3.3 General Packet Radio Service (GPRS)

GPRS as a technology is a parallel enhancement that is built alongside standard GSM and utilises standard GSM resources. From the Um interface point of view this means that the interface structure (timeslots, timing values and frame structures) remain unchanged, but the contents of timeslots are constructed differently.

Packet-switched traffic requires a different approach from the CS approach. Because of this, the whole protocol stack carrying GPRS traffic is completely different. GPRS introduces the Medium Access Control (MAC) and RLC (Radio Link Control) layers and some other vital layers for packet-switched traffic.

GPRS traffic uses its own channels (note the list is not exhaustive):

- PRACH (Packet Random Access Channel): the terminal sends a request to start packet-switched traffic through this channel.
- PACCH (Packet Associated Control Channel): works in a similar way to the SACCH in CS traffic.

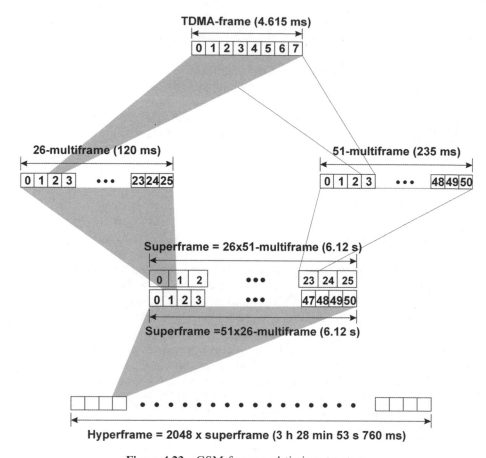

Figure 4.23 GSM frame and timing structure

- PAGCH (Packet Access Grant Channel): the same as AGCH in CS traffic; the network grants access to a packet-switched terminal.
- PPCH (Packet Paging Channel): through this channel the packet-switched terminal receives paging commands.
- PDTCH (Packet Data Traffic Channel): this channel is for actual packet traffic delivery, both uplink and downlink.

As a default, GPRS traffic is deemed second-priority traffic in the GSM radio interface; it is delivered when there is free space for it. In order to add a simultaneous number of GPRS transactions (and also taking into account the nature of packet-switched traffic) the GPRS contains mechanisms about how to set up many users over the same physical timeslot (i.e., sharing the same physical air interface resource).

There are times when we need to speed up GPRS traffic, in this case more timeslots are required for GPRS use. There are also services that cannot tolerate a break in the connection due to "Are you alive?" enquiries (e.g., banking services). To improve the

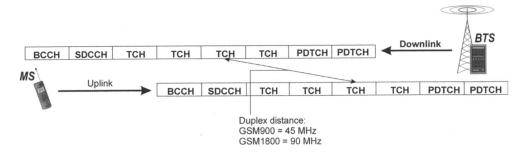

Figure 4.24 GSM/GPRS TRX with two dedicated PDTCHs

Table 4.6 GPRS coding schemes

Coding Scheme (CS)	Data rate (kb/s)
CS1	8.8
CS2	11.2
CS3	14.8
CS4	17.6

performance of these kinds of services, it is possible to dedicate air interface timeslots for GPRS use only. In this case these resources are not used by CS services at all.

Timewise and physically a PDTCH is the same size as a normal, full-rate TCH (Figure 4.24). GPRS, however, introduces various channel Coding Schemes (CSs) to be applied in the PDTCH. There are four CSs named "CS1–CS4" and using these the channel data rate varies as shown in Table 4.6.

Although a terminal supporting GPRS also supports CS1–CS4, the network must only support CS1, the other schemes being optional on the network side. In practice, CS1 and CS2 are in use both in the terminals and in the network. If an operator wishes to apply CS3 and CS4, it will require frequency resources and careful planning. The reason for this is that CS3 and CS4 work perfectly well in good conditions close to antennae. The farther away the terminal from the BS the lower the CS, since interference and other radio interface characteristics start to have an effect on the quality of the radio path.

4.3.4 Enhanced Data Rates for Global/GSM Evolution (EDGE)

As can be seen from the previous subsection, data rates are an issue with GPRS. EDGE contains answers to questions about how to boost 2G GPRS average data rates. Technically, there are three entities implementing EDGE:

- New modulation Octagonal Phase Shift Keying (8-PSK): this improves GMSK capacity markedly.

- New Modulation Coding Schemes (MCSs): new channel coding methods enable more bits in physical timeslots of the same size.
- Radio and transport network planning.

As in GPRS, the radio interface physical structure remains the same in sense of time and frames. The changes EDGE introduces occur within the timeslots. New MCSs enable data transfer rates as shown in Table 4.7.

MCS1–MCS4 are the same as CS1–CS4 in GPRS. The EDGE add-in is in CS5–CS9. The higher the data rate the less the system observes the bitstream (i.e., the error correction is minimized in the highest MCSs).

As result, the highest MCSs can be used in very good radio conditions, typically very

Table 4.7   EDGE Modulation Coding Schemes and data rates

MCS	Modulation	Data rate (kb/s)
MCS1	GMSK	8.8
MCS2	GMSK	11.2
MCS3	GMSK	14.8
MCS4	GMSK	17.6
MCS5	8-PSK	22.4
MCS6	8-PSK	27.2
MCS7	8-PSK	44.8
MCS8	8-PSK	54.4
MCS9	8-PSK	59.2

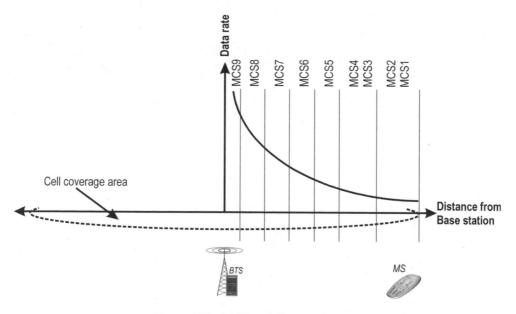

Figure 4.25   MCS and distance from BS

close to BS antennae. The lowest MCSs, MCS1 and MCS2, are available throughout the cell coverage.

If an operator wishes to offer higher average data rates requiring higher MCSs, such as MCS6 or MCS7, the radio network will experience marked changes. Higher MCS implementation throughout the network leads inevitably to smaller cell sizes. This in turn requires more cells in order to keep the coverage area the same. In other words, higher average data rates increase the number of network cells. The increase in the number of cells would be very high if the highest MCSs are available throughout the whole coverage area.

Another issue is the transport network. If user data rates coming and going through the radio interface increase, this must be taken into account in the transmission network. This phenomenon first occurs at MCS4, since the data rate MCS4 transfers exceeds the amount of data amount that one standard GSM Abis-channel (BTS–BSC interface) is able to carry.

4.4 WLAN Technology

In this section we introduce WLAN technology as one of the 3GPP complementary access candidates. IEEE 802.11 is the underlying specification family for the commercially available technology, often called "WLAN" or "Wi-Fi" (the Wireless Fidelity Alliance whose task is to promote and certify the interoperability and compatibility of IEEE 802.11-based commercial products). Although IEEE 802.11 radio technology was originally developed to replace the wire in Local Area Networks (LANs) and extend the Internet network, it is gradually becoming a complementary means of access to the UMTS network as well. Recently, 3GPP has also been working on the interworking aspects of UMTS and the 802.11 family to support hot spots via the wide-area-networking cellular network and vice versa.

4.4.1 Physical Technology

Originally approved in 1997, IEEE 802.11 standards have essentially focused on developing the radio and link layer for WLAN. In this respect, the IEEE 802.11 family defines a radio technology with three underlying physical alternatives. The radio specification includes diffused infrared, Direct Sequence Spread Spectrum (DSSS) and Frequency Hopping Spread Spectrum (FHSS) with three different hopping sets comprising 26 hopping sequences per set. Of them, the DSSS and the FHSS that operate within the 2.4-GHz Industrial, Scientific and Medical (ISM) band have so far had the most commercial significance (Figure 4.26). Due to the restrictions of the FHSS scheme for supporting high speed (higher than 2 Mb/s) the DSSS has practically predominated the physical-layer technique for higher bit-rate uses.

As in the traditional DSSS, when WLAN radio uses this scheme, the data stream is modulated and demodulated with a high-speed Pseudo-Random Numerical (PRN) sequence at the TRX, respectively. The 1-Mb/s data are encoded using Differential Binary Shift Keying (DBPSK). Also, a Differential Quadrature Phase Shift Keying (DQPSK) is used for the 2-Mb/s data rate.

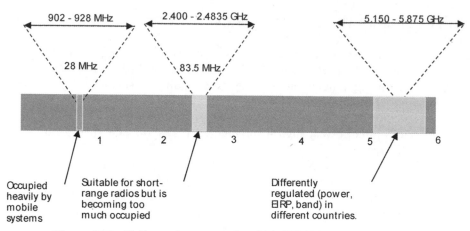

Figure 4.26   Unlicensed spectrum in which WLAN may also operate

Later, 802.11b started developing a standard for the physical layer for 2.4 GHz as well, but this time the specification included the so-called "Complementary Code Keying" (CCK) method in association with DSSS to reach a data rate up to 11 Mb/s. Later still, 802.11a specifications were developed. A new specification that introduces a new physical layer for the 5-GHz Unlicensed National Information Infrastructure (UNII) spectrum, operating in the 5.15–5.35-GHz and 5.725–5.825-GHz bands, was allocated in the United States and used the 20-MHz channel space. Recently, based on the results of WRC-03, it has been agreed that the 5,150–5,350-MHz and 5,470–5,725-MHz bands will be harmonised globally for 802.11.

The underlying technology for 802.11a is Orthogonal Frequency Division Multiple Access (OFDM), which brings more robustness against radio channel hazards. Particularly when equipped with convolutional coding and bit interleaving, OFDM is very efficient at overcoming the multipath propagation problem. In addition, 802.11a employs AMC. These, combined with BPSK and 64-quadrature Amplitude Modulation (QAM), make the physical layer capable of providing bit rates of up to 54 Mb/s. Operating at 2.4 GHz the 802.11g specification also makes use of OFDM with the Packet Binary Convolution Code (PBCC), further improving the robustness of radio. Table 4.8 summarises the very basic parameters of 802.11 specifications.

4.4.2 Medium Access Control

WLAN uses a Carrier Sense Multiple Access with Collision Avoidance (CSMA/CA) MAC solution layer that handles channel allocation, Protocol Data Unit (PDU) addressing, frame formatting, error checking, security, power management, synchronisation, and fragmentation and reassembly. The MAC layer can be in two modes, a priority/polling-based Point Coordination Function (PCF) and a contention-based Distributed Coordination Function (DCF). PCF mode provides a cell-wise network environment by allowing the Access Point (AP) to poll a number of terminals within the Basic Service Set (BSS) or the coverage unit area. On the other hand, DCF mode is the

Table 4.8 Key parameters of 802.11 radio technology

Parameter	802.11	802.11b	802.11a	802.11g
Frequency	2.4 GHz	2.4 GHz	5 GHz	2.4 GHz
Bitrate	1–2 Mb/s	1, 2, 5 and 11 Mb/s	6, 12, 24 and 54 Mb/s	54 Mb/s
Physical layer	IR, DSSS, FHSS	HR/DSSS equipped with CCK	OFDM	OFDM, PBCC
Timeslot	20 µs	20 µs	9 µs	20 or 9 µs
Modulation	DS-DBPSK, DS-DQPSK	DS-DBPSK, DS-DQPSK, PBCC	BPSK, QPSK, QAM	DS-DBPSK, ERP-OFDM, PBCC
Range	50–100 m	50–100 m	50–100 m	50–100 m

PBCC = Packet Binary Convolution Code; QAM = Quadrature Amplitude Modulation.

basic access method and supports data transport on the best effort manner and peer-to-peer *ad hoc*-type communications. DCF mode can also operate in association with PCF mode, providing a random access protocol similar to CSMA, equipped with Request-To-Send (RTS) and Clear-To-Send (CTS) messages for collision avoidance and with an ACK message returned by the receiver after successful transmission. This control message interaction between TRX devices is called "handshaking".

Figure 4.27 illustrates a basic handshaking mechanism that is employed, especially when proximity WLAN devices intend to establish a collision-free connection without the help of AP. In this case, device A sends an RTS message to device B to disover whether or not the data can be transmitted. On receiving that and having used its waiting time, device B replies with a CTS message. Assuming device C is free it also receives the CTS frame sent by device B. In this way, B informs both A of its readiness to receive the data and B not to disturb it. To defer transmission to a requested time, devicc C uses the Network Allocation Vector (NAV) parameter received in association with the CTS. As a result, connection is successfully established between A and B. It should be noted that if this mechanism is not used the connection set-up process might encounter the so-called "hidden effect". Suppose C was also trying to send data to B at the same time as A was trying. This would naturally lead to a collision at B. Although basic RTS–CTS handshaking may cause the so-called "exposed effect" (e.g., when C is prevented from transmitting to devices other than A or B), it nevertheless considerably helps alleviate the hidden problem.

In WLAN the carrier can be sensed at both the physical and MAC layer, where it is referred to as "Clear Channel Assignment" (CCA) and "virtual carrier sensing", respectively. In the former case only basic RTS–CTS is applied, while in the latter case, as shown in the figure, an extended RTS–CTS coupled with reservation time is employed. In addition to the fundamental features of the CSMA/CA and handshaking mechanism, the MAC uses different delay-based techniques and parameters: for example, InterFrame Spaces (IFSs) and NAV parameters for controlling priority access and successful transmission of the MAC Protocol Data Unit (MPDU). Being equipped

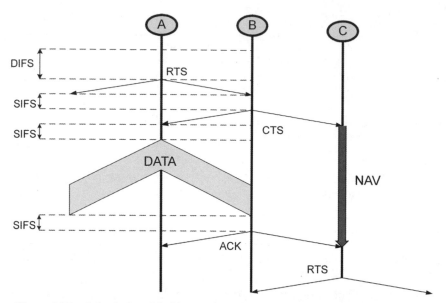

Figure 4.27   A basic handshaking and carrier-sensing procedure used in 802.11

with NAV and CCA for indicating channel elapsed time and a back-off period, the MAC uses three types of IFS:

- Short InterFrame Space (SIFS).
- Point coordination function InterFrame Space (PIFS).
- DCF InterFrame Space (DIFS).

The MAC supports three types of frames: management, control, and data. Of these management frames are used for handling terminal association and re-association with AP, timing and synchronisation as well as authentication and de-authentication. Control frames are used for RTS–CTS handshaking, positive ACK and alternating between the contention modes (CP and CFP). Data frames can be used in both CP and CFP modes, and can also be combined with polling and ACK during the CFP.

MAC frame format is of variable length and consists of payload and encryption/ decryption octets for optional Wired Equivalent Privacy (WEP) which, abreast with its recent enhancements, forms the basic security method used in 802.11. The WEP algo- rithm is basically used to encrypt the message exchange between 802.11 parties and, to some extent, for authenticating peers.

The IEEE standard 48-bit MAC address is used for source and destination device identification. Furthermore, Transmitter Address (TA), Receiver Address (RA) and Basic Service Set Identifier (BSSI) have been defined to help ease device mobility within the WLAN coverage. There is also a place for indicating the frame type as control, management or data. In addition, a Cyclic Redundancy Check (CRC) is used for error detection. Synchronisation is handled by utilising a timing beacon and

AP-referenced clock. The beacon is also used when alternating terminal transmission power between awake and doze states.

In order to ease the power consumption of the WLAN device, the 802.11 specification also introduces a power management mechanism. In this context, it defines two modes of operation: the so-called "Continuous Power Mode" (CPM) and "Power Save Mode" (PSM). In the former mode the device continuously gives information about its state to the AP or neighbours, while in the latter case the device only relays information about its wake-up periods. In this way the WLAN device is able to lower its power consumption markedly.

4.4.3 Network Formation

Regarding access architecture, the WLAN access network consists of an AP and a group of terminals that are under the direct control of the AP. These form a so-called "Basic Service Set" (BSS) also known as the Basic Service Area (BSA) of the access network. The main function of an AP is to establish a bridge between wireless and wired LANs while playing a master role for the network. Indeed, the AP resembles the BS in cellular networks. When an AP is present, terminals do not communicate on a proximity peer-to-peer basis but rather a centralised star network topology is formed. All communications between terminals or between a terminal and a wired network client go through the AP. Originally, APs were not planned to support terminal mobililily, but some degree of terminal portablility mobility is supported. Portable devices can roam between APs and in this way seamless local coverage is possible. As illustatred in Figure 4.28, a WLAN network operating in this configuration is said to be operating in "infrastructure mode".

In addition to infrastructure mode the IEEE 802.11 standard supports, to a certain extent, *ad hoc* configurations as well. In *ad hoc* mode fixed APs are not necessarily required. Instead, a promiscuous node is "elected" master of the network, forming a local network with other nodes as slave nodes. Any node can handle the logical function of the master. Thus, the nodes communicate directly with each other via a direct proximity radio link while sharing the given cell coverage area of the master. However, a single shared channel, lack of multihop routing, hidden and exposed effects, coupled with shortage in conveying secure and efficient Transmission Control Protocol (TCP) traffic over multi-hops, form the most severe limitations of WLAN when used for *ad hoc* network formation.

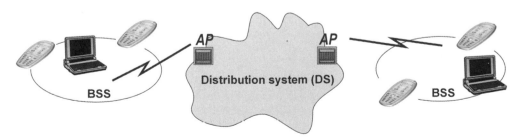

Figure 4.28   Basic architecture of the 802.11 access network

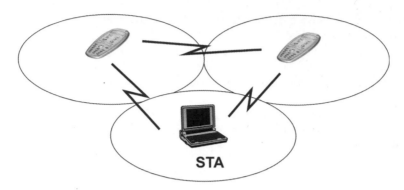

Figure 4.29 Independent Basic Service Set (IBSS) with a stations (STA) as the only building block

As WLANs have rapidly spread beyond their original territory of hot spots, new enhancements for residential and enterprise purposes have become urgent. This has led to the development of standards that mainly differentiate on the basis of the frequency used, data rate and applied modulation methods. Although interoperability has also been a matter of concern and covered in the standardisation, inevitably this further development has led to a situation in which some of the standards are not backward-compatible. As the unprecedented success of WLAN continues the basic standards encounter new challenges. As a result, standardisation has been continued to meet different requirements (e.g., related to QoS, security, Internet-working using an external network, radio resource measurement, handovers and fast roaming).

5

UMTS Radio Access Network

Siamäk Naghian and Heikki Kaaranen

Although the Universal Mobile Telecommunication System (UMTS) is rapidly evolving towards a fully multi-access network, the UMTS Terrestrial Access Network (UTRAN) forms its main Radio Access Network (RAN). The main task of the UTRAN is to create and maintain Radio Access Bearers (RABs) for communication between User Equipment (UE) and the Core Network (CN). With RAB the CN elements are given a rough idea about a fixed communication path to the UE, thus releasing them from the need to take care of radio communication aspects. Referring to the network architecture models presented in Chapter 1, the UTRAN makes certain parts of the Quality of Service (QoS) architecture happen independently (Figure 5.1).

The UTRAN is located between two open interfaces: Uu and Iu. From the bearer architecture point of view, the main task of the UTRAN is to provide a bearer service over these interfaces; in this respect the UTRAN controls the Uu interface, and bearer service provision in the Iu interface is done in cooperation with the CN.

The RAB fulfils the QoS requirements set by the CN. The handling of end-to-end QoS requirements in the CN and in the UE is the responsibility of Communication Management. These requirements are then mapped onto the RAB, which is "visible" to the Mobile Terminal (MT) and the CN. As just mentioned, the main task of the UTRAN is to create and maintain RABs such that the end-to-end QoS requirements are fulfilled in all respects.

One of the main ideas of this layered structure is to encapsulate the physical radio access; later it can be modified or replaced without changing the whole system. In addition, it is a known fact that the radio path is a very complex and continually changing transmission medium. This bearer architecture awards the principal role to the Radio Network Controller (RNC), since the RNC and the CN map the end-to-end QoS requirements over the Iu interface and the RNC takes care of satisfying QoS requirements over the radio path. These two bearers exist in the system because the Iu bearer is more stable in nature; the RAB experiences more changes during the connection. For example, one UE may have three continuously changing RABs maintained between itself and the RNC, while the RNC has only one Iu bearer for this connection. This kind of situation occurs with soft handovers, which are described in this chapter (see p. 115).

UMTS Networks Second Edition   H. Kaaranen, A. Ahtiainen, L. Laitinen, S. Naghian and V. Niemi
© 2005 John Wiley & Sons, Ltd   ISBN: 0-470-01103-3

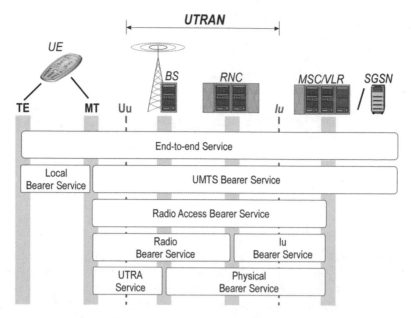

Figure 5.1   Bearer and QoS architecture in UTRAN

The physical basis of the end-to-end service provided by the Uu interface is the Universal Terrestrial Radio Access (UTRA) service. As explained in Chapter 4, UTRA is implemented using Wideband Code Division Multiple Access (WCDMA) radio technology and at the outset of UMTS the variant used was WCDMA-FDD (Frequency Division Duplex). From the QoS point of view UTRA contains mechanisms that show how end-to-end QoS requirements are mapped onto the physical radio path. Respectively, every Uu interface connection requires a terrestrial counterpart through the UTRAN. The equal terrestrial physical basis for the end-to-end service is the Physical Bearer Service (PBS). This could be implemented using various technologies, but in 3G Partnership Project (3GPP) R99 implementation the most probable alternative is Asynchronous Transfer Mode (ATM) over physical transmission media. Both ATM and some of the physical transmission alternatives are explained in Chapter 10. As the UTRAN evolves the ATM will have Internet Protocol (IP) as an alternative means of implementation (see Chapters 2 and 3).

Referring to the conceptual model presented in Chapter 1, the protocols that bring about the RB service belong to the access stratum (RB is defined as the service provided by the RLC layer for transfer of user data between the UE and the RNC). The protocols above the RB service belong to the non-access stratum and are responsible for the UMTS bearer service.

5.1 UTRAN Architecture

Figure 5.2 illustrates the UTRAN architecture at the network element level. The UTRAN consists of Radio Network Subsystems (RNSs) and each RNS contains

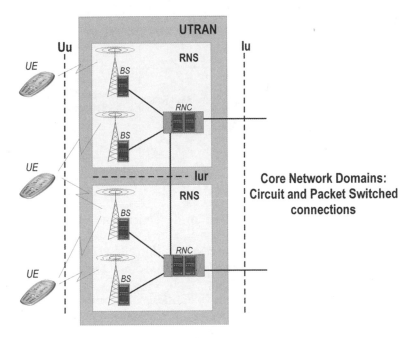

Figure 5.2   UTRAN general architecture

different numbers of Base Stations (BSs or, officially, Node Bs, which make the Uu interface a reality) and one RNC.

RNSs are separated from each other by the UMTS interface which is situated between RNC (Iur) interfaces that form connections between two RNCs. The Iur, which has been specified as an open interface, carries both signalling and traffic information.

5.2 Base Station (BS, Node B)

The Base Station (BS) is located between the Uu and UMTS interface which is further situated between RNC and BS (Iub) interfaces. Its main tasks are to effect physical implementation of the Uu interface and, towards the network, implementation of the Iub interface by utilising the protocol stacks specified for these interfaces. Regarding the Uu interface, the BS implements WCDMA radio access physical channels and transfers information from transport channels to the physical channels based on the arrangement determined by the RNC, as described in Chapter 4.

5.2.1 Base Station Structure

The BS can be considered as the radio edge of the UTRAN and, hence, its underlying task is to perform radio signal receiving and transmitting (Rx and Tx), signal filtering and amplifying, signal modulation and demodulation, and interfacing to the RAN.

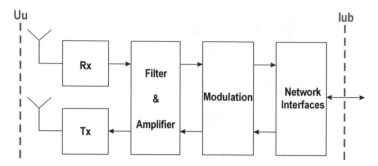

Figure 5.3   The basic structure of BS

The internal structure of the BS is very vendor-dependent, but basically consists of the components shown in Figure 5.3.

Its logical structure (i.e., how a BS is treated within the UTRAN) is generic. From the network point of view, the BS can be divided into several logical entities as shown in Figure 5.4.

On the Iub side, a BS is a collection of two entities: common transport and a number of Traffic Termination Points (TTPs). Common transport represents those transport channels that are common to all UE in the cell as well as those channels used for initial access. The common transport entity also contains one Node B control port used for

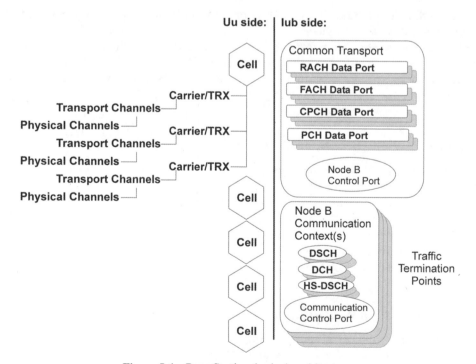

Figure 5.4   Base Station logical architecture

Operation and Maintenance (O&M) purposes. One TTP consists of a number of *Node B communication contexts*. Node B communication contexts in turn consist of all dedicated resources required when the UE is in dedicated mode. Thus, one Node B communication context may contain at least one Dedicated Channel (DCH), for example. The exception is the Downlink Shared Channel (DSCH) and High Speed DSCH (HS-DSCH), which also belong to the Node B communication context. From the point of view of UMTS network infrastructures, the BS can be considered to be a logical O&M entity that is subject to network management operations. In other words, this term describes the physical BS and its circumstances (i.e., the BS site).

From the point of view of the radio network and its control, the BS consists of several other logical entities called "cells". A cell is the smallest radio network entity having its own identification number (cell ID) that is publicly visible to the UE. When the radio network is configured, it is actually the cell(s) data that are changed. The term "sector" stands for the physical occurrence of the cell (i.e., radio coverage).

Every cell has one scrambling code, and the UE recognises a cell by two values: scrambling code (when logging on to a cell) and cell ID (for radio network topology). One cell may have several Transmitter–Receivers (TRXs, also called "carriers") under it. The TRX of the cell delivers the broadcast information towards the UE; that is, the Primary Common Control Physical Channel (P-CCPCH) containing the Broadcast Channel (BCH) information is transmitted here. One TRX maintains physical channels through the Uu interface and these carry the transport channels containing actual information, which may be either common or dedicated in nature.

A cell may consist a minimum of one TRX. The TRX is a physical part of the BS performing various functions, such as the conversion of data flows from the terrestrial Iub connection to the radio path and vice versa.

5.2.2 Modulation Method

From the system point of view, the used modulation method is of interest since it has a close relationship to overall system capacity and performance. WCDMA uses Quadrature Phase Shift Keying (QPSK), its variant—dual QPSK—as well as Quadrature Amplitude Modulation (16QAM) as its modulation methods. QAM coupled with Adaptive Modulation and Coding (AMC) and Hybrid Automatic Repeat Request (HARQ) were the most significant new features of WCDMA specified by 3GPP Release 5, in association with High Speed Downlink Packet Access (HSDPA) enhancements. While some channels, such as the downlink physical channels P-CCPCH, S-CCPCH, CPICH, AICH, AP-AICH, CSICH, CD/CA-ICH, PICH, PDSCH, HS-SCCH and downlink DPCH, use QPSK modulation, the downlink physical channel HS-PDSCH can use either QPSK or 16QAM depending on the requested bit rate and radio channel condition.

The QPSK modulation method expresses a single bit and its status using a different phase of the carrier. The bits in the modulation process are handled as pairs and, thus, this generates four possible 2-bit combinations that must be indicated.

As illustrated in Figure 5.5, when the QPSK modulation process is applied, the data stream is the first to be modulated (i.e., the physical channels are first converted from

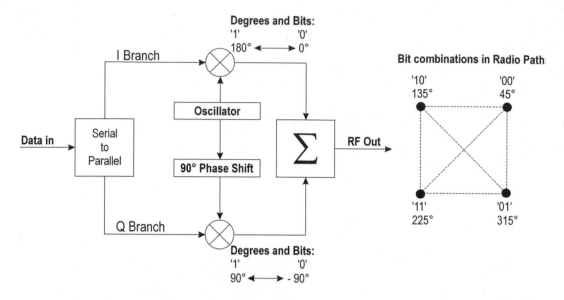

Figure 5.5 The main principle behind QPSK modulation

serial to parallel format). After this conversion, the modulation process divides the data stream into two branches: I and Q. In the I branch the bit having a value of "1" represents a +180° phase shift and a value of "0" means the phase of the carrier has not shifted. In the Q branch the bit having a value of "1" represents a +90° phase shift and a value of "0" stands for a −90° phase shift. These branches, I and Q, are fed by the oscillator, the I branch directly and the Q branch with a 90° shift. When the incoming data stream is combined with the oscillator output, the 2-bit combinations (one bit from the I branch and one from Q) represent the phase shifting shown on the right-hand side of the figure. Thin lines in the square indicate the "paths" along which the system can transfer itself from one state (phase) to another.

This system works fine, but certain 2-bit combinations are more difficult to handle. For instance, when the bit values change from "00" to "11" this means there has been a 180° phase shift, which is considered a very marked amplitude change. Very big amplitude changes cause problems, especially if the bandwidth used in the radio connection is wide. In this case the BS must have linear amplifiers in order to guarantee that amplitude changes are represented correctly throughout the bandwidth used.

To eliminate the problem of amplitude changing too rapidly, another variant of QPSK is used: the so-called dual Quadrature Phase Shift Keying (dual QPSK). As a result, instead of time-multiplexing, the Dedicated Physical Common Channel (DPCCH) and the Dedicated Physical Data Channels (DPDCHs) are I/Q code-multiplexed with the complex scrambling shown in Figure 5.6. This leads to a time delay close to 0.5 bits (chips) on the Q branch of the modulator. As shown in Figure 5.7 this prevents 180° phase changes and limits them to 90° steps. Based on this, the transition from the combination "11" to "00" is now "11"–"10"–"00" and takes place within the same time frame as in the QPSK.

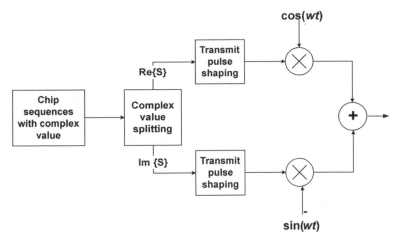

Figure 5.6   Uplink modulation in UTRAN

Therefore, the spectrum used for QPSK and dual QPSK are the same, but dual QPSK has a smoother signal. This allows amplifiers to also operate on non-linear operating areas without significant problems. This apparently simple operation has a significant impact, especially on terminal design, as it reduces the required Peak Average Ratio (PAR), resulting in amplifiers with higher power and, ultimately, cost-efficiency. Alternatively, conventional QPSK could have been used in both directions, but then the UE would have suffered power consumption problems and high prices: the linear power amplifier needs to be very accurate and, thus, is very expensive. Using dual QPSK these amplifying problems can be avoided. From the BS point of view, though, dual QPSK is not good enough because the signal the BS transmits must be very accurate and the UE must achieve accurate synchronisation. Using dual QPSK, these features are compromised and, thus, conventional QPSK is better for modulation performed on the transmission side of the BS.

In addition to QPSK and dual QPSK, WCDMA utilises QAM modulation as well. As was discussed in the chapter on HSDPA (Section 4.2), this modulation method was selected in Release 5 for use in association with HSDPA data enhancements in WCDMA. The reason for this transition was straightforward: the main goal of

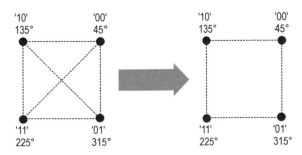

Figure 5.7   The result of combining I/Q code-multiplexing with uplink modulation

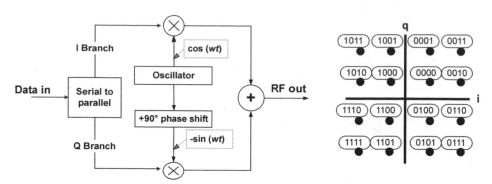

Figure 5.8   The basic process of QAM modulation and its 16QAM bit constellation

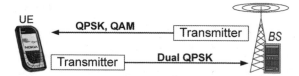

Figure 5.9   Used modulation methods and their direction in UTRAN

HSDPA was to achieve much higher bit rates than those pursued in 3GPP R4. The targeted bit rate was not efficiently achievable by employing QPSK, if at all. Unlike QPSK, which utilises phase information in the modulation process, QAM can benefit from both phase and amplitude information when forming the signal constellation. Thus, as can be seen in Figure 5.8, it improves data transmission by several times, depending on the modulation level employed. In QAM two branches of carriers, with a 90° phase difference, are used to transmit data over a given physical channel. Thanks to the 90° phase difference and the orthogonal carrier, each of these branches can be modulated independently and transmitted over the same channel and ultimately separated by demodulation at the terminal (e.g., in an HSDPA application).

QAM modulation efficiency coupled with AMC increases the WCDMA downlink bit rate by several orders of magnitude when compared with Release 4 data capability (DSCH). The modulation level of QAM can range up to 64QAM, but, due to cost, complexity and efficiency reasons, 16QAM was considered reasonable for a UTRAN/ HSDPA application.

Based on these facts, and as shown in Figure 5.9, WCDMA actually uses both QPSK variants, conventional QPSK in the downlink direction and dual QPSK in the uplink direction, as well as QAM in the downlink direction for high-bit-rate applications.

5.2.3 Receiver Technique

WCDMA utilises multipath propagation: in other words, the transmitted signal propagates in many different ways from the transmitter to the receiver. On the other hand, to gain better capacity in the radio network, the transmit powers of the UE (and

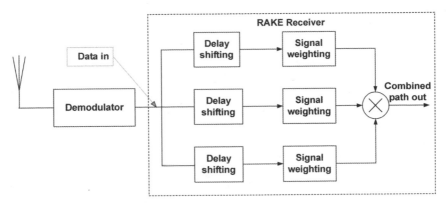

Figure 5.10   The basic principle behind the RAKE receiver

BSs) should be relatively small. This decreases interference in the radio interface and gives more space for other transmissions. This leads to a situation where it is very useful for both the UE and the BS to be able to "collect" the many weak-level signals indicative of the same transmission, thus helping to combine them together. This requires a special type of receiver. An example of this kind of arrangement is called RAKE (noted that RAKE is not an acronym, it is the real name for this type of receiver).

The purpose of the RAKE receiver is to improve the level of the received signal by exploiting the multipath propagation characteristics of the radio wave, as signals propagating through different paths have different attenuations. As shown in Figure 5.10, a basic RAKE receiver consists of a number of fingers, a combiner, a matched filter and a delay equaliser. Practically every finger can receive part of the transmitted signal, which can be either from the serving BS or from a neighbouring BS. Different branches of the received signal are combined in such a way that the phase and amplitude deviations of the branches compensate for each other, resulting in a signal with markedly higher signal strength than any individual signal branch. To distinguish multipath signal branches from each other, there must be a delay (within a given range) between consecutive branches.

In general, diversity techniques are efficient ways of overcoming the radio signal deterioration due to shadowing and fading (discussed in Chapter 3). In addition, utilising a diversity technique is a prerequisite for providing a soft-handover feature in cellular systems. Different diversity techniques may be used in mobile systems, such as time diversity, space diversity, frequency diversity, etc. In WCDMA technology, polarisation diversity is typically utilised for both uplink and downlink transmission. The purpose of multipath diversity is to resolve individual multipath components and combine them to obtain a sum signal component with better quality. Having used the RAKE receiver in both the UE and the BS makes it possible to capture and combine different branches of the desired signal and improve the quality of the ultimate signal or data stream for further processing. Maximum Ratio Combining (MRC) is the diversity algorithm used for signal processing. The responsibility for diversity is distributed in radio access between the UE, the BS and the RNC, depending on the link direction (uplink, downlink) and the position of the network element in the system architecture hierarchy.

5.2.4 Cell capacity

In WCDMA technology, all users share the common physical resource: a frequency band in 5-MHz slices. All users of the WCDMA TRX coexist on the frequency band at the same time and different transactions are recognised using spreading codes. In the first UMTS systems, the UTRAN used to employ the WCDMA-FDD variant. In this variant, the Uu interface's transmission directions use separate frequency bands. One of the most interesting questions (and one of the most confusing) for people is the capacity of the WCDMA TRX. In the Global System for Mobile Communications (GSM), TRX capacity calculation is a very straightforward procedure, but because the radio interface in WCDMA is handled differently and system capacity is limited by variable factors, the capacity of the WCDMA TRX is not very easy to determine.

Let us now outline a *rough* theoretical estimate of WCDMA TRX capacity, based on radio conditions. In order to simplify the issue, we must make some assumptions:

- All subscribers in the TRX coverage area are equally distributed so that they are situated at equal distances from the TRX antenna.
- The power level they use is the same and, thus, the interference they cause is at the same level.
- Subscribers in the TRX area use the same base band bit rate (i.e., the same symbol rates).

Under these circumstances a value called "processing gain" (G_p) can be defined. G_p is a relative indicator that provides information about the relationship between the whole bandwidth available (B_{RF}) and the base band bit rate $(B_{Information})$:

$$G_p = \frac{B_{RF}}{B_{Information}}$$

There is another way to express G_p by using chip and data rates:

$$G_p = \frac{\text{Chip rate}}{\text{Data rate}}$$

As a result both ways (when expressed in dB values) show an improvement in the Signal to Noise (S/N) ratio between the received signal and the output of the receiver.

Later on, we can see G_p is actually the same as the Spreading Factor (SF). Note that the base band bit rate discussed here is the one achieved after rate matching. In this process the original (user) bit rate is adjusted to the bearer bit rate. Bearer bit rates are fixed (e.g., 30 kb/s, 60 kb/s, 120 kb/s, 240 kb/s, 480 kb/s and 960 kb/s). The system chip rate is constant: 3.84 Mchip/s (38,400,000 chip/s). Hence, as an example a bearer having a bit rate of 30 kb/s will have an SF of 128:

$$G_p = \frac{38,400,000}{30,000} = 128 = SF$$

The power P required for information transfer in one channel is a multiple of the energy used per bit and the base band data rate:

$$P = E'_b \text{ Base band data rate}$$

On the other hand, it is known that the noise on one channel (N_{Channel}) that partially uses the whole bandwidth B_{RF} can be expressed as:

$$N_{\text{Channel}} = B'_{\text{RF}} N_0$$

where N_0 is noise spectral density (W/Hz).

Based on this, the S/N ratio is:

$$S/N = \frac{P}{N_{\text{Channel}}} = \frac{E'_b \text{ base band data rate}}{B'_{\text{RF}} N_0} = \frac{E_b/N_0}{G_p}$$

If we assume that there are X users in the TRX area and that the assumptions presented earlier in this chapter are applied, this would mean that there are (statistically thinking) $X - 1$ users causing interference to a single user. This also indicates the S/N ratio and, when expressed in a mathematical format, the outcome is the following equation:

$$S/N = \frac{P}{P'(X - 1)} = \frac{1}{X - 1}$$

If, later on, there are many users (say, tens of them), then the equation could be simplified:

$$S/N = \frac{1}{X - 1} \approx \frac{1}{X}$$

Now we have two different ways to calculate the S/N ratio:

$$\frac{E_b/N_0}{G_p} \approx \frac{1}{X} \quad \text{and} \quad X \approx \frac{G_p}{E_b/N_0}$$

This equation is a *very rough* expression and should be used for estimation purposes only. The "official" way to calculate TRX capacity has several more parameters that need to be taken into account.

Example Assume that the SF used in the cell is 128 and for these transactions the E_b/N_0 is 3 dB. How many users can the cell contain simultaneously when there is only one TRX available?

$$X \approx \frac{G_p}{E_b/N_0} = \frac{128}{3} \text{ dB} \approx \frac{128}{2} = 64 \text{ users}$$

This is the maximum number of users the TRX is able to handle, in theory, when taking intra-cell interference into account. In reality, the neighbouring cells produce inter-cell interference. If we assume inter-cell interference to be the same as intra-cell interference, then the user amount is split to 32 (64/2 = 32).

The E_b/N_0 relationship is an important point of interest. It is a constant-like numerical relationship that may have several values, which later are related to radio interface bearer bit rates. Thus, it can be stated that the E_b/N_0 relationship has a marked effect on the TRX/cell capacity as far as the number of simultaneous users is concerned. In the E_b/N_0 relationship, the following issues should be considered:

- N_0 is a type of local constant value, which also contains some receiver-specific values.
- E_b is a naturally changing value that depends on the following bullets.
- G_p/SF: the bigger the SF value the smaller the E_b.
- The higher the base band bit rate the bigger the E_b (this is a direct consequence from the previous point).
- Distance between the terminal and BS receiver: the longer the distance the bigger the E_b.
- Terminal motion speed: the higher the speed the bigger the E_b.

The above calculation is a rough way to estimate TRX capacity, but there are many other ways to do this. For further details, refer to the *WCDMA for UMTS* by Holma and Toskala (2001).

5.2.5 Control Functions in BS

Although the main functions of BS are related to radio signal transmitting and receiving, as well as other base-band-dependent processing functions, there are nevertheless other UTRAN control functions in which the BS is involved. These functions are partly limited to supporting such functions as fulfilling, collecting and filtering radio measurements and providing them to the RNC for the RAN to execute its control functions. On the other hand, there are BS function that have a central role. These include code-generating, power control executing and O&M, particularly in the network element or at the cell level. The O&M function in the BS includes both network-element-level (HW&SW) and UTRAN-level logical functions controlled by the RNC. The main network-element-level function, which is basically implementation-dependent, forms a basis for the Network Management System (NMS). The way this function is handled is highly vendor-dependent.

From the standpoint of Radio Resource Management (RRM), the BS is involved in power control (inner loop) and code generation, but due to the recent enhancements in Release 5 it is also involved in packet-scheduling. Therefore, as the UTRAN has evolved so it has increased the control function of the BS. Nevertheless, the role of the RNC still remains central in terms of overall control of the UTRAN. As this is currently being made a reality—in particular, in conjunction with the RRM—we will now describe these functions in more detail in association with other RNC functions.

5.3 Radio Network Controller (RNC)

The RNC is the switching and controlling element of the UTRAN. The RNC is located between the Iub and Iu interfaces. It also has a third interface called an "Iur" for inter-RNS connections. Implementation of the RNC is vendor-dependent, but some generic points can be highlighted, as illustrated in Figure 5.11.

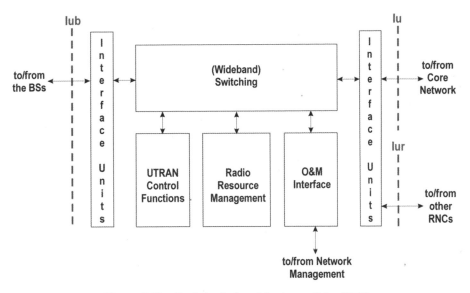

Figure 5.11   Basic logical architecture of the RNC

As explained earlier, the RNC sees the BS as two entities: common transport and a collection of Node B communication contexts. The RNC controlling these for a BS is called the *Controlling RNC* (CRNC).

As far as the bearers are concerned, the RNC is a switching point between the Iu bearer and the RB(s). One of the radio connections between the UE and the RNC that carry user data is an RB. The RB in turn is related to the UE context, which is a set of definitions required at the Iub interface to arrange both common and dedicated connections between the UE and the RNC. Since the UTRAN utilises macrodiversity, the UE may have several RBs between itself and the RNC. This situation is known as a soft handover and is discussed later on p. 115. The RNC holding the Iu bearer for some UE is called the *Serving RNC* (SRNC).

The third logical role the RNC can play is that of the *Drifting RNC* (DRNC). In this mode the RNC allocates the UE context to itself—the request to perform this activity comes from the SRNC through the Iur interface.

Both the SRNC and DRNC roles are functionalities that can change their physical location. When the UE moves in the network performing soft handovers, the radio connection of the UE will be accessed by an entirely different RNC from the SRNC that originally performed the first RAB set-up for this UE. In this case the SRNC functionality will be transferred to the RNC, which handles radio connection to the UE. This procedure is called "SRNC or SRNS relocation".

The whole functionality of RNC can be classified into two parts: UTRAN RRM and control functions. The RRM is a collection of algorithms used to guarantee the stability of the radio path and the QoS of radio connection by the efficient sharing and managing of radio resources. UTRAN control functions include all the functions that related to set-up, maintenance and release of RBs, including the support functions for RRM algorithms.

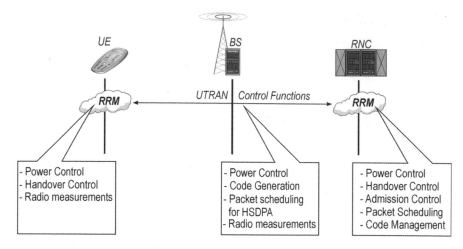

Figure 5.12   The main entities of Radio Resource Management

5.3.1 Radio Resource Management (RRM)

As explained in Chapter 1, RRM is a management responsibility of the UTRAN. RRM is located in the UE, the BS and the RNC inside the UTRAN. RRM makes use of various algorithms, which aim to stabilise the radio path enabling it to fulfil the QoS criteria set by the service using the radio path (Figure 5.12).

The RRM algorithms must deliver information over the radio path, which is called the "UTRA" service. The control protocol used for this purpose is the Radio Resource Control (RRC) protocol. UTRAN control functionalities are discussed later in this chapter. The RRC protocol is overviewed in Chapter 10.

The RRM algorithms we will shortly present are:

- Handover control.
- Power control.
- Admission Control (AC) and Packet Scheduling (PS).
- Code management.

5.3.1.1 RRM—Handover Control

Handover is one of the essential means to guarantee user mobility in a mobile communications network. Maintaining the traffic connection for a moving subscriber is made possible with the help of the *handover* function. The basic concept is simple: when the subscriber moves from the coverage area of one cell to another, a new connection with the target cell has to be set up and the connection with the old cell may be released. Controlling the handover mechanism is, however, a quite complicated issue in cellular systems—the Code Division Multiple Access (CDMA) system, in particular, adds some sophisticated ingredients to it.

5.3.1.1.1 *Reasons behind the Handover*

There are many reasons handover procedures are activated. The basic reason behind a handover crops up when the air interface connection no longer fulfils the desired criteria set for it and, thus, either the UE or the UTRAN initiates actions in order to improve the quality of the connection. In WCDMA, handover on-the-fly is used for circuit-switched calls. In the case of packet-switched calls, handovers are mainly achieved when neither the network nor the UE has any packet transfer activity.

Regardless of the sorts of handovers, they have a common denominator: the current QoS of the connection is neither acceptable nor optimised. The logic behind how the need for the handover is investigated is also quite common. The handover execution criteria depend mainly upon the handover strategy implemented in the system. However, most of the criteria behind handover activation may rely on signal quality, user mobility, traffic distribution, bandwidth and so forth.

Signal quality handover occurs when the quality or the strength of the radio signal falls below certain parameters specified by the RNC. Deterioration of the signal is detected by constant signal measurements carried out by both the UE and the BS. Signal quality handover may be applied for both uplink and downlink radio links.

Traffic handover occurs when the traffic capacity of a cell has reached its maximum or is approaching it. In such a case, the high load of a UE nearing the edges of a cell may be handed over to neighbouring cells with a smaller traffic load. By using this sort of handover, system load can be distributed more uniformly and the needed coverage and capacity can be adapted efficiently to meet the traffic demand within the network. Traffic handovers may rely on pre-emption or directed retry approaches.

The number of handovers is dependent in a straightforward way on the degree of UE mobility. If we assume that the UE keeps on moving in the same direction, then it can be said that the faster the UE is moving the more handovers it causes the UTRAN to undertake. To avoid undesirable handovers the UE moving at high speed may be handed over, for instance, from micro-cells to macro-cells. When the UE is moving slowly or not at all, on the other hand, it can be handed over from macro-cells to micro-cells to improve radio signal strength and avoid draining its battery.

The decision to perform a handover is always made by the RNC that is currently serving the subscriber, except for handovers for traffic reasons. In the latter case the Mobile service Switching Centre (MSC) may also make the decision. In addition to the above, there are many other reasons for handovers (e.g., a change of services).

5.3.1.1.2 *Handover Process*

Figure 5.13 illustrates a basic handover process consisting of three main phases: measurement phase, decision phase and execution phase. The overall handover process discussed here is specifically related to the WCDMA system. Nevertheless, as far as the basic principles are concerned, they are valid for any kind of cellular system.

Handover measurement provision is a pivotal task from the system performance standpoint: first, the signal strength of the radio channel may vary drastically due to fading and signal path loss, resulting from the cell environment and user mobility; second, an excess of measurement reports by UE or handover execution by the network increases overall signalling, which is undesirable.

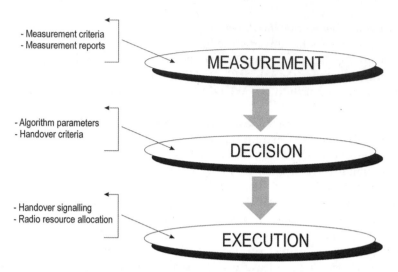

Figure 5.13   The basic process of handover

For handover purposes and during connection the UE continuously measures the signal strength of neighbouring cells, and reports the results to the network (to the radio access controller) in much the same way as the RNC in the WCDMA system.

According to the 3GPP TS 25.331, UE measurements may be grouped in different categories, depending on what the UE should measure. The different types of measurements are:

- Intra-frequency—including measurements of the strength of downlink physical channels (for signals with the same frequencies).
- Inter-frequency—including measurements of the strength of downlink physical channels (for signals with different frequencies).
- Inter-system—covering measurements of the strength of the downlink physical channels that belong to an access system other than the UTRAN (e.g., the GSM).
- Traffic volume—containing measurements of the uplink traffic volume.
- Quality—including measurements of quality parameters (e.g., the downlink transport block error rate).
- Internal—including measurements of UE transmission power and UE received signal level.

Measurements, on the other hand, may be triggered as a consequence of the following criteria:

- Change of best cell.
- Changes in Primary Common Pilot Channel (CPICH) signal level.
- Changes in P-CCPCH signal level
- Changes in Signal to Interference Ratio (SIR) level.
- Changes in Interference Signal Code Power (ISCP) level.

- Periodic reporting.
- Time-to-trigger.

So, the WCDMA specification provides various measurement criteria to support the handover mechanisms in the system. In order to improve system performance it is pivotal to select the most appropriate measurement procedure and measurement criteria as well as the filtering intervals to do with handover mechanisms. Handover signalling load can be optimised by fine-tuning the trade-off between handover criteria, handover measurements and the traffic model utilised in network planning.

The decision phase consists of assessment of the overall QoS of the connection and comparing it with the requested QoS attributes and estimates from neighbouring cells. Depending on the outcome of this comparison, the handover procedure may or may not be triggered.

The SRNC checks whether the values indicated in measurement reports trigger any criteria set. If they do, then it allows the handover to take place.

As far as handover decision-making is concerned, there are two main types:

- Network Evaluated HandOver (NEHO).
- Mobile Evaluated HandOver (MEHO).

In the case of NEHO, the network SRNC makes the handover decision, while in the MEHO approach the UE mainly prepares the handover decision. In the case of a combination of NEHO and MEHO, the decision is made jointly by the SRNC and the UE.

Note that, even in a MEHO, the final decision about handover execution is made by the SRNC. The reason is that the RNC is responsible for the overall RRM of the system and, thus, it is aware of the overall load of the system and other necessary information needed for handover execution.

Handover decision-making is based on the measurements reported by the UE and the BS as well as the criteria set by the handover algorithm. The handover algorithms, as such, are not subject to standardisation; rather, they are implementation-dependent aspects of the system. Therefore, advanced handover algorithms may be utilised freely, based on the available parameters in association with the network element measurement capabilities, traffic distribution, network planning, network infrastructure and overall traffic strategy applied by operators.

The general principles underpinning a handover algorithm are presented in Figure 5.14. In this example it is assumed that the decision-making criteria of the algorithm are based on the pilot signal strength reported by the UE. The following terms and parameters are used in this handover algorithm example:

- *Upper threshold*: the level at which the signal strength of the connection is at the maximum acceptable level with respect to the requested QoS.
- *Lower threshold*: the level at which the signal strength of the connection is at the minimum acceptable level to satisfy the required QoS. Thus, the signal strength of the connection should not fall below it.

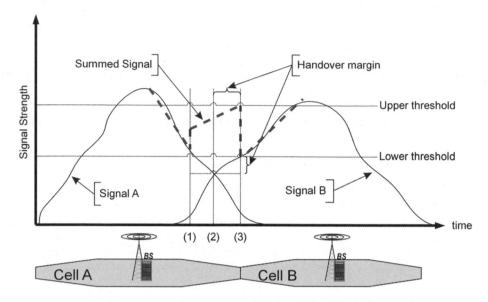

Figure 5.14   Basic principle underlying a handover algorithm

- *Handover margin*: a predefined parameter which is set at the point where the signal strength of the neighbouring cell (B) has started to exceed the signal strength of current cell (A) by a certain amount and/or for a certain time.
- *Active set*: a list of the signal branches (cells) through which the UE has simultaneous connection to the UTRAN.

Let us assume that a UE camping in cell A is moving towards cell B. As the UE is moving towards cell B the pilot signal A—to which the UE currently has a connection—deteriorates, approaching the lower threshold as shown in Figure 5.14. This may result in the triggering of handover whenever the following steps can be distinguished:

1. The strength of signal A becomes equal to the defined lower threshold. On the other hand, based on UE measurements the RNC recognises that there is already a neighbouring signal (signal B in Figure 5.14) available with adequate strength to improve the quality of the connection. Therefore, it adds signal B to the active set. Upon this event, the UE has two simultaneous connections to the UTRAN and, hence, it benefits from the summed signal, which consists of signals A and B.
2. At this point the quality of signal B starts to become better than signal A. Therefore, the RNC keeps this point as the starting point for handover margin calculation.
3. The strength of signal B becomes equal or better than the defined lower threshold. Thus, its strength is adequate to satisfy the required QoS of the connection. On the other hand, the strength of the summed signal exceeds the defined upper threshold, causing additional interference to the system. As a result, the RNC deletes signal A from the active set.

Note that the size of the active set may vary, but usually it ranges from 1 to 3 signals. In this example the size of the active set is 2 between event (1) and (3).

Because the direction of UE motion varies randomly it is possible for it to return to cell A immediately after the first handover. This results in the so-called *ping-pong* effect, which is harmful to the system in terms of capacity and overall performance. Using the handover margin or hysteresis parameter is the best way to avoid undesired handovers, which cause additional signalling load to the UTRAN.

5.3.1.1.3 Handover Types

Depending on the diversity used with handover mechanisms, they can be categorised in three ways: hard handover, soft handover and softer handover. Hard handover can be further divided into intra-frequency and inter-frequency. All of these mechanisms are provided by the UMTS system.

If the old connection during the handover process is released before making the new connection it is called a *hard handover*. Therefore, there are not only lack of simultaneous signals but also a very short cut in the connection, which is not distinguishable for the mobile user.

In the case of *inter-frequency* hard handover the carrier frequency of the new radio access is different from the old carrier frequency to which the UE was connected. On the other hand, if the new carrier, through which the UE is accessed after the handover procedure, is the same as the original carrier, then there is an *intra-frequency* handover in question.

Figures 5.15 and 5.16 show hard handover situations in which neighbouring BSs may transmit with the same frequency or with a different frequency, respectively.

In Figure 5.15, the neighbouring RNC is not connected to the Iur interface due to the radio network planning strategy or for transmission reasons and, hence, inter-RNC soft handover is not possible. Under these circumstances, intra-frequency hard handover is

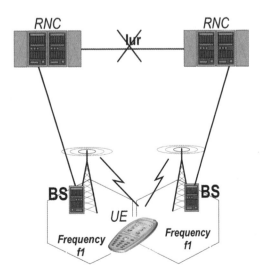

Figure 5.15   Inter-frequency hard handover

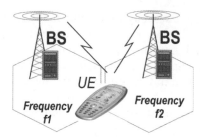

Figure 5.16   Inter-frequency hard handover

the only handover to support seamless radio access connection and subscriber mobility from the old BS to the new BS. In fact, this leads to an inter-RNC handover event, in which the MSC is also involved.

Generally, the frequency reuse factor is one for WCDMA, meaning that all BSs transmit on the same frequency and all items of UE share a common frequency within the network. This does not mean, however, that other frequency reuse cannot be utilised in WCDMA at all. Therefore, if different carriers are allocated to the cells for some other reason, inter-frequency handover is required to ensure there is a handover path from one cell to another cell in the cell cluster.

Inter-frequency handover also occurs in the Hierarchical Cell Structure (HCS) network between separate cell layers (e.g., between macro-cells and micro-cells) which use different carrier frequencies within the same coverage area. In this case inter-frequency handover is used not only because the UE would otherwise lose its connection to the network, but also in order to increase the system performance in terms of capacity and QoS. Inter-frequency hard handover is always a NEHO.

Moreover, inter-frequency handover may happen between two different RANs, such as between the GSM and WCDMA. In this context it can also be called an "inter-system handover" (Figure 5.17). An inter-system handover is always a type of inter-frequency handover, since different frequencies are used in different systems.

The possibility of carrying out an inter-system handover is enabled in WCDMA by a special functioning mode, *compressed mode* (also known as *slotted mode*). From the WCDMA point of view, the SF value of the channel can be reduced when the UE is in slotted mode. As a consequence, connection to the radio interface uses only a fraction of the space in the WCDMA frame slot. The rest of the slot can be used for other purposes by the UE (e.g., to measure surrounding GSM cells). In other words, this

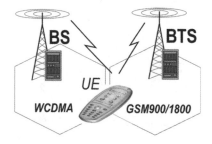

Figure 5.17   Inter-system handover

mechanism is a way of implementing the GSM/UMTS interoperability requirement in UTRAN. In addition, slotted mode can be achieved by reducing the data rate using higher layer controlling and by reducing the symbol rate in association with physical layer multiplexing. When the UE uses the Uu interface in this mode, the contents of the WCDMA frame are "compressed" a little in order to open a time window through which the UE is able to view and decode GSM Broadcast Control Channel (BCCH) information. Additionally, both the WCDMA RAN and GSM Base Station System (BSS) must be able to send each other's identity information along BCCHs so that the UE is able to decode properly.

The inter-system WCDMA and GSM handover can be applied in areas where WCDMA and GSM systems coexist. Inter-system handover is required to complement the coverage areas of each other in order to ensure continuity of services. Inter-system handover can also be used to control the load between GSM and WCDMA systems, when the coverage area of the two systems overlap each other. Other inter-system handover reasons might include the service requested by the UE and the user's subscription profile.

Inter-system handover is a NEHO. However, the UE must support inter-system handover completely. The RNC recognises the possibility of inter-system handover based on the configuration of the radio network which mainly depends on neighbouring cell definitions and other control parameters. The same applies for the BSC at the GSM RAN end.

Unlike in hard handover, if a new connection is established before the old connection is released, then the handover is called a *soft handover*. In the WCDMA system, the majority of handovers are intra-frequency soft handovers. As is illustrated in Figure 5.18, the neighbouring BS involved in a soft handover event transmits the same frequency.

Soft handover is performed between two cells belonging to different BSs but not necessarily to the same RNC. In any case the RNC involved in a soft handover must coordinate the execution of soft handover over the Iur interface. In a soft handover event both source and target cells have the same frequency. In the case of a circuit-switched call, the terminal actually performs soft handovers almost all the time if the radio network environment has small cells. There are several variations of soft handover, including softer and soft–softer handovers.

A softer handover is a handover as a result of which a new signal is either added to or deleted from the active set, or replaced by a stronger signal within the different sectors that are under the control of the same BS (Figure 5.19).

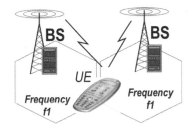

Figure 5.18   Intra-frequency soft handover

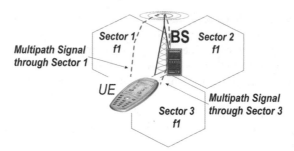

Figure 5.19 Intra-frequency softer handover

In a softer handover the BS transmits through one sector but receives from more than one sector. In this case the UE has active uplink radio connections with the network through more than one sector populating the same BS.

When soft and softer handovers occur simultaneously, the term soft–softer handover is usually used. A soft–softer handover may occur, for instance, in association with an inter-RNC handover, at the same time as an inter-sector signal is added to the UE's active set and a new signal is added via another cell controlled by another RNC.

As far as soft handover and active sets are concerned, there are two terms describing the handling of the multipath components: microdiversity and macrodiversity.

Microdiversity means the situation where propagating multipath components are combined in the BS as shown in Figure 5.20. WCDMA utilises multipath propagation. This means that the BS RAKE receiver, which was introduced in Section 5.2.3, is able to determine, differentiate and combine the signals received from the radio path. In reality, a signal sent to the radio path is reflected from the ground, water, buildings, etc. and at the receiving end can be "seen" as many copies, all of them reaching the receiver at a slightly different phase and time. Microdiversity functionality at the BS level combines the different signal paths received from one cell, and, in the case of a BS with many sectors, the outcome from different sectors is also referred to as a softer handover.

Because the UE can use cells belonging to different BSs or even different RNCs, macrodiversity functionality also exists at the RNC level. However, the means of

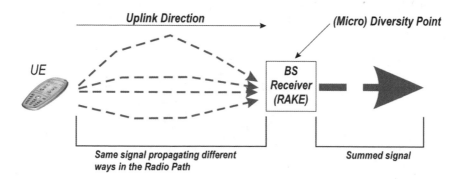

Figure 5.20 Microdiversity at the Base Station

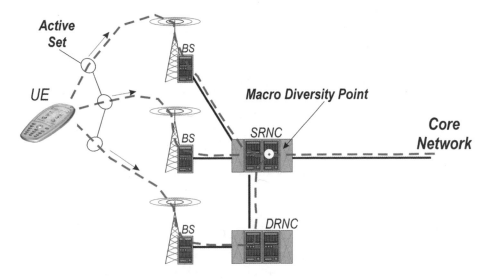

Figure 5.21   Macrodiversity at the RNC

combining the signals is quite different from the microdiversity case at the BS, because there is no RAKE receiver at RNC. Therefore, other approaches, like the quality of data flow, may be utilised to combine or select the desired data stream. Figure 5.21 presents a case in which the UE has a three-cell active set in use, one of which is connected to another RNC. In this case, the BSs first combine the signal concerning their own radio paths and, then, combining of the data stream is done at the RNC level.

As far as soft and softer handovers are concerned, the idea is that subjective call quality will be better when the "final" signal is constructed from several sources (i.e., multipath propagation). In the GSM, subjective call quality depends on the transmission power used: put roughly, the greater the power the better the quality. In WCDMA, terminals cannot use much power because transmission levels that are too high will start blocking other users; thus, the best way to gain improved subjective call quality is to utilise multipath propagation.

In conclusion it can be stated that soft and softer handovers do in fact consume radio access capacity because the UE is occupying more than one radio link connection in the Uu interface. On the other hand, the added capacity gained from interference reduction is greater and, hence, system capacity is actually increased as long as soft and softer handovers are used reasonably. Another point worthy of mention concerning soft and softer handovers is that a softer handover utilises transport network capacity better than a soft handover; when summing is done at the BS level and macrodiversity is not in use, the system enjoys the advantages of multipath propagation and simultaneously minimises transmission resources. For this reason WCDMA BS sites are typically always expanded by sectorising first. When sectorising is no longer possible, new WCDMA carriers are implemented with a new frequency.

From the system structural architecture standpoint, the UMTS network supports the following types of handovers:

- Intra-BS/inter-cell handover (softer handover).
- Inter-BS handover, including hard and soft handovers.
- Inter-RNC handover, including hard, soft and soft–softer handovers.
- Inter-MSC handover.
- Inter-SGSN (Serving GPRS Support Node) handover.
- Inter-system handover.

5.3.1.2 RRM—Power Control

Power control is an essential feature of any CDMA-based cellular system. Without utilising an accurate power control mechanism, these systems cannot operate. In the following subsections we first describe why power control is so essential for these cellular systems and outline the main factors underlying this fact. We then describe the kinds of power control mechanisms utilised in WCDMA-FDD radio access.

The main reasons for implementing power control are the near–far problem (see p. 119), the interference-dependent capacity of WCDMA and the limited power source of the UE. Unlike Frequency Division Multiple Access (FDMA) and Time Division Multiple Access (TDMA), both of which are bandwidth-limited multi-accesses, WCDMA is an interference-limited multiple access. In FDMA and TDMA power control is applied to reduce inter-cell interference within the cellular system which arises from frequency reuse, while in WCDMA systems the purpose of power control is mainly to reduce intra-cell interference. Meeting these targets requires optimisation of radio transmission power (i.e., that the power of every transmitter is adjusted to the level required to meet the requested QoS). Determining the transmission power level is, however, a very sophisticated task due to unpredictable variation of the radio channel.

Whatever the radio environment, power received should be at an acceptable level (e.g., at the BS for the uplink to support the requested QoS). The target of power control is to adjust the power to the desired level without any unnecessary increase in UE transmit power. This ensures that transmit power is just within the required level (neither higher nor less), taking into account the existing interference in the system.

The influence of the multipath propagation characteristics and the technical characteristics of the WCDMA system (e.g., simultaneous bandwidth sharing and near–far phenomena) has the effect of power control being essential for the WCDMA system, to overcome the drawbacks caused by the radio environment and the nature of electromagnetic waves. Without power control such phenomena as fading and interference will drive down system stability and ultimately degrade its performance dramatically.

Maximising system capacity is an invaluable asset for both advanced cellular technology suppliers and cellular network operators. System capacity is maximised if the transmitted power of each terminal is controlled such that its signal arrives at the BS with the minimum required SIR. If a terminal's signal arrives at the BS with a power value that is too low, then the required QoS for radio connection cannot be met. If the received power value is too high, then, although the performance of this terminal is good, interference to all the other terminal transmitters sharing the channel is increased and may result in unacceptable performance for other users, unless their number is reduced.

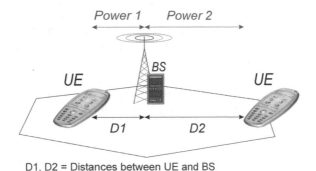

Figure 5.22   Near–far effect in the CDMA system

Due to the fact that the total bandwidth in the WCDMA system is shared simultaneously, other users can be experienced as a noise-like interference for a specific user. When the power control mechanism is missing or operates imperfectly, common sharing of the bandwidth creates a severe problem, called the "near–far" effect. In near–far situations the signal of the terminal that is close to the serving BS may dominate the signal of those terminals that are distant from the same BS. Figure 5.22 illustrates a situation in which the near–far problem could occur. The main factors that cause the near–far problem include the path loss variation of simultaneous users at different distances from the BS, the fading variation and other signal power variation of users caused by radio wave propagation mechanisms (described in Chapter 3).

In WCDMA the near–far effect can be mitigated by applying power control mechanisms, diversity techniques, soft handovers, multi-user receivers and, more generally, near–far resistance receivers. Because of the crucial drawback of the near–far effect on the performance of the WCDMA system, its mitigation is one of the pivotal purposes of power control mechanisms. These mechanisms have a considerable impact on WCDMA system capacity.

5.3.1.2.1 Basic Approaches Used in Power Control

For the reasons just outlined, it is relatively easy to determine that the optimal situation in the uplink case, from the BS receiver point of view, is that the power representing one UE's signal should always be equal to another UE's signal regardless of their distance from the BS. If so, the SIR will be optimal and the BS receiver able to decode the maximum number of transmissions. In reality, however, the radio channel is extremely unstable and radio services requested vary for different users (even for the same user and during the same radio connection). Therefore, the transmission power of UE should be controlled very accurately by utilising efficient mechanisms.

To achieve this, power control has been thoroughly investigated and, as a consequence, many power control algorithms have been developed since the advent of the CDMA scheme. These include distributed, centralised, synchronous, asynchronous, iterative and non-iterative algorithms. Most current algorithms either utilise SIR or transmit power as a reference point in the power control decision-making process.

The primary principle of Centralised Power Control (CPC) schemes is that they keep the overall power control mechanism centralised. As a result they require a central controller, which should have knowledge of all the radio connections in the RAN.

In contrast to CPC methods, distributed power control methods do not utilise a central controller. Instead, they distribute the controlling mechanism within the RAN and toward its edge. This feature makes them of special interest. CPC approaches bring about added complexity, latency and network vulnerability. The main advantage of a distributed power control algorithm is that it can respond more adaptively to a variable QoS, which is greatly important for cellular systems with packet-based transmission characteristics, like WCDMA.

5.3.1.2.2 Power Control Mechanism in UTRAN (WCDMA-FDD)

In WCDMA, power control is employed in both uplink and downlink directions. Downlink power control is basically for minimising interference with other cells and compensating for the interference from other cells, as well as achieving an acceptable SIR. However, power control for the downlink is not as vital as it is for the uplink. It is still implemented for the downlink, because it improves system performance by controlling interference from other cells.

The main target of uplink power control is to mitigate the near–far problem by making the transmission power level received from all terminals as equal as possible at the home cell for the same QoS. Therefore, uplink power control is used for fine-tuning terminal transmission power, resulting in mitigation of intra-cell interference and the near -far effect. Note that the power control mechanism specified for WCDMA is, in principle, a distributed approach.

The power control mechanisms used in the GSM are clearly inadequate to guarantee this situation in WCDMA and, thus, WCDMA takes a different approach to the matter. In the GSM, power control is applied to the connection once or twice per second, but, due to its critical nature in WCDMA, the power used in the connection is adjusted 1,500 times per second (i.e., the power control cycle is repeated for each radio frame in association with the DCH). Therefore, power adjustment steps are considerably faster than in the GSM.

To manage power control properly in WCDMA, the system uses two different power control mechanisms, as defined in Figure 5.23. These power control mechanisms are:

- Open Loop Power Control (OLPC).
- Closed Loop Power Control (CLPC), including inner and outer loop power control mechanisms.

By applying all these different power control mechanisms together, the UTRAN benefits from the advantages of CPC as well by overlaying the inner CLPC with the outer CLPC mechanism in order to keep the target SIR at an acceptable level.

5.3.1.2.3 Open Loop Power Control (OLPC)

In OLPC, which is basically used for uplink power adjusting, the UE adjusts its transmission power based on an estimate of the received signal level from the BS

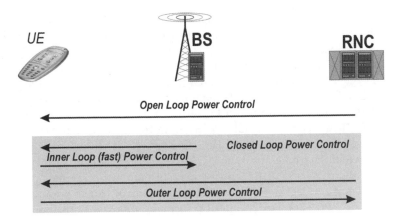

Figure 5.23   The main power control mechanisms employed in WCDMA

CPICH when the UE is in idle mode and prior to Physical Random Access Channel (PRACH) transmission. In addition, the UE receives information about the allowed power parameters from the cell BCCH when in idle mode. The UE evaluates the path loss, and from this together with figures received from the BCCH and the UE it is able to estimate an appropriate power level to initialise the connection.

Figure 5.24 illustrates the OLPC as applied for the uplink case. In this process, the UE estimates the strength of the transmission signal by measuring the received power level of the pilot signal from the BS in the downlink, and adjusts its transmission power level in a way that is inversely proportional to the pilot signal power level. Consequently, the stronger the received pilot signal the lower the UE power transmitted.

In the case of WCDMA-FDD, OLPC alone is insufficient to adjust UE transmission power, because the fading characteristics of the radio channel vary rapidly and independently between uplink and downlink. Therefore, to compensate for the rapid changes in signal strength a CLPC mechanism is also needed. Nevertheless, OLPC is useful for determining the initial value of transmitted power and for mitigating drawbacks of the log-normal-distributed path loss and shadowing.

5.3.1.2.4 Closed Loop Power Control (CLPC)
CLPC is utilised to adjust transmission power when the radio connection has already been established. Its principal purpose is to compensate for the effect of rapid changes in radio signal strength and, hence, it needs to be fast enough to respond to these changes.

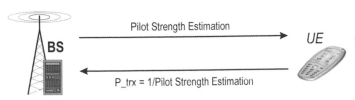

Figure 5.24   Open Loop Power Control for uplink

Figure 5.25 illustrates the basic uplink CLPC mechanism specified for WCDMA. In this case the BS commands the UE to either increase or decrease its transmission power using a cycle of 1.5 kHz (1,500 times per second) by 1-, 2- or 3-dB step sizes. The decision to increase or decrease power is based on the received SIR as estimated by the BS. When the BS receives the UE signal it compares the signal strength with the predefined threshold value at the BS. If the UE transmission power exceeds the threshold value, the BS sends a Transmission Power Command (TPC) to the UE to decrease its signal power. If the received signal is lower than the threshold target the BS sends a command to the UE to increase its transmission power. It should be emphasised that various measurement parameters, such as SIR, signal strength, Frame Error Ratio (FER) and Bit Error Ratio (BER), can be used to compare the quality of the power received and to make a decision about whether to control transmission power or not.

Note also that CLPC is utilised to adjust transmission power in the downlink as well. In the case of downlink CLPC, the roles of the BS and the UE are interchanged; that is, the UE compares the received signal strength from the BS with a predefined threshold and sends the TPC to the BS to adjust its transmission power accordingly.

In WCDMA, the CLPC mechanism consists of inner loop and outer loop variants. What we have so far described is related to the inner loop variant; this is the fastest loop in the WCDMA power control mechanism and, hence, it is occasionally called "fast power control".

Another variant of the CLPC is the OLPC mechanism. The main target of OLPC is to keep the target SIR for the uplink inner loop power control mechanism at a satisfactory quality level. Thanks to macrodiversity, the RNC is aware of current radio connection conditions and quality and, therefore, is able to define the allowed power levels of the cell and target SIR to be used by the BS when determining TPCs. To maintain the quality of the radio connection, the RNC uses this power control method to adjust the target SIR and keep any variation in the quality of the connection in check. By doing this, the network is able to compensate for changes in radio interface propagation conditions and to achieve the maximum target quality for BER connection and FER observation. In fact, OLPC fine-tunes the performance of inner loop power control.

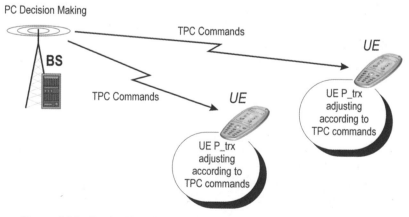

Figure 5.25   Basic Closed Loop Power Control mechanism in uplink

Together, the OLPC and CLPC mechanisms have a considerable impact on the terminal's battery life and overall system capacity in any cellular system (particularly in CDMA-based mobile systems).

5.3.1.2.5 Power Control in Specific Cases

In addition to the ordinary power control mechanisms used in WCDMA, there are additional approaches to cope with specific cases. These include controlling the transmission power associated with soft-handover mode, Site Selection Diversity (SSDT) and compressed mode (slotted mode).

In soft-handover mode, the transmission power of the UE is adjusted based on selection of the most suitable power control command from those TPCs that it receives from the different BSs to which it has simultaneous radio links (Figure 5.26). In this case and because the UE receives more than one TPC command from the different BSs independently, the received TPC commands may differ from each other. This may result from the fact that power control commands are not efficiently protected against errors or, simply, it may result from the network environment. This leads to a conflict situation for the UE. The basic approach to resolve this problem is that, if at least one of the TPC commands calls for a decrease in transmission power, then the UE decreases its power. In soft-handover mode the UE can utilise a threshold for detecting reliable commands, from which it can decide whether to increase or decrease its transmission power.

SSDT is another special power control solution. The principle behind SSDT is that the BS with the strongest signal is dynamically chosen as the only transmitting BS (Figure 5.27). Then, the other BSs, to which the UE is simultaneously radio-connected, turn off their DPDCHs. Therefore, the transmit power is adjusted based on the power control commands of the BS with the strongest signal. It seems that this method reduces the downlink interference generated while the UE is in soft-handover mode.

Yet, in compressed mode the transmission and reception of the BS and UE are stopped for a predefined period to allow time for inter-frequency radio measurements to be carried out (e.g., in conjunction with an inter-system handover event). As a result there is also a break in operation of the transmission power-adjusting mechanism.

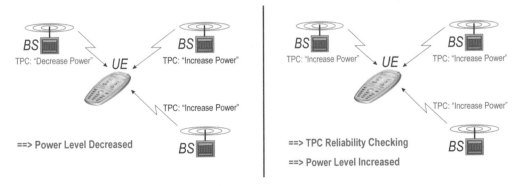

Figure 5.26 Basic principle underlying the WCDMA power control mechanism in soft-handover mode

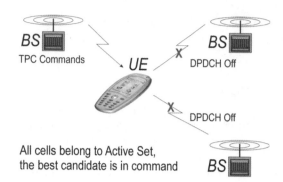

Figure 5.27   Basic principle underlying the Site Selection Diversity Technique (SSDT)

In this case the receiver of the power control commands (e.g., the UE for the uplink case) is allowed to increase or decrease its transmit power with larger step sizes so that it can reach the desirable SIR level as fast as possible.

5.3.1.3 RRM—Admission Control (AC) and Packet Scheduler (PS)

WCDMA radio access has its very own limiting factors, some are a result of the underlying principles used in it and others are environment-dependent. The most important and at the same time the most difficult limitation to control is the interference that occurs due to the radio path effect. Due to the nature and basic characteristics of WCDMA, every UE accessing the network generates a signal. This signal can also be interpreted as interference from another UE's point of view. When the WCDMA cellular network is eventually planned, one of the basic criteria behind the planning will be to define an acceptable interference level with which the network is expected to function correctly. Defining a planning threshold to fit the actual level of signals the UE transmits is a challenging task and may set practical limits for radio interface capacity.

To be more specific, a predefined SIR value will have to be used in this context. Based on radio network planning the network is, in theory, stable, as long as the predefined SIR level is not exceeded within the cell. This effectively means that, at the BS receiver, the interference and the signal must have different power levels in order to be able to extract one signal (code) from other signals using the same carrier. If the power distance between interfering components and the signal is too small the BS will no longer be able to extract an individual signal (code) from the carrier. Every UE having a bearer active through the cell "consumes" a part of the SIR and the cell is used up to its maximum level when the BS receiver is unable to extract the signal(s) from the carrier.

The main task of AC is to estimate whether a new call can have access to the system without sacrificing the bearer requirements of existing calls (Figure 5.28). Thus, the AC algorithm should predict the load of the cell if the new call is admitted. Note that the availability of terrestrial transmission resources is also verified. Based on the AC algorithm, the RNC either grants or rejects access.

Theoretically, it is possible to discover whether the SIR or interference margin has a direct relationship to cell load. If we express "cell load with load factor" as a parameter

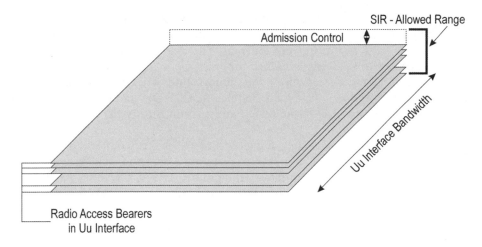

Figure 5.28 Main principle underlying Admission Control (AC)

that expresses cell percentage load as shown in Figure 5.29 and denote the interference margin as I, this leads to the following equation:

$$I = 10 \times \log\left(\frac{1}{1 - \text{Load factor}}\right)$$

When we lump interference margins calculated with different load factor values together we arrive at the results shown in Figure 5.29.

Based on the graph it is fairly easy to show that interference in a cell will be very difficult to control when cell load exceeds 70%. This is why the WCDMA radio network is normally dimensioned with expected capacity equivalent to a load factor value of 0.5 (50%); this value has a safety margin in it and the network will most likely operate steadily.

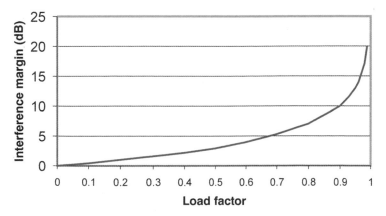

Figure 5.29 Interface margin as a function of cell load

Due to the nature of traffic in the UMTS system both Real Time (RT) and Non-Real Time (NRT) traffic should be controlled and balanced carefully. The ways of controlling RT and NRT traffic are very different. The main difference is that for an RT (circuit call) service an RAB is maintained continuously in a dedicated state. In the case of an NRT service (packet connection), the channel state varies as a consequence of RB bit-rate variation as well as system load. So, bit-rate variation and the bursty characteristic of packet connection should be taken into account during the packet connection admission process.

Figure 5.30 illustrates the main characteristics of packet connection based on a WWW browsing session. During WWW packet connection the user typically sends a WWW address to the network for data-fetching purposes. As a response, the network downloads data, typically a HyperText Markup Language (HTML) page, from the desired address. The desired address may also lead to document/file download. At the next step, most users spend a certain amount of time studying the information, called "reading time". During reading time there is no need for AC to keep the RAB in a dedicated state and radio resources should then be used for other purposes (e.g., other incoming and outgoing circuit- or packet-switched connections). Meanwhile, the information from the packet connection session is kept at an upper level. So, when the user wants to have another page downloaded, subsequent to reading time, the precondition for the call is still available.

Admission control is responsible for the handling of packet connections with bursty traffic (i.e., traffic having a very random arrival time), the number of packet calls per session, reading time and the number of packets within a call. Therefore, AC must utilise very sophisticated traffic models and statistical approaches to optimally control the requested RAB(s). This can be done by scheduling NRT RABs, and accepting,

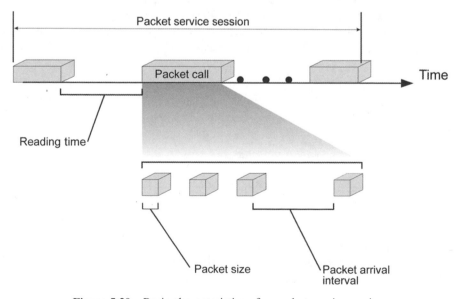

Figure 5.30   Basic characteristics of a packet service session

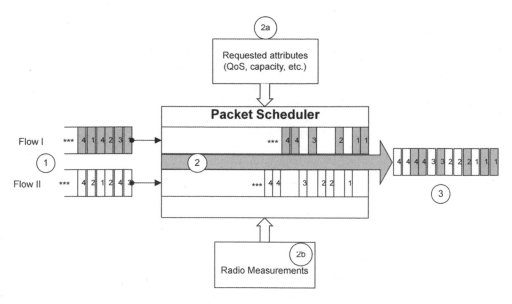

Figure 5.31 General principles underlying a Packet Scheduler (PS)

queuing or rejecting RT RABs. It is the responsibility of AC to maintain and control the QoS of accepted RABs and their influence on the overall performance of the UTRAN.

Figure 5.31 illustrates a basic packet-scheduling process. As shown, the main principles and steps in packet-scheduling can be summarised as follows:

1. The PS in the RNC (or in the BS for HSDPA) receives data flows from the radio channels related to one or more radio links. The data flow may have already been prioritised entirely or partially based on the requested attributes and link conditions of each original radio connection. Alternatively, and if necessary (based on the current state of the radio link, etc.), even that part of the scheduling function should be reconsidered in the PS.

2. The data packets are picked up from and preferably aligned on the original priority pattern of the packets or whole flows. The connection attributes are provided by upper layer protocols (2b). When mapped to the current link state and to the state of radio and physical resources, the contents of the requested QoS pattern may or may not be met. In the former case, the primary arrangement is preserved and possibly matched to the appropriate radio channel based on the nature and capability of the radio channel as well as its QoS requirements. However, if there is no room to handle the predefined QoS, both QoS and mapping to a suitable radio channel have to be carried out again. If so, the original QoS requirements, current link state and radio resource state constitute a new basis for reconsidering packet priorities, forming the final QoS figures as well as their matching to underlying radio channels.

3. The final scheduled and prioritised packets of the flows are released to the appropriate link(s) (i.e., a link(s) that has the best QoS considerations).

5.3.1.4 RRM—Code Management

The channelisation codes and scrambling codes used in Uu interface connections are managed by the RNC. In principle, the BS could manage them at the cell level, but then the system may behave unstably (e.g., there may be soft handovers at the RAN level) when the RNC is otherwise controlling the radio resources. When the codes are managed by the RNC, it is also easier to allocate Iub data ports for multipath connections. However, actual code generation is mainly done at the BS.

The Uu interface requires two kinds of codes for proper functionality: some codes must correlate with each other to a certain extent; others must be orthogonal or do not correlate at all. Every cell uses one scrambling code; the UE is able to distinguish between cells by recognising this code. Beneath every scrambling code the RNC has a set of channelisation codes. This set is the same under every scrambling code. The BCH information is coded with a scrambling code value and, thus, the UE must find the correct scrambling code value first in order to access the cell. When a connection between the UE and the network is established, the channels used must be separated; channelisation codes are used for this purpose. The information sent over the Uu interface is spread using one spreading code per channel and the spreading code used is the scrambling code $\times$ the channelisation code.

The various types and ways of using codes and their impact on the overall performance of the WCDMA system imply that intelligent mechanisms are employed to control code allocation and re-allocation. We can confirm this is true because the available codes are limited (especially for downlink): in the case of downlink, a total of $2^{18} - 1 = 262,143$ scrambling codes, numbered $0 \ldots 262,142$ can be generated. However, not all the scrambling codes are used. Scrambling codes are divided into 512 sets each having 1 primary scrambling code and 15 secondary scrambling codes (Figure 5.32).

Primary scrambling codes consist of scrambling codes $n = 16 * i$ where $i = 0, \ldots, 511$. The ith set of secondary scrambling codes consists of scrambling codes $16 * i + k$, where $k = 1, \ldots, 15$. Between each primary scrambling code and the 15 secondary scrambling codes one-to-one mapping is done in a such a way that the ith primary scrambling code corresponds to the ith set of scrambling codes.

Hence, according to the above, there are $k = 0, 1, \ldots, 8,191$ scrambling codes available. Based on 3GPP TS 25.213, each of these codes is associated with an even-numbered alternative scrambling code and an odd-numbered alternative scrambling code, which can be used for compressed frames. The even-numbered alternative scrambling code corresponding to scrambling code k is scrambling code number $k + 8,192$, while the odd-numbered alternative scrambling code corresponding to scrambling code k is scrambling code number $k + 16,384$. The set of primary scrambling codes is further divided into 64 scrambling code groups, each consisting of 8 primary scrambling codes. The jth scrambling code group consists of primary scrambling codes $16 * 8 * j + 16 * k$, where $j = 0, \ldots, 63$ and $k = 0, \ldots, 7$.

Each cell is allocated one and only one primary scrambling code. The primary CCPCH (Common Control Physical Channel) is always transmitted using the primary scrambling code. Hence, the other downlink physical channels can be trans-

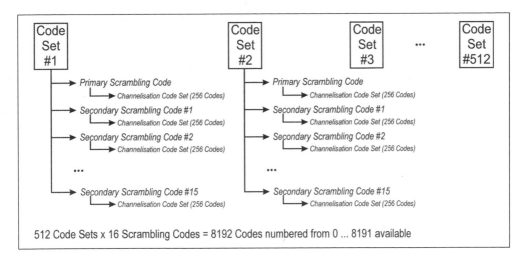

Figure 5.32 Scrambling codes and channelisation codes, their amounts and relationship

mitted with either the primary scrambling code or a secondary scrambling code from the set associated with the primary scrambling code of the cell.

For the uplink case the situation is different because there are millions of uplink scrambling codes available. The specified number of uplink scrambling codes for WCDMA is 2^{24}. All uplink channels have to use either short or long scrambling codes, except the PRACH, for which only the long scrambling code is used. Therefore, uplink code allocation is not as crucial as downlink code allocation in WCDMA.

Channelisation codes are used for channel separation in both uplink and downlink directions, have different SF values and, thus, also different symbol rates. There are 256 channelisation codes available and the SF indicates how many of these codes are used in the connection. Thus, the greater the SF value the better the channelisation codes are utilised and the radio resources are used. In the case of high user data rates the SF gets a relatively small value. This leads to a situation in which high user data rates consume more air interface code capacity.

Channelisation codes are by their very nature *orthogonal* or, at least, they have orthogonal properties. "Orthogonality" in this case means that the channelisation codes in the 256-member code list are selected in such a way that they interfere with each other as little as possible. This is necessary to get good channel separation. On the other hand, the code used for user and cell separation (scrambling code) must have good correlation properties. Channelisation codes do have these, and this is the basic reason that both scrambling codes and channelisation codes are used.

Every WCDMA cell normally uses one downlink scrambling code that is locally unique and that acts basically like a cell ID—cell IDs are given to the UE for network identity recognition purposes and locally unique scrambling codes are for correct radio wave recognition, respectively. The characteristics of this scrambling code are pseudo-random (i.e., it is not always orthogonal). Under this scrambling code the cell has a set of channelisation codes, which are orthogonal in nature and used for channel separation purposes.

5.3.2 UTRAN Control Functions

In order for the UTRAN to control and manage RABs, essential for provision of the RAB service, it should perform other functions in addition to RRM algorithms. These can be classified as:

- System information broadcasting.
- Random access and signalling bearer set-up.
- RAB management.
- UTRAN security functions.
- UTRAN-level Mobility Management (MM).
- Database-handling.
- UE positioning.

5.3.2.1 System Information Broadcasting

An important function of the RNC is to handle the system information task. System information is used to maintain both the radio connections between the UE and the UTRAN and also control the overall operation of the UTRAN. The RNC broadcasts the system information elements to assist the UTRAN's controlling functions by providing the UE with the essential data needed when communicating with the UTRAN (e.g., radio measurement criteria, paging occasion indication, radio path information, assistance data for positioning purposes, etc.). The system information can be received by the UE both in idle mode and any connected mode when it has been identified by the UTRAN. System information services may also be used (e.g., by the CN to provide broadcast services). The RNC utilises point-to-multipoint system information broadcasting to keep the UE in touch with the UTRAN when necessary.

From the protocol architecture point of view, system information broadcast functionality is part of the RRC and is terminated at the RNC.

Figure 5.33 illustrates the system information structure, which together with segmentation class information is used by the RNC as the basis for controlling system information segmentation and scheduling. As shown, the system information elements are organised based on the Master Information Block (MIB), two optional Scheduling Blocks (SBs) and System Information Blocks (SIBs), which contains actual system information. The 3GPP specification TS 25.331 defines up to 17 different types of SIBs, some of which may also contain sub-SIBs. An MIB contains reference and scheduling information about the number of SIBs in a cell. It may also contain reference and scheduling information about one or two SIBs, which in turn gives reference and scheduling information about further SIBs. Therefore, only the MIB and SIBs can contain scheduling information for an SIB.

Because SIBs are characterised differently and grouped specifically in terms of their repetition rate and criticality the RNC can use them for different purposes (e.g., it uses SIB#1 to inform the UE about timers and counters to be used in idle and connected mode). Other examples are SIB#2 and SIB#3 which the RNC uses to inform the UE about UTRAN-level MM, cell selection and re-selection, respectively.

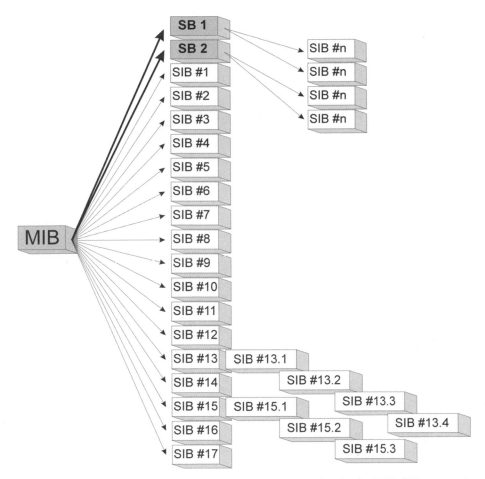

Figure 5.33 Basic structure of system information in the UTRAN

The RNC may handle the constructing process together with the BS (e.g., when modifying system information scheduling). Nevertheless, the controlling function of system information belongs entirely to the RNC. Once the RNC has structured the system information data it forwards the information to the BS to be broadcast over the air interface to the UE. Based on the air interface situation the BS informs the RNC about the ability to broadcast system information on the Uu interface (Figure 5.34).

5.3.2.2 Initial Access and Signalling Connection Management

Before the RNC can map any requested RABs to the RB, it needs to create the signalling connection between the UE and the CN. In this context, 3GPP TR 25.990 defines the *signalling connection* as an acknowledgement-mode link between the UE and the CN to transfer higher layer information between the entities in the Non Access

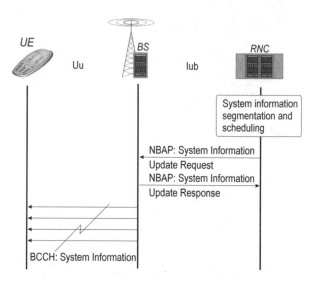

Figure 5.34   System information broadcasting

Stratum (NAS). In order to do this the RNC uses RRC connection services in creating
the *Signalling Radio Bearer* (SRB) between the UE and the UTRAN to provide the
transferring services for signalling connection. Once the RNC has created the SRB, it
uses the first RLC RBs to convert and transfer the signalling connection through the
UTRAN to the UE. (Remember: RB is defined as the service provided by the RLC
layer for transfer of user data between the UE and the RNC.) However, primary RBs
have also used been defined for signalling connection purposes (Figure 5.35).

The whole process starts when the UE enters idle mode by turning on its power. After
power-on, the UE attempts to make contact with the UTRAN. The UE looks for a
suitable cell in the UTRAN and chooses the cell to provide available services and tunes
in to its control channel. This is known as "camping on the cell". Camping includes a
cell search, in which the UE looks for a cell and determines the downlink scrambling
code and frame synchronisation of that cell. Cell search is done over several steps.

During the first step the UE uses the Synchronisation Channel's (SCH) primary
synchronisation code, which is common to all cells, to bring about slot synchronisation
with a cell. During the second step, the UE uses the SCH's secondary synchronisation
code to bring about frame synchronisation and identify the code group of the cell found
in the first step. During the third and last step, the UE determines the exact primary
scrambling code used by the cell found within the UTRAN. If the UE identifies the
primary scrambling code, then the P-CCPCH can be detected and the system and cell-
specific broadcast information defined by the RNC can be read.

In this regard the UE should be aware of the way system information is structured
and scheduled by the RNC (i.e., on the basis of MIBs, SBs and SIBs). If the Public
Land Mobile Network (PLMN) identity found from the MIB on the BCH matches the
PLMN (or list of PLMNs) that the UE is searching for, it can continue to read the
remaining BCH information, such as parameters for the configuration of the common
physical channels in the cell (including PRACH and secondary CCPCH). Otherwise,

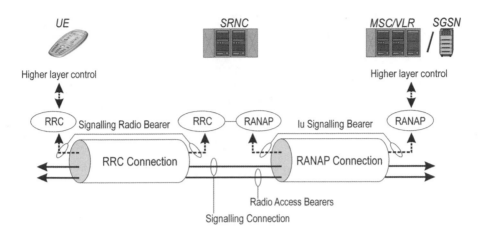

Figure 5.35 Relation between SRB, RB, signalling connection, RAB and the role of RNC

the UE may store the identity of the found PLMN for possible future use and restart the cell search mechanism.

The first cell search for a PLMN is normally the most difficult for the UE, since it has to scan through a number of scrambling codes to find the correct one. Once the UE obtains the necessary information to capture the BS controlled by the corresponding RNC it can request *initial access* to the UTRAN, resulting in a transition from idle mode to connected mode.

Up to this point the RNC has been involved in the controlling function in terms of system information broadcasting and controlling the related BS to which the UE was trying to get radio connection. Once this has been carried out, the RNC participates actively in controlling how access is provided by considering the context of the RRC connection set-up message requested by the UE via the Random Access Channel (RACH). In this regard, the RNC plays a central role in controlling radio connection by checking the UE identity, the reason for the requested RRC connection and UE capability, which helps it in allocating the initial signalling RB for the UE. Based on this information the RNC decides whether or not to allocate the signalling RB to the UE, which is used to carry the rest of the signalling for initial access and to provide all the services thereafter. In case the RNC does not accept the access request the UE can restart the initial access within a predefined time span.

Irrespective of the direction from which the higher layer service is requested, the RRC connection requested for creating the SRB is always started by the UE, as shown in Figure 5.36. When the RNC receives the RRC connection set-up request, it sets up a radio link over the Iub interface to the BS with which the UE is going to have radio connection. In this context, *radio link* is defined as the logical association between a single UE and a single UTRAN access point, and the physical act of making it a reality comprises one or more RB transmissions. If this step is successful, then the RNC informs the UE that the RRC connection is set up and the UE responds to the RNC concerning completion of RRC connection.

The SRB is now ready between the RNC and the UE and, hence, the RNC can convert and transfer signalling connections and RABs between the UE and the CN.

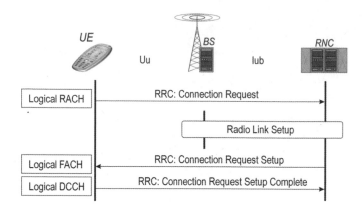

Figure 5.36   Radio Resource Control (RRC) connection set-up

It should be mentioned that, regardless of how many signalling connections and RABs exist between the UE and the CN, there is only one RRC connection used by the RNC to control and transfer them between the UE and the UTRAN. It is the function of the RNC to reconfigure the lower layer services needed, based on the number of signalling connections and RABs that need to be transferred.

5.3.2.3 Radio Bearer Management

Once the SRB is established between the radio network and the UE, as just described, it is time for the RNC to map the requested RAB(s) onto the radio bearer for transfer between the UTRAN and the UE. In this way the RNC creates the impression that there is a fixed bearer between the CN and the UE.

After the requested RABs have been negotiated between the UE and the CN via the signalling connection, it is forwarded to the RNC for further action. In this regard the main function of the RNC is to analyse the attributes of the requested RAB(s), evaluate the radio resources needed for these, activate or reconfigure the radio channels by using lower layer services and map the requested RAB(s) to the radio bearer. Therefore, the RNC covers all of the control functions related to the supporting of the RAB(s) as well as configuring the available radio resources for that purpose.

As shown in Figure 5.35, the RAB is carried within the RRC connection between the RNC and the UE over the radio interface and within the Radio Access Network Application Part (RANAP) protocol connection between the RNC and the CN over the Iu interface. In this association the RNC acts as a protocol converter between the RAN and the CN. The RNC maps the requested RABs onto the radio bearers by using current radio resource information and controls the lower layer services.

To optimise the use of radio resources as well as sharing the bandwidth and network physical resources possessed by different operators, the UTRAN can also support the CN distribution function for NAS messages. To accomplish this, the RRC protocol transparently transfers the CN messages within the access network by using the direct transfer procedure. When this happens, a CN-specific indicator is inserted in these messages and the distribution function entities in the UE and the SRNC use the indicator to direct the messages to the appropriate CN and vice versa. In this way appropriate MM in the UE domain and the appropriate CN domain is reached.

5.3.2.4 UTRAN Security

Radio connection security is another important function handled by the RNC. The RNC is involved in both integrity-checking and ciphering mechanisms. The former is used to protect the signalling connection between the UE and the UTRAN over the radio interface and the latter is used for protecting the user data transferred between the UE and the UTRAN. The RNC ciphers the signalling and user data by using predefined integrity and ciphering algorithms. In this regard it needs to generate random numbers and maintain time-dependent counter values for integrity-checking of signalling messages. It also verifies and deciphers received messages by using the same algorithms. Algorithms and other UMTS-security-related issues are thoroughly addressed in Chapter 9.

5.3.2.5 UTRAN-level Mobility Management (MM)

UTRAN-level MM refers to those functions the RNC handles in order to keep the UE in touch with UTRAN radio cells, taking into account the user's mobility within the UTRAN and the type of traffic or RAB(s) it is using.

As mentioned in Section 5.3.1, the nature of traffic in the UMTS network environment is substantially different from the traditional circuit-switched type of traffic. This means that the UTRAN is able to share radio resources for RABs having a different QoS. Therefore, more sophisticated mechanisms need to be utilised to exploit radio resources efficiently and to meet the diverging QoS requirements of the RABs as effectively as possible. To respond to this demand, the RRM algorithms must be accompanied by more adaptive MM than was the case for 2G.

As a result, the concepts of RRC-state transition and hierarchical MM have been specified for the UMTS, including UTRAN-level MM, which is a much newer system compared with the GSM MM, which was taken care of by the CN subsystem. In this approach, the UE can have different states during radio connection, depending on the type of connection it has to the UTRAN as well as the motion speed of the UE. It is the responsibility of the RNC to control UE states by considering UE mobility, the requested RAB(s) and its variation in terms of bit rate.

The cornerstones of UTRAN-level mobility are based on the concept of cell, UTRAN Registration Area (URA), UTRAN Radio Network Temporary Identifier (U-RNTI) and the RRC-state transition model. In addition, the primary purpose of defining the different logical roles for the RNC and specifying the Iur interface was also to support UTRAN's internal MM.

A *URA* is defined as an area covered by a number of cells. It is only internally known in the UTRAN and, hence, is not visible to the CN. This means that whenever a UE has an RRC connection its location is known by the UTRAN at the accuracy level of one URA. Every time the UE enters a new URA it has to perform a URA-updating procedure. Having the interface between RNCs supports the URA procedure, which in principle can cover different RNC areas.

Based on the 3GPP specification TS 25.401, *Radio Network Temporary Identities* (RNTIs) are used as UE identifiers within the UTRAN and in signalling messages between the UE and the UTRAN. There are four types of RNTI used to handle UTRAN's internal mobility:

- SRNC RNTI (S-RNTI).
- Drift RNC RNTI (D-RNTI).
- Cell RNTI (C-RNTI).
- UTRAN RNTI (U-RNTI).

S-RNTI is allocated at the same time as RRC connection set-up. This is carried out by the SRNC with which the UE has an RRC connection. By this means the UE can identify itself to the SRNC and also the SRNC can reach the UE.

D-RNTI is allocated by a Drift RNC (DRNC) at the same time as context establishment. It is used to handle UE connection and context over the Iur interface.

C-RNTI, which is allocated when the UE accesses a new cell, is a CRNC-specific identifier that is used to identify the CRNC with which the UE has an RRC connection. By this means the UE can identify itself to the CRNC and also the CRNC can reach the UE. This identifier is used when the UE is in cell Forward Access Channel (FACH) state.

U-RNTI is, on the other hand, allocated to a UE having an RRC connection and identifies the UE within the UTRAN. It is used as a UE identifier for initial cell access (at cell change) when an RRC connection exists for this UE and for UTRAN-originated paging (including associated response messages).

Allocating and handling these identifiers is primarily the responsibility of the RNC.

Having defined the RNC roles, UE identifiers and the concept of URA, let us turn to the question of how to combine this with the state transition in order to handle UTRAN's internal mobility and RRM. The main principles underlying the way in which RRC-state transitions are handled are shown in Figure 5.37 and can be summarised as follows:

- *No radio connection*: the UE location is known only by the CN (within the accuracy of the actual location area and/or routing area, which are CN-level location area concepts that we will describe in Chapter 6). This means that location information is stored in the network in response to the latest MM activity the UE has performed with the CN.
- *Radio connection over common channels*: if radio connection uses common channels (e.g., FACH and CPCH), the location of the UE is known within the accuracy of a cell. This location information is updated by means of the RRC procedure called "cell update". This procedure can be used when there are low-bit-rate data to be transferred between the UTRAN and the UE.
- *Radio connection over DCHs*: in this case the UTRAN has allocated dedicated resources for connection—one DPDCH and one DPCCH is the minimum, but there may be a number of DPDCHs depending on the bandwidth used. The

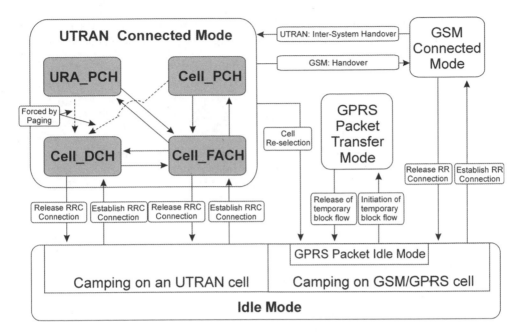

Figure 5.37 RRC-state automation including GSM/GPRS connection alternatives

location of the UE is known at the cell level. Depending on the type of connection, different RRC procedures can occur. If a dedicated connection is used to carry a service using the highest QoS class (e.g., a circuit-switched voice call) the UTRAN and the UE perform handovers. If the service the radio connection carries uses a lower QoS class (e.g., Web-surfing) that allows buffering and delays, handovers may not be completed as such. Instead, the UE uses the cell update procedure to inform the UTRAN about its location (i.e., the place for delivery of data over the radio connection).

- *Radio connection in cell Paging Channel (PCH) state*: once the UE is in cell FACH or cell DCH state, but there are no data to transfer, the UE state is changed to the cell PCH state so that the UE can monitor paging occasions based on predefined Discontinuous Reception (DRX) cycles and, hence, hear the paging channel. In this state the location of the UE is known to an accuracy of a single cell, which is sometimes called the "home cell".

- *Radio connection in URA PCH state*: once the UE is in cell FACH or cell DCH state and the amount of data for transfer between the UE and the UTRAN is not considerable, or UE mobility is high, then its state can be transferred to URA PCH state to avoid periodic cell update and to release dedicated radio resources, respectively. In this case the location of the UE is known only at the URA level and, hence, to obtain cell-level location accuracy the UE should be paged by the UTRAN/RNC. This is the case, for example, in the context of the mobile positioning process when cell-based positioning is used. In this state the UTRAN benefits from having the Iur interface between the RNCs and the URA.

- *Idle mode*: RRC idle mode is equivalent to the state in which the UE and the UTRAN do not have any radio connection (e.g., the UE is switched off). In this RRC state the network has no valid information concerning the location of the UE.

5.3.2.6 Database Handling

Like the GSM's Base Station Controller (BSC), the RNC holds a store of information called a "radio network database". This database is where cell information is stored. It is in this database that cell information about the cells the RNC controls is stored. The RNC then sends this cell-related information to the correct cell, which further distributes it by broadcasting it over the Uu interface towards the UE. The radio network database contains a great deal of cell-related information and the codes used in the cell are only part of this information.

The information in a radio network database can roughly be classified as:

- Cell ID information: codes, cell ID number, location area ID and routing area ID.
- Power control information: allowed power levels in uplink and downlink directions within the cell coverage area.
- Handover-related information: connection quality and traffic-related parameters triggering the handover process for the UE.
- Environmental information: neighbouring cell information (for use by both the GSM and WCDMA). These cell lists are delivered to the UE, which then undertakes radio environment measurements as the preliminary work for handovers.

5.3.2.7 UE Positioning

Another important function carried out by the RNC is control of the UE-positioning mechanism in the UTRAN. In this association it selects the appropriate positioning method and controls how the positioning method is carried out within the UTRAN and in the UE. It also coordinates the UTRAN resources involved in the positioning of the UE.

With network-based positioning methods the RNC calculates an estimate of the location and later indicates the accuracy achieved. It also controls a number of Location Measurement Units/BSs (LMUs/BSs) for the purpose of obtaining radio measurements to locate or help to locate the UE.

Since UE-positioning is seen as a value-added service in UMTS networks, we have described it more thoroughly in Chapter 8.

6

UMTS Core Network

Heikki Kaaranen and Miikka Poikselkä

The Universal Mobile Telecommunication System (UMTS) Core Network (CN) can be seen as the basic platform for all communication services provided to UMTS subscribers. The basic communication services include switching of circuit-switched calls and routing of packet data. The 3G Partnership Project (3GPP) R5 also introduces a new subsystem called the "IP Multimedia Subsystem" (IMS). The IMS opens up the Internet Protocol (IP)-based service world for mobile use by seamlessly integrating the mobile world and the Internet world and providing sophisticated service mechanisms to be used in the context of mobile communications.

The CN maps end-to-end Quality of Service (QoS) requirements to the UMTS bearer service. When inter-connecting with other networks, QoS requirements also need to be mapped onto the available external bearer service. The gateway role of the UMTS CN in creating an end-to-end service path is illustrated in Figure 6.1. The external bearer does not fall within the scope of UMTS system specifications and this may create some local problems if the QoS requirements to be satisfied between the UMTS and external network do not match.

Between the Mobile Termination (MT) and the CN the QoS is provided by the radio access bearer. The radio access bearer hides QoS handling over the radio path from the CN. Within the CN, QoS requirements are mapped to its own bearer service, which in turn is carried by backbone bearers on top of the underlying physical bearer service. A challenge to CN implementation is that the operator is pretty much free to choose how to implement physical backbone bearers. These bearers rely on the physical transmission technologies used between CN nodes. Typical transmission technologies, like PDH and SDH, with Pulse Code Modulation (PCM) channelling or with Asynchronous Transfer Mode (ATM) cell-switching are used. In 3GPP R5 the emphasis is on replacing these technologies by the Internet Protocol (IP) wherever and whenever possible, since making this transport network uniform simplifies the functionality of higher protocol layers.

The UMTS represents a kind of philosophy for use in production of a universal core that is able to handle a wide set of different radio accesses. Looking back at the network evolution discussed in Chapter 2, we see there are three types of recognised radio accesses as far as 3GPP R5 is concerned: WCDMA/HSDPA, GSM/EDGE and,

UMTS Networks Second Edition   H. Kaaranen, A. Ahtiainen, L. Laitinen, S. Naghian and V. Niemi
© 2005 John Wiley & Sons, Ltd   ISBN: 0-470-01103-3

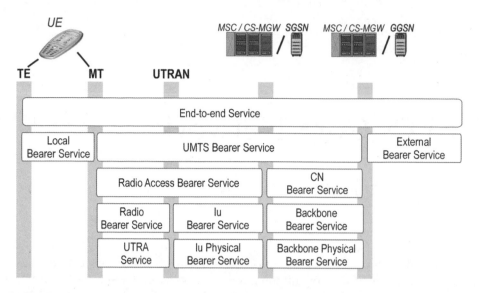

Figure 6.1   Bearer and Quality of Service (QoS) architecture in the Core Network (CN)

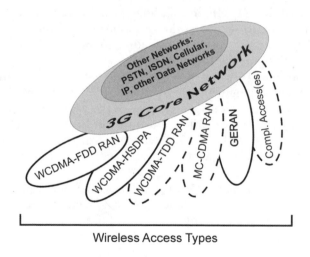

Figure 6.2   Universal core for wireless access

possibly, complementary access. Of these, WCDMA/HSPDA and GSM/EDGE are implemented, while complementary access is under study. The core part of the UMTS network does not evolve in as straightforward a way as the radio network due to both the CN's traditional infrastructure basis and its advanced technologies, which may have a number of different impacts on the evolution of the core part of the UMTS. Figure 6.2 shows the conceptual nature of the UMTS CN: the radio accesses drawn as continuous lines are the ones used at the outset and the others are regarded as access candidates as time goes by.

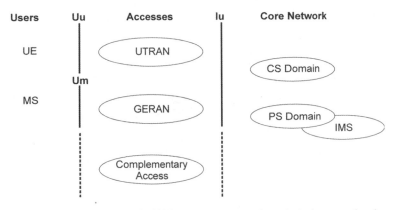

Figure 6.3   Core Network (CN) structure on a domain/subsystem level

6.1 UMTS Core Network Architecture

3GPP R99 introduced new mechanisms and capacity increases for the access network side. Starting with 3GPP R4 and its actual realisation in 3GPP R5 the CN has undergone major changes. In this chapter we will introduce, albeit briefly, the main characteristics of 3GPP R5.

As shown in Figure 6.3 the UMTS CN consists of equipment entities called "domains" and "subsystems" whose purpose is to describe the traffic characteristics the equipment takes care of. Based on this division, the UMTS CN contains the following entities:

- Circuit Switched (CS) domain.
- Packet Switched (PS) domain.
- IP Multimedia Subsystem (IMS).
- BroadCast (BC) domain.

What is the difference between a domain and a subsystem as far as the CN is concerned? The CN domain is an entity directly interfacing one or more access networks. This interface is called "Iu". Due to the nature of traffic and to identify the domain, the Iu is very often subscripted: for example, Iu_{CS} is the interface between an access network and the CS domain and delivers CS traffic; Iu_{PS} is the interface for PS traffic purposes; and Iu_{BC} is the interface that carries a broadcast/multicast type of traffic. CN subsystems do not have a direct Iu-type interface with access networks. Instead, they utilise other, separately defined interfaces to connect themselves to one or more CN domains.

Figure 6.4 is not exhaustive but aims to illustrate the most important interfaces within the UMTS CN. For a complete presentation, see 3GPP TS23.002, version 5.12.0.

In this figure, the bold lines indicate user traffic (user plane) and thinner lines indicate signalling connections (control plane). As far as the CN is concerned, there are some items that need to be pointed out:

- The connections drawn in the figure represent logical, direct connections. In reality, however, the connections have other ways of connecting due to transport network solutions.

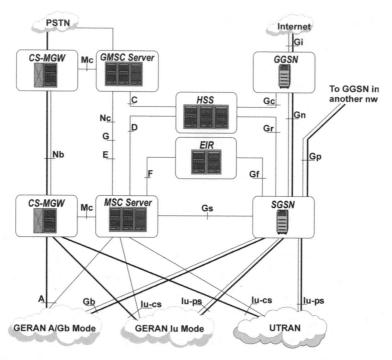

Figure 6.4 Core network (CN) configuration supporting Circuit Switched (CS) and Packet Switched (PS) traffic

- The CS Media Gateway (CS-MGW) and the Gateway Mobile Services Switching Centre (GMSC) server can be combined into one physical entity. In this case the entity is simply called the "GMSC".
- If the CS domain structure follows 3GPP R99, the CS-MGW and MSC Server could be combined into one physical entity. In this case the entity is called the MSC/VLR (Visitor Location Register).
- If the Serving GPRS Support Node (SGSN) and MSC/VLR are combined into one physical entity it is called the UMSC (UMTS MSC).

Sections 6.1.1–6.1.3 handle CN-domain-related issues. The IMS is handled in Sections 6.4–6.6.

CN management tasks and control duties with related issues, like identities and addressing, are handled in Section 6.2.

6.1.1 Core Network Entities that Are Common to All Domains and Subsystems

In addition to domains the CN contains some functionalities that are common to all CN domains and subsystems. These common functionalities are mainly collected in an entity called the "Home Subscriber Server" (HSS).

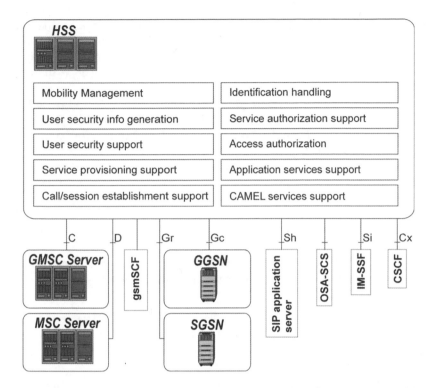

Figure 6.5 Logical diagram about Home Subscriber Server (HSS) functionalities and interfaces to Core Network (CN) domains

If we look at Figure 6.5, we see that the BC domain is not included. Although it has been defined to be part of the CN, its implementation with the 3G network is for further study.

We can see from Figure 6.5 that the majority of HSS functionalities are the ones that have existed in the network for a long time. They used to be taken care of by separate elements: the Home Location Register (HLR) and the Authentication Centre (AuC). In the 3GPP R5 architecture, the HLR and AuC are considered *HSS subsets*, but they still provide the same functionalities:

- Mobility Management (MM) functionality supports user mobility through the CS domain, PS domain and IMS. In this role the HSS, for instance, stores addressing information that can pinpoint the user/terminal location within the MM hierarchy.
- User security information generation, user security support and access authorization: these functionalities are mainly taken care of by the AuC subset, which sends signals to the CN domains and subsystems through the HLR subset.
- Service-provisioning support: the HSS provides access to the service profile data that are used within the CS domain, PS domain and/or IMS application services and Customised Applications for Mobile network Enhanced Logic (CAMEL) services support. The HSS communicates with the Session Initiation Protocol (SIP)

Application Server (AS) and the Open Service Architecture/Service Capability Server (OSA/SCS) to support application services in the IM CN subsystem. It also communicates with the IM-SSF to support CAMEL services related to the IM CN subsystem and with the GSM SCF to support CAMEL services in the CS domain and PS domain.

- Call/session establishment support: the HSS supports the call and/or session establishment procedures in the CS domain, PS domain and the IMS. For terminating traffic, it provides information about which call and/or session control entity is currently hosting the user.
- Identification handling: the HSS provides the appropriate relations among all the identifiers uniquely determining the user in the system: IMSI and MSISDNs for the CS domain; IMSI, MSISDNs and IP addresses for the PS domain; private identity and public identities for the IM CN subsystem. These items are discussed in more detail in Section 6.2.1.1.
- Service authorisation support: the HSS provides the basic authorisation for MT call/session establishment and service invocation. Furthermore, it updates the appropriate serving entities with the relevant information related to the services to be provided to the user.

In addition to the HSS, the Equipment Identity Register (EIR) is a functionality common to all domains and subsystems. The EIR stores information about end-user equipment and the status of this equipment. To do this, it makes use of three "lists": put roughly, the white list contains information about approved, normal terminal equipment; the black list stores information about stolen equipment; and the grey list contains serial number information about suspect equipment. Of these lists, the black and grey ones are normally implemented—it is unusual for the white list to be used. The EIR maintains these lists and provides information about user equipment to the CN Domain on request. If the EIR indicates that the terminal equipment is blacklisted, the CN domain refuses to deliver traffic to and from that terminal. In case the terminal equipment is on the grey list, the traffic will be delivered but some trace activity reporting may occur.

6.1.2 CS Domain

The CS domain is included in 3GPP R5 because the network must support CS services for backward compatibility. 3GPP R99 introduced CS domain structure that it directly inherited from the Global System for Mobile Communication (GSM). In 3GPP R4 the CS domain structure was given an alternative implementation method, where an operator had the ability to fine-tune CS domain control and traffic delivery capacity separately (Figure 6.6).

The aim of CS-MGW–MSC server division is to separate the control and user plane from each other within the CS domain. This introduces scalability to the system, since a single MSC server could control many CS-MGWs. Another advantage of this distributed CS domain architecture is that it opens up the possibilities for user plane geographical optimisation. For instance, an operator could locate CS-MGWs freely within its network and, by proper routing arrangements, it will be possible to arrange

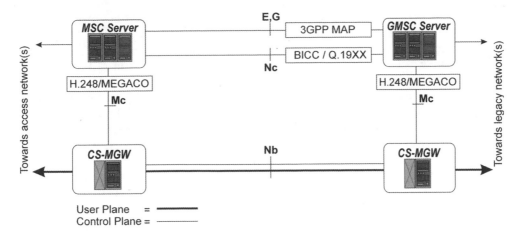

Figure 6.6 UMTS Core Network Circuit Switched (CN CS) domain with distributed Mobile Switching Centre (MSC) functionality (3GPP R4)

things in such a way that the user plane goes through the network geographically in the shortest possible way. The CS-MGW may also contain various conversion packages, which would give the operator the possibility of considering optimised transport network arrangements. For example, using the CS-MGW concept the operator could convert the CS domain backbone to use IP instead of other transport network mechanisms between the access network edge CS-MGW and the legacy Public Switched Telephone Network (PSTN) edge gateway.

The 3GPP-R4-distributed CS domain architecture defines MSC division, where call control functionality and the VLR are brought into an entity called the "MSC server". Respectively, user plane connectivity and related items (e.g., network inter-working) are brought into an entity called the "Media Gateway" (MGW). The CN as a whole contains all kinds of gateways and, thus, it is recommended to add the lettering "CS" in front of MGW to make it crystal clear that we are speaking about the Circuit Switched domain Media Gateway (CS-MGW).

When MSC server–CS-MGW separation is implemented, it opens up a new interface within the CS domain. That interface, Mc, uses the ITU-T H.248-defined Media Gateway Control Protocol (MGCP) for its purposes. H.248 only forms the basis for the information transfer mechanisms in this interface—the complete implementation contains various 3GPP-specific extensions. The Mc interface carries both *call-independent* and *call-dependent* H.248 transactions. The call-independent transactions in this interface contain mechanisms that control the way the CS-MGW imparts its functioning state to the MSC Server. Call-dependent transactions, in turn, can be visualised as "envelopes" transferring the control plane information that is either coming from a user through an access network or from a legacy network. Both call-independent and call-dependent transactions were first described in 3GPP TS29.232 version 5.0.0.

The Nc interface carries network–network call control information. In principle, any call control protocol is suitable for this purpose, as long as the protocol supports a call bearer and its control flow separation. For this purpose the 3GPP adopted a control protocol called "Bearer Independent Call Control" (BICC). To be exact, BICC is not a unique protocol; instead, it is a combination of various packages defined to be used together. These packages are mainly defined in ITU-T specification Q.1950 "Bearer independent call bearer control protocol".

The Nb interface carries both the user plane and the so-called transport network control plane. At the user plane end the Nb interface contains suitable frame protocols and other mechanisms for user data transfer. According to the relevant specifications, the Nb interface can be implemented either using ATM or IP transport. Both transport options were first covered in 3GPP TS29.414 version 5.0.0.

Of course, MSC servers need to communicate with each other. A couple of situations that could initiate this communication include MSC–MSC handover in GSM/EDGE Radio Access Network (GERAN) or the serving Radio Network Controller (RNC) relocation procedure occurring in UMTS Terrestrial Access Network (UTRAN), respectively. In these situations the control of user traffic moves from one MSC server to another and at the same time the CS-MGW on the edge of access network will change as well. These are the reasons MSC servers have Mobile Application Protocol (MAP) interfaces E and G, which transfer MM and other related information between MSC servers. For a detailed description of MAP see 3GPP TS29.002.

Maybe surprisingly, the 3GPP R5 CS domain does not need to be implemented according to 3GPP R4 directives. Another alternative is to continue with 3GPP R99 implementation at the CS end of the network. In this case the CS domain is directly inherited from the GSM world and follows GSM's traditional functionality. The advantage of this approach is that the need to invest in the network is smaller. However, there are also some drawbacks. By keeping the 3GPP R99 architecture within the CS domain the operator may lose the possibility of scalability. In addition, optimal user plane routing is not feasible with this solution, where the control and user plane are not so strictly separated from each other.

6.1.3 PS Domain

The two main elements of the PS domain are types of mobile network-specific servers: Serving GPRS Support Node (SGSN) and Gateway GPRS Support Node (GGSN).

SGSN contains the location registration function, which maintains data needed for originating and terminating packet data transfer. These data are subscription information containing the International Mobile Subscriber Identity (IMSI, see Section 6.2.1.1), various temporary identities, location information (see Section 6.2.1.2), Packet Data Protocol (PDP) addresses (de facto but not necessarily IP addresses), subscripted QoS (see Chapter 8) and so on.

The tool for data transfer within the PS domain is called the "PDP context" (see Section 6.2.2.2). In order to transfer data, the SGSN must know with which GGSN the active PDP context of a certain end-user exists. It is for this purpose that the SGSN stores the GGSN address for each active PDP context. Note that one SGSN may have active PDP contexts going through numerous GGSNs.

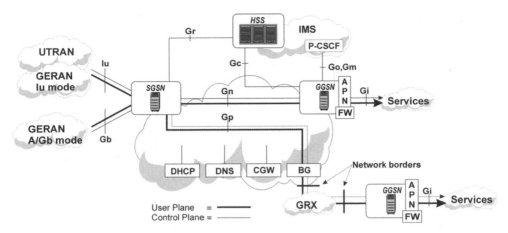

Figure 6.7   Rough guide to the Packet Switched (PS) domain structure

The GGSN also holds some data about the subscriber. These data may also contain the IMSI number, PDP addresses, location information and information about the SGSN that the subscriber has registered.

As far as the PS domain architecture is concerned, the SGSN and GGSN as such are insufficient. Packet traffic require additional elements/functionalitics for addressing, security and charging. Figure 6.7 aims to illustrate the most relevant functionalities within the PS domain.

For security reasons operators now use dynamic address allocation for end-users. These addresses can be allocated in many ways, but the normal way to do this is to use Dynamic Host Configuration Protocol (DHCP) functionality/server. Depending on the operator's configuration, the DHCP allocates either IPv4 or IPv6 addresses for the end-user's terminal equipment.

Actually, the PS domain is, in a way, a sophisticated intranet. In order to address the various elcments within this intranet, the Domain Name Server (DNS) is needed. The DNS within the PS domain is responsible for addressing PS domain elements. For example, when an SGSN establishes traffic to a certain GGSN, the SGSN requests the required GGSN address from the DNS.

When a user has gained a dynamically allocated address and the connection has been established between the SGSN and GGSN, the user is ready to access services that are made accessible by the operator. Service access is arranged through Access Point Names (APNs) which can be freely defined but very often are service-specific. For instance, one APN could be the "Internet" and through this APN the user is able to start Internet-browsing. Another APN could be, say, the "WAP" and this APN leads the end-user to browse WAP menus made available by the operator. One GGSN may contain tens of thousands of APN definitions: they could be company/corporate-specific, they could lead to any place, any network, etc. If the operator does not want to have this kind of access control, a so-called "wild card" APN can be brought into use. In this case end-user preferences as such are allowed and the operator just provides the connection.

Since security is an issue, the GGSN has a FireWall (FW) facility integrated. Every connection to and from the PS domain is done through the FW in order to guarantee security for end-user traffic.

There are many networks that contain a PS domain and roaming between these networks is a most vital issue as far as business is concerned. The PS domain contains a separate functionality in order to enable roaming and to make an interconnection between two PS domains belonging to separate networks. This functionality is called the "Border Gateway" (BG). GPRS Roaming Exchange (GRX) is a concept designed and implemented for General Packet Radio Service (GPRS) roaming purposes.

For charging data collection purposes the PS domain contains a separate functionality called the "Charging Gateway" (CGW). The CGW collects charging data from PS domain elements and relays them to the billing centre to be post-processed. Charging is also the main factor behind some GRX roaming arrangements. A very typical way of doing this is when a user is visiting a GPRS-capable network: the GGSN for GPRS connection is arranged from the home network of the user. By doing this the home network operator is in a position to collect charging data related to this GPRS connection. This arrangement also relinquishes control about APNs to the home network operator. Referring to the APN explanation above, this "home network GGSN" arrangement does not allow wild card APNs. If a visited network GGSN was used during roaming, wild card APNs are allowed, respectively.

As Figure 6.7 states, the PS domain maintains various connections. First, it maintains the IuPS interface towards access networks. Through this interface UTRAN and GERAN are connected. When GERAN is connected to the network in this way, it is said that the network uses *GERAN Iu mode*. There is still a possibility to use a frame-relay-based Gb interface for GERAN connection. In this case it is said that the network uses *GERAN Gb mode*. UTRAN is restricted to using the Iu interface for PS domain connections. Possible complementary accesses and their interconnection mechanisms are under study.

Second, the PS domain has a connection to CN common functionalities, like HSS and EIR. Through these connections the PS domain handles information related to the tasks presented in Section 6.1.1.

The PS domain is the network platform for sophisticated multimedia services enabled and maintained by the IMS. Thus, the PS domain contains interfaces towards the IMS. The IMS and its architecture are explained in Section 6.4.

6.2 CN Management Tasks and Control Duties

The previous section provided a short overview of the architectural aspects of the CN. In this section we will follow a slightly different approach: we will study the role of the CN through its management tasks and control duties.

As shown in Figure 6.8, as far as Communication Management (CM) is concerned, the two main tasks are connection management and session management. Connection management is the management task responsible for CS transactions and related issues, session management could be considered to be its counterpart at the PS end of the network. The control protocols carrying CM information, which deal with call and

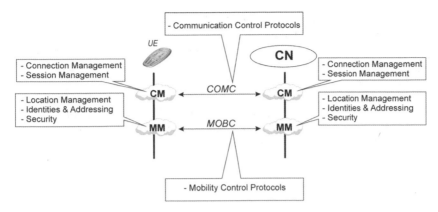

Figure 6.8   Core network (CN) management tasks and control duties

session control, are referred to here as a set of Communication Control (COMC) protocols.

The MM task covers the management of User Equipment (UE) locations together with their identities and addresses, related issues (security is also considered a part of MM). Security is discussed in Chapter 9 in more detail. The control protocols supporting execution of MM tasks are referred to as Mobility Control (MOBC) protocols.

6.2.1 Mobility Management (MM)

With worldwide 2G cellular network mobility is here to stay in communication networks. Understanding the essence of mobility makes the mobile network design significantly different—though more complex as well—from fixed communications and creates a lot of potential for provision of completely new kinds of services to end-users.

Let us first clarify the difference between two basic concepts related to user mobility:

- Location.
- Position.

The term "location" is used to refer to the location of the end-user (and his or her terminal) within the logical structure of the network. The identifiable elements within such a logical structure are the cells and areas (composed of groups of cells). Please note that the word "area" does not need to refer to a set of geographically neighbouring cells, but is simply used by the network operator for network operation purposes.

The term "position", on the other hand, refers to the geographical position of the end-user (and his or her terminal) within the coverage area of the network. The geographical position is given as a pair of standardised coordinates. In the most elementary case, when no geographical position can be determined, the position may be given as a cell identity, from which the position can be derived (e.g., as the geographical coordinates of the BS site controlling that cell).

Although both location and position deal with the whereabouts of this user, the answers are used by the UMTS network in a completely different manner. Location information is used by the network itself to reach end-users whenever there is a communication service activity addressed to them. Position information is determined by the UMTS network when requested by some external service (e.g., an emergency call centre). Although position information may well be life-critical to the end-user making the emergency call, location information is "life-critical" to the network itself in being able to provide services to mobile users in an uninterrupted manner.

Note that the primary purpose of positioning is to support application-oriented services—position information could also be utilised internally by the network. Examples of internal applications are position-aided handover and network-planning optimisation.

Mobile positioning as a service enabler is introduced among other services in Chapter 8.

Another key service created by the MM is *roaming*. The MM functions inside a single Public Land Mobile Network (PLMN) allowing a UMTS user to move freely within the coverage area of that single PLMN. Roaming is a capability that also enables users to move from one PLMN to another operated by a different operator company and, possibly, even in a different country. For the purpose of roaming many of the interfaces defined to be used in the CN are inter-operator interfaces, which are used by the CN elements in visited serving networks to retrieve subscriber location and subscription information from their home networks.

As stated, MM needs a kind of logical, relative hierarchy for its functioning. In addition to this structure, MM handles identities (permanent and temporary) and the addressing information of the subscribers and their terminals as well as those of the network elements involved. In the specifications, security aspects are also counted as part of MM.

6.2.1.1 Identities and Addressing of Users and Their Terminals

Unlike in fixed networks, the UMTS network requires the use of many kinds of numbers and identities for different purposes. In fixed networks the location of the subscriber and the equipment is, like the name says, fixed and this in turn makes many issues constant. When the location of the subscriber is not fixed, the fixed manner of numbering is no longer valid. The purposes of the different identities used in the UMTS can be summarised as follows:

- *Unique identity*: this is used to provide a globally unique identity for a subscriber. This value acts as a primary search key for all registers holding subscriber information; it is also used as a basis for charging purposes.
- *Service separation*: especially in the case of mobile terminating transactions the service going to be used must be recognised. This is done by using an identity having a relationship to the unique identity of the subscriber.
- *Routing purposes*: some special arrangements are required to perform transaction routing; this is not fixed to any network and country borders.

- *Security*: security is a very important topic in the cellular environment and this is why additional identities are generated to improve the privacy of users. Basically, these security-related identities are optional, but it is strongly recommended to use them anyway.

6.2.1.1.1 International Mobile Subscriber Identity (IMSI)

The unique identity for the mobile subscriber is called the *International Mobile Subscriber Identity* (IMSI). The IMSI consists of three parts:

$$IMSI = MCC + MNC + MSN$$

where MCC = the mobile country code (three digits), MNC = the mobile network code (2–3 digits) and MSN = the mobile subscriber number (9–10 digits). This number is stored in the SIM card (USIM). The IMSI acts as a unique database search key in the HLR, VLR, AuC and SGSN. This number follows the ITU-T specification E.214 on numbering. When the mobile user is roaming outside the home network, the visited serving network is able to recognise the home network by requesting the UE to provide this number. Because the IMSI is a unique database key for the subscriber's HLR located in the subscriber's home network, the HLR is able to return the subscriber profile and other information when requested by the IMSI. The same procedure is applied for security information requests from the home network.

6.2.1.1.2 Mobile Subscriber ISDN Number (MSISDN) and PDP Context Address

The IMSI number is used for exact subscriber identification. The *Mobile Subscriber ISDN Number* (MSISDN) is then used for service separation. Because one subscriber may have several services provisioned and activated, this number acts as a separator between them. For instance, the mobile user may have one MSISDN number for a speech service, another MSISDN number for a fax and so on. In the case of

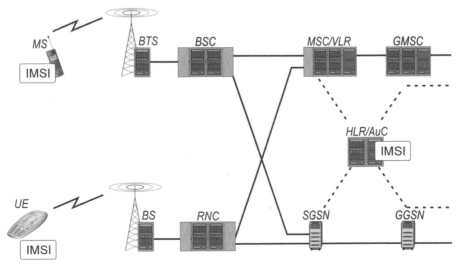

Figure 6.9   International Mobile Subscriber Identity (IMSI)

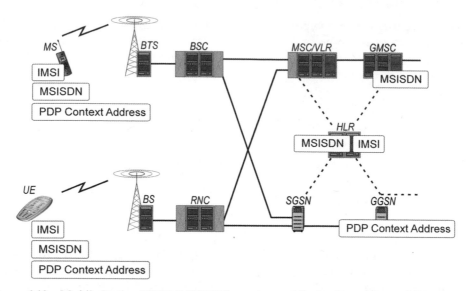

Figure 6.10 Mobile Station ISDN (MSISDN) number and Packet Data Prococol (PDP) context address

mobile-originated transactions the MSISDN is not required for service separation because indication of the service is provided within the CM message(s) during transaction establishment. In the mobile-terminated direction different MSISDN numbers are required for different services because the surrounding networks are not necessarily able to provide the service information by other means. The MSISDN consists of three parts:

$$MSISDN = CC + NDC + SN$$

where CC = the country code (1–3 digits), NDC = the national destination code (1–3 digits) and SN = the subscriber number. This number format follows the ITU-T specification E.164 on numbering. Very often this number is called a "directory number" or just simply a "subscriber number".

The functional PS counterpart for MSISDN is the *PDP context address*, which is an IP address of the mobile user. The PDP context address can be either dynamic or static. If it is dynamic, it is created when a packet session is created. If it is static, it has been defined in the HLR. If the PDP context address is static it behaves like an MSISDN at the CS end of the network.

6.2.1.1.3 Mobile Subscriber Roaming Number (MSRN) and Handover Number (HON)

The *Mobile Subscriber Roaming Number* (MSRN) is used for call-routing purposes. The format of the MSRN is the same as MSISDN (i.e., it consists of three parts, CC, NDC, SN and follows the E.164 numbering specification).

MSRN is used in mobile-terminated call path connection between the GMSC and serving MSC/VLR. This is possible because the allocated MSRN number recognises

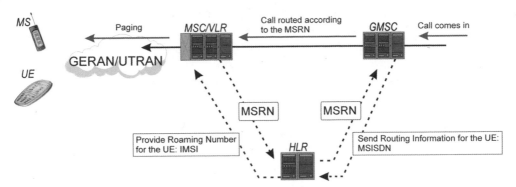

Figure 6.11   Mobile Subscriber Roaming Number (MSRN)

the country, network and the network element within the network. The "subscriber part" of MSRN is for subscriber recognition. The MSRN is also used for call path connection between two MSC/VLRs in the case of MSC–MSC handover. In this context the MSRN is often called the *Handover Number* (HON).

6.2.1.1.4 Temporary Mobile Subscriber Identity (TMSI) and P-TMSI

For security reasons it is very important that the unique identity IMSI is transferred to non-ciphered mode as seldom as possible. For this purpose, the UMTS system makes use of the *Temporary Mobile Subscriber Identity* (TMSI) instead of the original IMSI. The PS domain of the CN allocates similar temporary identities for the same purpose. In order to separate these from the TMSI, they are called *Packet Temporary Mobile Subscriber Identities* (P-TMSIs).

The TMSI and P-TMSI are random format numbers, which have limited validity time and validity area. TMSI numbers are allocated by the VLR and are valid until the UE performs the next transaction. The P-TMSI is allocated by the SGSN and is valid over the SGSN area. The P-TMSI changes when the UE carries out a routing area update.

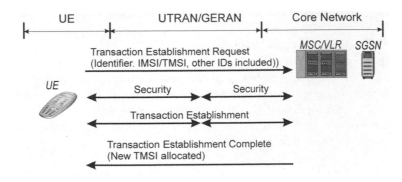

Figure 6.12   Allocation of Temporary Mobile Subscriber Identity (TMSI)

Figure 6.13 International Mobile Equipment Identity (IMEI)

6.2.1.1.5 *International Mobile Equipment Identity*

Two slightly different numbers are defined for mobile equipment identification purposes: the International Mobile Equipment Identity (IMEI) and its extension, the International Mobile Equipment Identity and Software Version (IMEISV). Both of these numbers are handled by the EIR. The network procedures are similar for both numbers: the UE provides either of these numbers upon request and the network verifies the status of the number with the EIR. The structures of the number are as described in Figures 6.13 and 6.14.

Figure 6.14 International Mobile Equipment Identity and Software Version (IMEISV) number

Both of these numbers have common parts: the Type Allocation Code (TAC) defines the manufacturer and type of phone; and the Serial NumbeR (SNR) uniquely defines a piece of equipment within the TAC. The defined length of the IMEI is 15 digits and the remaining digit is spare, which the UE fills with the value "0" when sending it to the network. The network takes the 14 most significant digits of the IMEI and computes the check digit value in order to verify transmission correctness.

The IMEISV is two digits longer as a result of the IMEI containing the software version number of the hardware expressed by 2 digits. Either usage of these numbers is allowed: the UE may send the IMEI or IMEISV but not both.

It is also worth noting that the IMEI remains unchanged as it is the unique identity of the piece of hardware. In the case of IMEISV, the SVN part of the number will change in the context of software updates but the rest of the number remains unchanged.

6.2.1.1.6 *IMS—Home Network Domain Name*

The IMS home network domain name follows the naming structure as defined for Internet use but contains mobile-network-specific parts. To make a separation between a mobile network and an Internet namespace, the IMS home network domain name contains parts of an IMSI number (see Section 6.2.1.1.1).

When, for instance, the UE wants to ascertain the home network domain name, the procedure involves the following steps:

1. The MNC and MCC are ascertained from the IMSI number.
2. The home network domain name always starts with the label "ims".

3. The home network domain name always ends with the label "3gppnetwork.org".
4. The home network domain name is formed by joining parts (2) and (3) together as, for instance, "ims.mnc < MNC > .mcc < MCC > .3gppnetwork.org".

As an example, suppose we have an IMSI number 244 182 123123123. In this IMSI, MCC = 244, MNC = 182 and the Mobile Subscriber Identification Number (MSIN) = 123123123. Using this example IMSI, we can resolve the IMS home network domain name as:

ims.mnc182.mcc244.3gppnetwork.org

As explained in Section 6.2.1.1.1, in UMTS-based networks the MNC comprises 3 digits. However, GSM-based networks mostly use 2-digit MNC values. If the MNC of the network is 2 digits, an additional "0' is added in front of the MNC in the home network domain name.

6.2.1.1.7 IMS—Private User Identity
The IMS requires private user identities for its functionality. These follow the Internet naming structure: *username@realm*. A private username could in principle be anything, but in practice UMTS usernames are derived from an IMSI number, since an IMSI constitutes a good means of providing a unique and confidential identity for the user.

If we use the same IMSI number as in the previous example, the private user identity will be as follows:

IMSI	244 182 123123123
MCC	244
MNC	182
MSIN	123123123
Realm	ims.mnc182.mcc244.3ppnetwork.org
Username	IMSI

Using these values, the private user identity in this example will be:

244182123123123@ims.mnc182.mcc244.3gppnetwork.org

A private user identity looks very much like an IMSI and behaves in part similarly. It is stored in an IMS Identity Module (ISIM) and has the following main characteristics:

- It is contained in all registration requests passed from the UE to the network (IMS).
- The IMS holds the registration and deregistration status of private user identities.
- It is permanently allocated and valid as long as subscriptions are maintained.
- The UE is not able to modify its private user identity under any circumstances.
- The HSS stores private user identities.

6.2.1.1.8 IMS—Public User Identity

A private user identity can also be used for purposes that are internal to the network. For reachability and addressing reasons a user must clearly have a public user identity. Since we are talking about a mobile user, he or she must be reachable in two ways: through the Internet using the Internet's addressing style or through conventional mobile equipment using an E.164 MSISDN type of address (see Section 6.2.1.1.2).

An Internet-style public user identity is actually a SIP URI. For example, its format looks like:

sip:firstname.lastname@operator.com

A telephone-style public user identity follows the standard MSISDN number definition as stated above. Suppose that the MSISDN number is +358 66 1231234. When it is expressed as a Tel URL it looks like:

tel: + 358661231234

It is also possible to indicate what kind of numbering plan the Tel URL follows. The example shown above expresses telephone numbers according to the global numbering plan. If a local numbering plan is used (country code with the network code possibly left out) the scope and owner of the numbering plan must be indicated within the Tel URL.

A public user identity has the following main characteristics:

- One of two possible formats: SIP URI or Tel URL.
- At least one public user identity is stored in the ISIM.
- It must be registered before it can be used with IMS sessions and services.

One user could have many public user identities.

6.2.1.2 Location Structures and Their Identities

In addition to the addressing and identities of the subscribers and their terminals the MM requires the network to have a logical structure. This structure can be represented as logical parts of the access network. Thus, these logical entities act like a "map" for MM procedures and their parameterisation. The UMTS basically contains four logical definitions:

- Location Area (LA).
- Routing Area (RA).
- UTRAN Registration Area (URA).
- Cell.

In the CN CS domain the LA is the area where the UE can freely move without performing a location update procedure. The LA consists of cells: the minimum is one cell and the maximum is all the cells under one VLR. In a location update procedure the location of the UE is updated in the VLR with LA accuracy. This

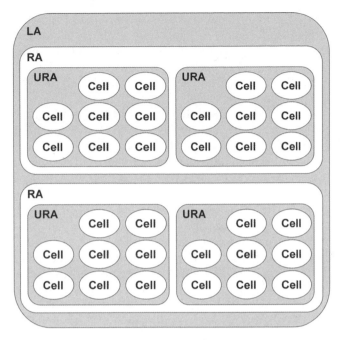

Figure 6.15   Mobility Management (MM) logical entities and their relationships

information is needed for mobile-terminated calls; to get this information the VLR pages the desired UE in the LA where the UE last performed a location update.

Note that in all other respects (other than the VLR) the LA does not have any hardware constraints. For instance, one RNC may have several LAs or one LA may cover several RNCs. Every LA is uniquely identified with a *Location Area Identity* (LAI). The LAI consists of the following parts:

$$LAI = MCC + MNC + LA \text{ code}$$

where the MCC and MNC have the same format as in the IMSI number. The LA code is just a number identifying an LA. The LAI is a globally unique number, and within the same network the same LA code should clearly not be repeated as a single VLR cannot handle duplicate LA codes. The UE listens to the LAI(s) from the Broadcast Channel (BCH). The content of this transport channel is cell-specific and is filled by the RNC.

Like the CN CS domain, the PS domain has its own location registration procedure based on an RA. An RA is very similar to an LA (i.e., it is the area where the UE may move without performing an RA update). On the other hand, an RA is a kind of "subset" of LA: one LA may have several RAs within it but not vice versa. In addition, one RA cannot belong to two LAs.

The CS and PS domains may have an optional Gs interface between them (VLR and SGSN) through which these nodes may exchange location information. Since the UMTS must interoperate with the GSM, the UMTS CN also supports features available in the GSM. One of these features is a combined LA/RA update, where the GSM terminal carries out update requests and sends them in the first place to the SGSN. If an optional Gs interface is available, the SGSN also uses this interface to request the VLR to update LA registration. In a normal UMTS network (i.e., without this option) the combined LA/RA update is not available and the UE has to register its location to both CN domains separately.

In the GSM network, MM is completely handled between the terminal and the NSS. In the UMTS the UTRAN is partially involved in MM and, therefore, contains a local mobility registration: this is called a *UTRAN Registration Area* (URA) and is discussed in Chapter 5. Although this sounds like a small change it causes marked changes in SGSN internal structure and, because of this, the 3G SGSN actually contains both 2G SGSN and 3G SGSN functionality. In the UMTS the SGSN carries tunnelled IP traffic to/from a UE according to the URA identity. In 2G the SGSN terminates tunnelled IP traffic and relays it over the 2G-specific Gb link.

Because a URA is defined in a very similar way to an LA or an RA, it does not have, in principle, any limitations insofar as network elements are concerned. In practice, it seems that the relationship between a URA and Radio Network Subsystems (RNSs) is more or less fixed. On the other hand, the URA is, in a way, a logical definition, which combines traffic routing and Radio Resource Control (RRC). In routing, the URA addressing entity points towards the access domain and in RRC the terminal has states indicating location accuracy and traffic reception ability. This is visible in the RRC-state model that was briefly presented in Chapter 5.

The smallest "building block" used for these MM logical entities is the "cell". Basically, the CN needs to be aware of sets of cells (i.e., areas) rather than individual cells. The cell in the access domain is the smallest entity that has its own publicly visible identity, called the *Cell ID* (CI). Like the LA code the CI is just a number that should be unique within the network. To globally separate cells from one another, the identity must be expanded, and in this case it is called a *Cell Global Identity* (CGI). The CGI has the following format:

$$CGI = MNC + MCC + LA\ code + CI$$

The CGI value covers the country of the network (MCC), the network within a country (MNC), the LA in the network and finally the cell number within the network. This information is distributed to the UE by the UTRAN functionality for system information broadcasting.

6.2.1.3 Network-level Identifiers Common to the CN and Access Network(s)

This section briefly introduces some identifiers that are transferred across the Iu interface and, thus, are common to the CN and the UTRAN. Some of these identities are also used when interconnecting with other networks.

As shown in Figure 6.16 every PLMN has its own, unique ID value called the PLMN-id. This value consists of two parameters: the MCC and the MNC. These are equivalent to the same parameters used in the IMSI.

$$PLMN\text{-}id = MCC + MNC$$

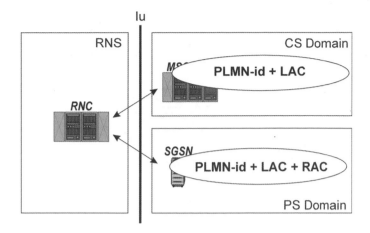

Figure 6.16   Core Network (CN) domain identifier

A PLMN-id is used for many purposes, since it is a very handy way to create a globally unique value that can be transferred between networks. For example, both the CGI and the LAI start with a PLMN-id.

Within a network the RNS must be able to locate the CN domain edge node that maintains the Iu interface. A CN domain identifier is used for this purpose. CN edge domain information is needed when a connection is established over the Iu interface and when a Serving RNS (SRNS) functionality is relocated (i.e., the bearers allocated for connection(s) over the Iu interface are changed so that another RNS can be used).

A CN CS domain identifier consists of a PLMN-id and a Location Area Code (LAC), while a CN PS domain identifier contains a PLMN-id, an LAC and a Routing Area Code (RAC).

Within a CN there may be a need to identify one of its elements globally. This can be done using a CN identifier (CN-id). This also sets theoretical limits for the number of CN element per network. A CN-id consists of two values: a PLMN-id and a numerical integer value between 0 and 4,095:

$$\text{Global CN-id} = \text{PLMN-id} + \langle 0 \dots 4{,}095 \rangle$$

Note that this identifier is *not* the same as an element address. A CN-id is a globally unique ordinal number and, in addition to this, CN elements may or may not have MSISDN-type addresses used for Signalling Connection Control Part (SCCP) routing purposes.

An RNC identifier (RNC-id) follows exactly the same format as a CN-id and is used in RNC elements and in Base Station Controllers (BSCs) when GERAN is the network employed in Iu mode:

$$\text{Global RNC-id} = \text{PLMN-id} + \langle 0 \dots 4{,}095 \rangle$$

A Service Area Identifier (SAI) identifies an area consisting of one or more cells belonging to the same LA and is recognised by all CN domains. It can be used to indicate the location of a UE to the CN domain. Figure 6.17 shows a sample SAI.

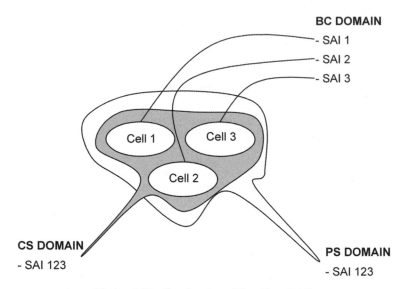

Figure 6.17   Service Area Identifier (SAI)

By studying Figure 6.17 we can see that the following characteristics and limitations apply:

- In both PS and CS domains a service area may consist of more than one cell. In the BC domain one service area is always one cell.
- A cell can have a maximum of two SAIs defined. In this case one is used for CS and PS domains and the other for the BC domain.
- The format of SAI is PLMN-id + LAC + SAC, where SAC stands for Service Area Code.

Due to licensing and implementation costs a commercial need for network sharing has been under discussion in many countries. Actually, network sharing is one of the topics in 3GPP R6 specification work currently under review. A network can be shared in many ways, but if sharing occurs there must be a way to identify which part of the network is shared. So far, a Shared Network Area Identifier (SNAI) has been introduced for this purpose.

A Shared Network Area (SNA) consists of LAs that are redefined to be an SNA. It is through this area that UEs gain access to different operator networks. An SNAI, or SNA-id, contains a PLMN-id and Shared Network Area Code (SNAC):

$$SNA\text{-}id = PLMN\text{-}id + SNAC$$

Since a PLMN-id is included in an SNA-id, the SNA-id is globally unique in nature.

6.2.1.4 Mobility Management State Model

The presence of a packet connection and its management brings in a new dimension as far as MM is concerned; for packet connections the MM has a state model. In CS

UE: MSC/VLR:

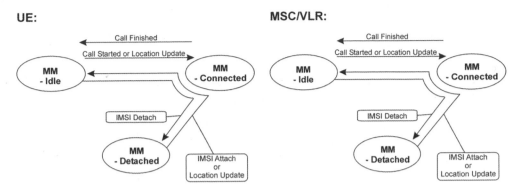

Figure 6.18   Circuit Switched (CS) Mobility Management (MM) state model

connections basically the same kind of model exists but it is rarely used because CS connection behaviour does not require this kind of state model. As already indicated in Figure 6.18, the abbreviation MM stands for CS MM, while PMM refers to PS MM.

6.2.1.4.1 MM States in CS Mode

From the MM point of view, a terminal's network connection may have three states: MM-detached, MM-idle and MM-connected. These MM states indicate how accurately the terminal location is known when compared with the logical structure presented in Figure 6.15. In the MM-detached state the network is not aware of the terminal/subscriber at all (i.e., the MM state when the terminal is switched off). In the MM-idle state the network knows the location of the terminal/subscriber down to the accuracy of an LA. In the MM-connected state the network knows the location of the terminal down to the accuracy of a cell.

The situation described in Figure 6.18 is similar to both the GSM Network Subsystem (NSS) and the UMTS CS domain (after 3GPP R99). When a subscriber switches his or her terminal on, the terminal performs either an IMSI attach or location update procedure. Accordingly, when the subscriber switches his or her terminal off the MM state changes from MM-idle to MM-detached. IMSI attach is performed if the LAI recognised by the camped cell is the same as the one the terminal has in its UMTS Service Identity Module (USIM). If the LAI received from the network differs from the one stored, the terminal performs location update to update and possibly register its new location within the CN CS domain and HLR. In either case the MM state changes as follows: MM-detached → MM-connected → MM-idle. The point we are trying to make is that when the terminal performs either of these briefly described procedures, the network is momentarily aware of the terminal location down to the accuracy of a cell. Both of these procedures cause the CS domain to "wake up" by delivering the same type of message containing originating cell information and the reason for the transaction. In the GSM this message is called a "CM service request" and in the UMTS it is called a "UE initial message".

When a subscriber is active (MM-idle, terminal switched on) the MM state is toggled between MM-idle and MM-connected states according to the use of the terminal. To put it simply: when a call starts the MM state goes from MM-idle → MM-connected and when the call is finished the MM state goes from MM-connected → MM-idle.

6.2.1.4.2 Mobility Management States in Packet Switched Mode

For PS connections the situation is different in terms of PMM procedures: PMM (Packet Mobility Management) states are the same but the triggers that allow movement from one state to another are different (Figure 6.19).

When the PMM state is PMM-detached the network does not have any valid routing information available for PS connection. If the PMM state needs to be changed an option is to perform packet IMSI attach. This procedure takes place whenever a terminal supporting PS operation mode is switched on. A confusing issue here is that this so-called packet attach is a very different procedure from its counterpart at the CS end of the network. Packet IMSI attach as a procedure is a relatively heavy signalling procedure reminiscent of location update. With packet IMSI attach the valid routing information for PS connection is "created" in every node involving PS connections: SGSN and GGSN. In addition to this, the subscriber profile is requested from the HLR and, possibly, old routing information is eliminated.

In PMM-connected state, data can be transferred between the terminal and the network: the SGSN knows the valid routing information for packet transfer down to the accuracy of the routing address of the actual SRNC. In PMM-idle state the location is known down to the accuracy of an RA identity. In PMM-idle state a paging procedure is needed to reach the terminal (e.g., for signalling).

From the end-user point of view, a PS mobile connection is often described as "being always on"; on the other hand, a packet call is said to be like many short CS calls. Both of these statements contain some truth but they are not quite exact as such. From the network point of view, a PS mobile connection gives the illusion of "being always on". This illusion is created by the PMM-connected and PMM-idle states. In PMM-idle state both the network and the terminal hold valid routing information and they are ready for packet data transfer but they are *not* able to transfer packets in this state since there is no connection present through the access network.

When the subscriber switches his or her packet-transfer-capable terminal off, MM reverts to the PMM-detached state and the routing information possibly present in the network nodes is no longer valid. If, for one reason or another, errors occur in the

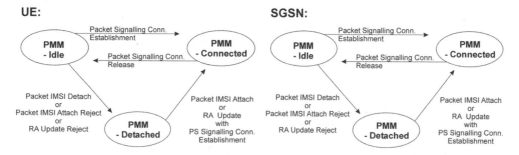

Figure 6.19 Packet Switched (PS) Mobility Management (MM) state model

context of packet IMSI attach or RA update, the MM state may return to PMM-detached. ·

6.2.2 Communication Management (CM)

In this subsection we briefly describe the main functions of CM in terms of both CS and PS communications. From a CS standpoint, the functionality is referred to as CM. From a PS standpoint, the entire functionality is referred to as Session Management (SM). These functionalities are covered by presenting the main phases of a connection or SM process as well as the entities.

6.2.2.1 Connection Management for Circuit Switching

Connection management is a high-level name describing the functions required for incoming and outgoing transaction handling within a switch. Generally speaking, the switch should perform three activities before a CS transaction can be connected. These activities are number analysis, routing and charging. Connection management can functionally be divided into three phases, which the transaction attempt must pass in order to perform through-connection (Figure 6.20).

Number analysis is a collection of rules on how the incoming transaction should be handled. The number of the subscriber who initiated the transaction is called the "calling number" and the number to which the transaction should be connected is the "called number". Number analysis investigates both of these numbers and makes decisions based on the rules defined. Number analysis is performed in both phases of connection management. In Phase I the switch checks whether the called number is obtainable and whether any restriction, such as call-barring, is to be applied to the calling number.

In Phase II the system concentrates on the called number. The nature of the transaction is investigated to ascertain whether it is an international or national call and whether there is any routing rule defined for the called number? In addition, the system checks whether the transaction requires any inter-working equipment (like a modem) to be connected and whether the transaction is chargeable or not. Also, the statistics for this transaction are initiated in this phase.

As a successful result of connection management Phase II, the system knows where the transaction attempt should be connected. This connection and channel selection procedure is called "routing". When the correct destination for the transaction is known the system starts to set up channel(s)/bandwidth towards the desired destination by using, for instance, ISDN User Part (ISUP) signalling protocol. During the transaction the switch stores statistical information about the transaction and its connection and collects charging information (if the transaction was judged to be chargeable). When the transaction is finished, connection management Phase III takes care of releasing all the resources related to the transaction.

In fixed networks every call is treated as an entity at both ends. In cellular networks the term "call" can be interpreted in many ways. Every "call" consists of call legs and each leg thus defines a part of a call.

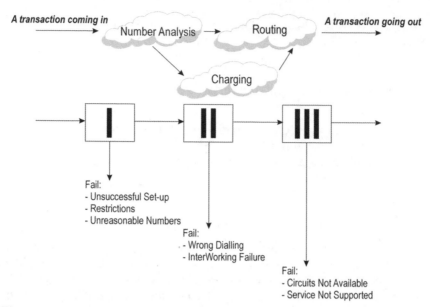

Figure 6.20   Circuit Switched (CS) connection management—connection diagram

From the connection management point of view, every call consists of at least two legs. There are four legs available: MOC (Mobile Originated Call), MTC (Mobile Terminated Call), POC (PSTN Originated Call) and PTC (PSTN Terminated Call). As Figure 6.21 indicates, connection management is actually a distributed functionality and, depending on which element is in question, different parts of call control are used. If a serving MSC/VLR is in question, it handles MOC and MTC call legs; and if the GMSC is in question, it handles POC and PTC legs. Call control is able to receive and create these call legs and also to determine whether any additional functionality is required based on the leg and call type. The most important additional facilities are network inter-working and charging.

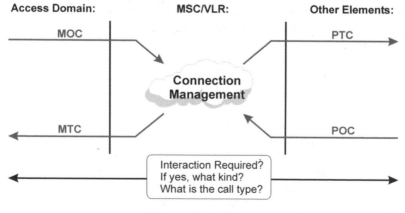

Figure 6.21   Call legs

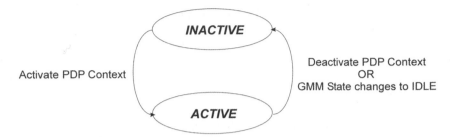

Figure 6.22 Session Management (SM) state model

Call control recognises the call type and based on this it decides further actions. The basic CS call types are:

- Normal call (voice).
- Emergency call.
- Data call (fax included).

In the MOC leg, this call type is included in the "CM service request message" (GSM) and "UE initial message" (UMTS with UTRAN). In the POC leg, the call type is "hidden" in the called party address (B Number); as explained in the context of MM, the service to be used is recognised in the terminating direction by the MSISDN number. In both MTC and PTC legs, connection management determines whether any kind of interaction is required between the call legs.

6.2.2.2 Session Management (SM) for Packet Communication

In the PS domain, packet connections are called "sessions" and they are established and managed by an entity called Session Management (SM).

SM as a logical entity has two main states: *inactive* and *active*. In the inactive state, packet data transfer is not possible at all and also routing information (if it exists) is not valid. In the active state, packet data transfer is possible and all valid routing information is present and defined.

The protocol used for packet data transfer during an active session is PDP. The design of the CN PS domain allows many different PDP protocols to be used. The most obvious case is to use IP as a PDP, but other protocols like X.25 could be supported, though these cases will rarely occur.

The SM handles packet session attributes as contexts and the term used here is *PDP context*. The PDP context contains all parameters describing packet data connection by means of end point addresses and QoS. For example, a PDP context holds such information as allocated IP addresses, connection type and related network element addresses. When SM is active (i.e., a PDP content exists), the user also has an IP address. From the service point of view, one PDP context is set up for one PS service with a certain QoS class. Thus, for instance, Web-surfing and streaming of video over packet connection have their own PDP contexts.

As defined, the UMTS uses the following QoS classes:

- Conversational class.
- Streaming class.
- Interactive class.
- Background class.

These classes are discussed in more detail in Chapter 8.

The PDP context is defined in the UE and the GGSN; and when it exists it contains all relevant parameters defining characteristics for packet connection, as described earlier. The PDP context can be activated, deactivated or modified.

Activation of the PDP context causes the SM to change its state from inactive to active. This SM state change in turn means that the UE forming a packet session with the network holds valid, allocated address information, and that the characteristics of packet connection are defined (e.g., QoS to be used). When the PDP context has been activated the UE and the network are able to establish a bearer for data transfer.

Deactivation of the PDP context causes the SM to change its state to inactive. When the SM state is inactive, the address information and packet session information the UE and the network may have is no longer valid. Thus, the UE and the network are not in a position to arrange any connections to transfer user data flows.

When the SM state is active and a PDP context exists, the PDP context can be modified. In this modification process the UE and the network renegotiate the packet session characteristics. A typical topic for this renegotiation is the QoS class of the packet session.

As shown in Figure 6.23, SM is a high-level entity and its activity depends on lower level entities: PMM and RRC. If RRC and PMM states are not suitable for the active packet session, the PDP context is deactivated and the SM state is changed from active to inactive. This kind of situation occurs, for example, when the RRC changes its state from connected to idle. This state change triggers the PMM to change its state from connected to idle and, as far as the SM state model in Figure 6.22 is concerned, the PMM state change triggers the SM to change its state from active to inactive.

The purpose of SM is to create the illusion of continuity in the connection to the end user, and this must be done in an effective way by saving network resources whenever possible; for instance, the Radio Access Bearers (RABs) carrying user data are established when required, and when there is nothing to transfer the RABs are cleared but the packet signalling connection still remains. Figure 5.19 shows how the different management and controlling entities within the CN and UTRAN change their states during an example packet data flow. This example describes a situation in which the subscriber turns his or her terminal on and IMSI attach is performed. After IMSI attach the subscriber sends some packet data to the network and the service provider sends some data packets to the terminal. This procedure is repeated and then, after some time, the network sends some packet data to the UE. Finally, the terminal is switched off. This example data flow can be realised, for example, during a WAP browsing session.

While the terminal is switched off there are no activities going on between the terminal and network. When it is switched on, an IMSI attach procedure is executed and the UE becomes identified by the network. The establishment of a signalling

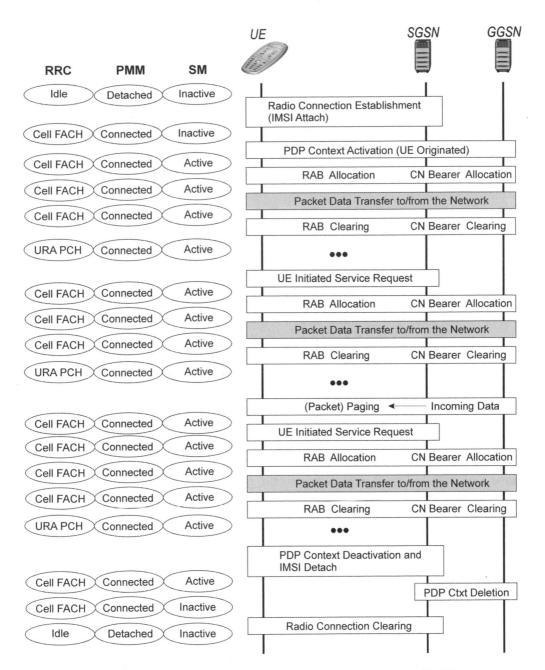

Figure 6.23 Radio Resource Control (RRC), Packet Mobile Management (PMM) and Session Management (SM) states during a packet flow (example)

connection brings the RRC state from "idle" to "connected cell (FACH)" (where FACH is the Forward Access Channel). At the same time the PMM state changes from "detached" to "connected" and the network now has valid information about the location of the subscriber. The SGSN gets this information in a "UE initial message" containing information about the desired activity. When the IMSI has been attached the UE initiates "PDP context activation". During this procedure the UE and the network negotiate the desired packet connection characteristics (e.g., the QoS class). The result of the "PDP context activation" procedure is that the SM state changes from inactive to active.

If there are any packet data to be transferred a bearer capable of carrying user data flows is established. The SGSN starts the RAB allocation procedure over UTRAN and the CN bearer is established between the SGSN and the GGSN. The network is now able to transfer packet data to and from the UE. The RRC state is used to optimise UTRAN resources; when the RRC is in the "connected cell FACH" state, only a small amount of packet data can be transferred over the Iu interface. In this RRC state the UE does not have a dedicated connection with the network. If the amount of packet data amount was larger, a dedicated channel would be allocated to the UE and the RRC state would be "connected cell DCH" (where DCH is the Dedicated Channel). When the packet data have been transferred the RAB and CN bearers carrying the user data flow are cleared but the PDP context is preserved. In addition, the signalling bearers forming the signalling connection between the UE and the network are maintained in this example. When the data bearers are cleared the RRC connection state changes to "connected URA PCH" (where PCH is the Paging Channel) to save UE resources. In this RRC state the network does not know the exact location of the UE, and if the network desires to communicate with the UE the UE must be paged.

If after a while the UE again desires to send packet data to the network it sends a "service request" message to the network and the state of the RRC connection changes once again to "connected cell FACH". The "service request" triggers the network to allocate RAB and CN bearers. Note that these bearers are allocated according to the parameterisation included in the PDP context. When these bearers have been established the packet data is transferred to and from the UE. Upon completion of packet data transfer the bearers carrying the user data flow are cleared, but the PDP context still remains active and the signalling connection is also maintained.

In the case of UE-terminated packet data the GGSN initiates the procedure by sending a data packet to the SGSN currently serving the addressed UE. The receipt of this data packet causes the SGSN to send packet paging to the desired UE. Packet paging forces the UE to change the RRC state from "connected URA PCH" to "connected cell FACH" and the UE sends a "service request" message to the network. Since SM is in active state the network is able to allocate RAB and CN bearers according to the correct, negotiated parameters describing the packet connection. When the bearers have been allocated, the packet data are transferred to the UE, and if the UE has packet data to be sent they are sent to the network. When the packet data have been transferred the RAB and CN bearers carrying the user data flow are cleared and only signalling bearers remain.

When the subscriber switches his or her terminal off, "PDP context deactivation" takes place. This procedure removes any address information stored in the network concerning the packet connection and the PDP context is deleted. This causes SM to change its state from active to inactive and packet data transfer is no longer possible. Since the UE switches itself off, the signalling connection is no longer required and is thus released. As a consequence, the PMM state goes to "detached" and the RRC state changes to "idle".

6.3 Charging, Billing and Accounting

This section provides a short overview of the charging, billing and accounting mechanisms used and their possible use in UMTS networks. But, first, these terms and their meanings should be clarified:

- *Charging* is a collection of procedures generating charging data. These procedures are located in CN elements. Charging (i.e., the identification of collected data) is specified on a common level in UMTS specifications.
- *Billing* is a procedure that post-processes charging data and, as a result, produces a bill for the end-user. Billing as such is beyond the scope of the UMTS specification. Instead of specifications, local laws and marketing practices regulate billing.
- *Accounting* is a common name for charging data collected over a predefined time period. The difference between billing and charging is that in the former accounting information is collected from the connections *between* operators or various commercial bodies. Hence, accounting has nothing directly to do with end-users. Another point is that, as far as telecommunication is concerned, accounting and charging are different issues, but in the context of the Internet these two terms are often used synonymously.

6.3.1 Charging and Accounting

As a result of its historical background and the inherited nature of UMTS networks, the network must support three triggers for end-user traffic identification and charging purposes. These three triggers are:

- *Time-based*: the system collects information about transaction duration: when it started, when it finished and how long it took.
- *Quantity-based*: the system collects information about the number of bits transferred during the transaction.
- *Quality-based*: the system collects information about the quality criteria used during the transaction. The quality profile/criteria of the transaction is called the "Quality of Service" (QoS). The QoS, its parameters and mechanisms are briefly described in Chapter 8.

Of these, the first was traditionally used in CS environments (e.g., PSTNs). The other two were introduced in cellular networks a couple of years ago in the context of the GPRS. Since the UMTS offers a number of transaction possibilities, these triggers may not be adequate. Note also that as the networking model has evolved it has become more complex and there are a number of commercial bodies involved. It is for these reasons that the charging requirements can be shortlisted (list is not exhaustive, though):

- Charging must be able to be applied separately for each medium type (voice, video, data) within a session and also for each used service (call, streaming video, file download, etc.).
- Charging must be able to be applied separately for the various QoS levels allocated for the medium or services within a session.
- It must be possible to charge each leg of a session or a call separately. This includes incoming and outgoing legs and any forwarded/redirected legs. The legs mentioned here are logical legs (i.e., not necessarily identical to the actual signal and traffic flow).
- Charging can be based on the access method used (i.e., 2G, 3G or complementary access). On the other hand, the operator may select access-independent charging and charge for actual service usage.
- It must be possible for the home network to charge its customers while roaming in the same way as when they are at home. For example, if duration-based charging is used for streaming music in the home network, then it must be possible to apply the same principle when the user is roaming.
- It must be possible for operators to have the option of applying charging mechanisms that are used in the GSM/GPRS, such as duration of a voice call, the amount of data transmitted (e.g., for streaming, file download, browsing) and for an event (one-off charge).
- It must be possible for charging to be applied based on location, presence, push services, etc.
- It must be possible to charge using pre-pay, post-pay, advice of charge and third-party charging techniques.
- It must be possible for the home network to apply different tariffs to national calls and short messages established/sent by their subscribers while roaming in their home PLMN, irrespective of whether or not the called subscriber's Home PLMN equals the calling subscriber's home PLMN, rather than on the called subscriber's MSISDN.

When these requirements are combined in a commercial environment, the result will be like that shown in Figure 6.24.

As can be seen, there are many commercial bodies involved. So far in this section, we have concentrated on "retail charging" and identified its requirements. Retail charging is identical to the term "charging" we defined at the beginning of this section. The other charging types identified in Figure 6.24 are basically accounting (i.e., they are not directly visible to end-users).

A typical accounting case is "wholesale charging" where a virtual operator/service provider buys network capacity and sells this capacity to its subscribers. This business is becoming more common nowadays. In many countries the authorities regulate prices used in accounting interfaces to guarantee competition. A good example of this is the case in Finland where local authorities ensure that mobile network operators have

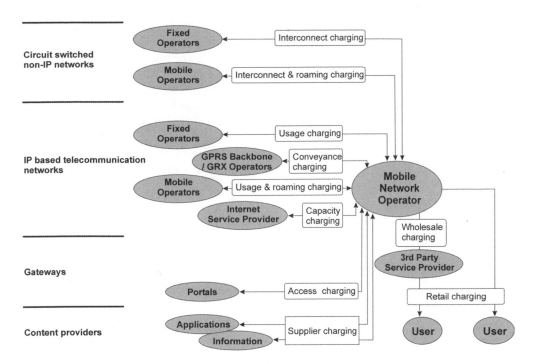

Figure 6.24   Charging types as defined in 3GPP TS 22.115

common and equal pricing for so-called virtual operators (i.e., any virtual operator can in principle buy network capacity from any mobile network operator).

In addition to "wholesale charging" the mobile network operator has altogether four further accounting interfaces: non-IP based telecommunication networks, IP-based telecommunication networks, various gateway types of services and content providers. Non-IP based telecommunication networks are mostly CS in nature and the interconnection towards these networks can be done in two ways. If the CN contains a CS domain, it interconnects directly using methods already defined in 3GPP R99. This means the control plane is taken care of by SS7 and ISUP types of signalling and the user plane is a timeslot(s) on a PCM trunk. If the CS domain follows 3GPP R4 implementation, the user plane is interconnected through CS-MGW elements and any related signalling is handled by means of signalling gateway functionality. In both cases the collected accounting information shows CS resource usage and charging is mainly based on call duration. For IP-based telecommunication networks the accounting information contains session information: what kinds of sessions and what transactions were made during the session? As far as gateways (portals) are concerned, the collected accounting information contains accessing figures (i.e., how often a portal is used). While, for content providers, the accounting information describes how many times a certain content is accessed, when it was accessed and by whom.

CS transactions and their charging will remain as it is and no major changes are expected in this area. PS communication in turn will change and its charging will be very challenging. PS communication is handled using the IMS, whose architectural aspects are handled in Section 6.4. From the charging and accounting point of view the IMS will involve the majority of these briefly presented accounting interfaces as well as retail charging.

Table 6.1 presents the various available charging options. In the table, A, B and C are parties to multimedia transactions performed through the IMS. As stated, the table presents charging *options*, rather than real charging cases. Finally, collected charging and accounting data is up to the IMS operator as is final billing, but the options presented in the table should be supported in the IMS configuration.

A multimedia session consists of media *components*. According to the requirements illustrated at the beginning of this section, the network must be able to recognise various media components and handle charging and accounting data accordingly.

The recognised media components are:

- Voice.
- Real-time audio.
- Streaming audio.
- Real-time video.
- Streaming video.
- Data download/upload.
- Interactive data (e.g., Web-browsing).
- Messaging (SMS, MMS type).
- Email.
- Data stream with unspecified content; this is where the network operator acts as a "bit pipe" and charging is based on quantity (bits transferred) and quality (QoS profile used in the transaction).

Table 6.2 summarises the available charging mechanism and type options for different media components.

6.3.2 Billing

Billing is not part of the scope of UMTS specifications: it is a separate process implemented using separate equipment foreign to the UMTS network. The basis of billing is the charging and accounting data collected from the operator's own network and, possibly, received from other networks.

Billing as a process is regulated by local authorities and laws and, thus, billing principles vary from country to country. In addition to charging and accounting information the billing process requires other source data:

- *Subscription information*. This defines the business relationship between the operator/ service provider and end-user: provisioned services, QoS profiles, identities, etc.
- *Interconnection/Service Level Agreement (SLA)*. This defines the business relationship between operators, service providers and content providers: bandwidth, QoS guarantees, O&M-related issues, etc.
- *Pricing policy*. This is defined by the operator/service provider and is regulated by

Table 6.1 Charging options available in multimedia sessions

No.	Connection	Description	Charging options required
1	A sets up a session with B	A simple connection between two subscribers or a subscriber and a service (e.g., voicemail)	A pays for the session set-up with B A pays for the session resource with B B pays for the session resource with A
2	A sets up a session with B	A simple connection where B is a "toll-free" (800)-type service	B pays for the session set-up B pays for the session resource A pays for part of the session resource (i.e., allowing split charging between A & B)
3	A requests a session with B, B redirects to C	This is redirection. The connection path is not set up to B from A, instead A is told to set up a connection direct to C	A pays for the session set-up with B A pays for the session resource with C C pays for the session resource with A A pays for the session resource as though it were with B and B pays for the session resource with C as though it came from B
4	A requests a session with B, B forwards it to C	This is normal forwarding as in the GSM. The connection path is A to B's home network and B's home network to C	A pays for the session set-up with B A pays for the session resource as though it were with B and B pays for the session resource with C
5	A sets up sessions with multiple parties (multi-party)	Connections to multiple parties are initiated by A	A pays for the set-up of each session A pays for each of the session resources with each of the called parties Each of the called parties pays for the session resource with A

(*continued*)

Table 6.1 (*cont.*)

6	A has a multi-party session where the individual parties set up the session with A	Each party pays for the session set-up with A
	The multiple parties in the session initiate the session with A	A pays for the session resource with the multiple parties
		The individual parties in the session each pay for the session resource with A
7	A is in a session with B, then puts B on hold to set up a session with C, then returns to B after dropping C	A pays for each session set-up with B & C
	A still has a connection with B while also in a session with C. The session with C is terminated after the session with C is terminated	A pays for the session resource with B & C
		B & C pay for the session resource with A
8	A is in a session with B then answers a session request from C while keeping B on hold	A pays for the session set-up with B
	A still has a connection with B while also in a session with C. The session with C is terminated after the session with C is terminated	C pays for the session set-up with A
		A pays for the session resource with B & C
		B & C pay for the session resource with A
9	A sets up a session with B who is roaming in another network	A pays for the session set-up with B
	The connection is made from A to B's home network and then forwarded to B in the visited network (normal GSM mechanism). Alternatively, A is redirected directly to B in the	A pays for the session resource as though it were with B in his home network and B pays for the session resource from his home network with the visited network
		A pays for the session resource with B in the visited network
		B pays for the session resource with A

Table 6.2 Charging mechanism and type options

Component	Charging mechanism options	Charging-type options
Voice	Charging principles as described in Table 6.1	Charging by duration of session Charging by QoS requested and/or delivered One-off set-up charge
Real time audio and video	Charging principles as described in Table 6.1	Charging by duration of session Charging by QoS requested and/or delivered One-off set-up charge
Streaming audio and video	Charged to the initiator of the request Charged to the sender of the audio or video	Charging by duration of session Charging by volume of data, optionally QoS-differentiated One-off set-up charge
Data (upload or download)	Charged to the initiator of the request Charged to the sender of the data	Charging by duration of session Charging by volume of data, optionally QoS-differentiated One-off set-up charge
Interactive data	Charged to the initiator of the session	Charging by duration of session Charging by volume of data, optionally QoS-differentiated One-off set-up charge
Messaging (SMS, MMS type)	Charged to the initiator of the message Charged to the recipient of the message	Charging by event (e.g., SMS) Charging by volume of data
Unspecified content (data stream)	Charged to the initiator of the session Charged to all parties involved	Charging by duration of session Charging by volume of data (sent & received), optionally QoS-differentiated One-off set-up charge

the marketplace and in some cases by authorities: price per call, price per session, price per media component, monthly fees, reduced tariffs, etc.

The function of billing is to combine these items with charging and accounting data and to produce bills for end-users and other related parties.

6.4 IP Multimedia Subsystem (IMS)

The previous sections in this chapter covered CN domain structures and CN management tasks and controlling duties. This section will cover the final ingredient of the whole architecture, the IP Multimedia Subsystem (IMS), which enables applications in mobile devices to establish peer-to-peer connections.

People have a natural need to share experiences: share what they see, share things they do, share emotions. Nowadays people have plain telephony to talk to each other, the Multimedia Messaging Service (MMS) to send pictures and voice clips and the possibility of browsing web pages using their terminals. Some may think that this is sufficient, but there is a wave of other multimedia communication services coming, such as interactive gaming, interactive web services, application sharing, video communication, rich messaging, presence and group communication (e.g., Push to talk over Cellular, or PoC). Clearly, many of these new services will be exercised simultaneously.

UMTS networks bring flexible IP bearers and excellent data capabilities to terminals using the GPRS, Enhanced Data for GSM Evolution (EDGE) and Wideband Code Division Multiple Access (WCDMA) networks. However, the network lacks a mechanism to connect terminals using IP. This is where the IMS comes in. As shown in Figure 6.25 the IMS introduces multimedia session control using SIP in the PS domain; this allows users to establish connections with various ASs and, especially, to use the IP-based services between the terminals.

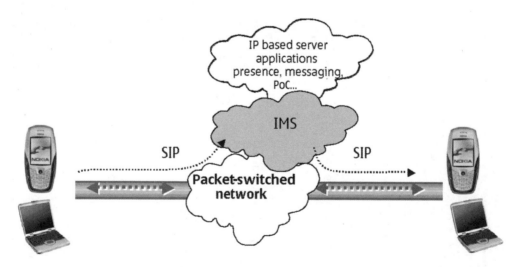

Figure 6.25 IMS brings multimedia session control in the Packet Switched (PS) domain

In this section we introduce the IMS. We will explain IMS design principles and you will also get a grip on the building blocks of the IMS, how different functions are connected and the key protocols of the IMS. However, this chapter is intentionally light on examples and protocol behaviour. You will find a detailed and complete description of the IMS in the book *IMS IP Multimedia Concepts and Services in the Mobile Domain.*

6.5 IP Multimedia Subsystem Fundamentals

There is a set of basic requirements that guides the way in which the IMS architecture has been created and how it should evolve in the future. The following ten issues form the baseline for the IMS architecture:

- IP connectivity.
- Access independence.
- Layered design.
- Quality of Service (QoS).
- IP policy control.
- Secure communication.
- Charging.
- Possibility to roam.
- Interworking with other networks.
- Service development and service control for IP-based applications.

As the name "IP Multimedia Subsystem" implies, a fundamental requirement is that a terminal has to have IP connectivity to access it. Peer-to-peer applications require end-to-end reachability and this connectivity is easiest attainable with IP version 6 (IPv6) because IPv6 does not have address shortage. Therefore, the 3GPP has arranged matters so that the IMS exclusively supports IPv6 [3GPP TS 23.221]. However, early IMS implementations and deployments may use IP version 4 (IPv4). There exists a study report that contains guidelines and recommendations if IPv4 is used to access the IMS [3GPP TR 23.981]. IP connectivity can be obtained either from the home network or the visited network when a user is roaming. The leftmost part of Figure 6.26 presents an option in which the UE has obtained an IP address from a visited network. In the UMTS network this means that the Radio Access Network (RAN), SGSN and GGSN are located in the visited network. The rightmost part of Figure 6.26 presents an option in which the UE has obtained an IP address from a home network. In the UMTS network this means that the RAN and SGSN are located in the visited network. Obviously, when a user is located in the home network all necessary functions are in the home network and IP connectivity is obtained in that network.

Although this is a book about UMTS networks it is important to realise that the IMS is designed to be access-independent so that IMS services can be provided over any IP connectivity networks (e.g., GPRS, WLAN, broadband access x-Digital Subscriber Line). In fact the first IMS release, Release 5, is tied to UMTS because the only possible

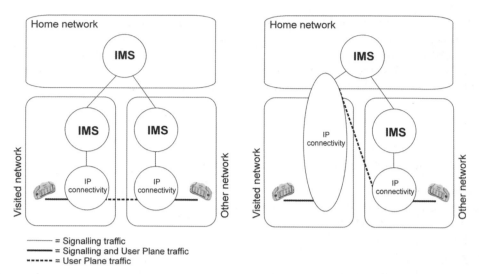

Figure 6.26 IP Multimedia Subsystem (IMS) connectivity options when a user is roaming

IP connectivity access network is GPRS. However, the specifications were corrected in
the second release, Release 6, in such a way that all other accesses are possible as well.

In addition to access independence the IMS architecture is based on a layered
approach. This means that transport and bearer services are separated from the IMS
signalling network and session management services. Further services are run on top of
the IMS signalling network (Figure 6.27 shows the design). The layered approach aims
at a minimum dependence between layers. A benefit is that it facilitates the addition of
new access networks to the system later on. Wireless Local Area Network (WLAN)
access to the IMS, in 3GPP Release 6, will test how well the layering has been done.
Other accesses may follow (e.g., fixed broadband). The layered approach increases the
importance of the application layer. When applications are isolated and common
functionalities can be provided by the underlying IMS network the same applications
can run on the UE using diverse access types.

Quality of Service (QoS) plays an important role in IP networks, as it is commonly
known that delays tend to be high and variable, packets arrive out of order and some
packets are completely lost on the public Internet. This paradigm must be corrected in
the IMS otherwise end-users will not make good use of the new, fancy IMS services. Via
the IMS, the UE negotiates its capabilities and expresses its QoS requirements during a
SIP session set-up or session modification procedure. The UE is able to negotiate such
parameters as: media type, direction of traffic, media bit rate, packet size, packet
transport frequency. After negotiating the parameters at the application level, the
UE is able to map information to UMTS QoS parameters and is able to reserve suitable
resources from the RAN and GPRS network. Chapter 8 in this book contains information about QoS classification, attributes and related network mechanisms. We have just
described how UEs could guarantee QoS between the UE and the GGSN for completing end-to-end QoS. In addition, we need the connections between the GGSNs to
provide the necessary QoS. This could be achieved by SLAs between operators.

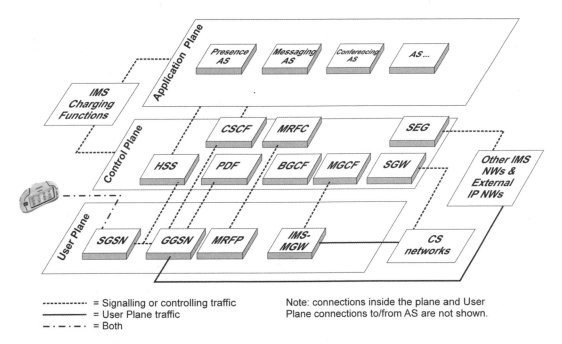

----------- = Signalling or controlling traffic   Note: connections inside the plane and User
——————— = User Plane traffic   Plane connections to/from AS are not shown.
— · — · — · = Both

Figure 6.27 IP Multimedia Subsystem (IMS) and layering architecture

IP policy control means the capability to authorise and control the usage of bearer traffic intended for IMS media, based on the signalling parameters at the IMS session. This requires interaction between the GPRS network and the IMS:

- The policy control element is able to verify that values negotiated in SIP signalling are used when activating bearers for media traffic. This allows an operator to verify that its bearer resources are used for that particular peer-to-peer connection as was negotiated in SIP signalling.
- The policy control element controls when media traffic between the end points of a SIP session can start or stop.
- The policy control element is able to receive notifications from the GPRS network about modification, suspension or deactivation of the PDP context(s) of a user associated with a SIP session.

Security is a fundamental requirement in every part of the UTMS network and the IMS is not an exception. The IMS has its own authentication and authorisation mechanisms between the UE and the IMS network in addition to the GPRS network procedures. Moreover, the integrity and *optional* confidentiality of SIP messages is provided between the UE and the IMS network and between IMS network elements regardless of the underlying RAN and GPRS network. The security issues in the UMTS environment are described in detail in Chapter 9.

From an operator or service provider perspective, the ability to charge users is a must in any network. The IMS architecture allows different charging models to be used. This includes, say, capability to charge just the calling party or charge both the calling party and the called party based on resources used at the transport level. In the latter case the calling party could be charged entirely for an IMS-level session: that is, it is possible to use different charging schemes at the transport and IMS level. However, an operator might be interested in correlating charging information generated at the transport and IMS (service and content) charging levels. This capability is provided if an operator utilises a policy control reference point. As IMS sessions may include multiple media components (e.g., audio and video), it is required that the IMS provide a means for charging per media component. This would allow the possibility of charging the called party if he or she adds a new media component in a session. It is also required that different IMS networks are able to exchange information about the charging to be applied to a current session.

From a user point of view, it is important to gain access to his or her services regardless of geographical location. Roaming is a feature that allows the use of services when the user is not geographically located in the service area of the home network. Three different types of roaming models can be identified: GPRS roaming, IMS roaming and IMS CS roaming. GPRS roaming entails the capability of accessing the IMS when a visited network provides the RAN and SGSN, and the home network provides the GGSN and IMS. The IMS roaming model refers to a network configuration in which the visited network provides the RAN, SGSN, GGSN and the IMS entry point (i.e., P-CSCF) and the home network provides the remaining IMS functionalities. Roaming between the IMS and the CS CN domain refers to inter-domain roaming between the IMS and CS CN. This means that if a user is not reachable in the IMS, then the IMS routes the session towards the CS CN and vice versa.

It is foreseen that different types of networks will co-exist for many years; therefore, inter-working with legacy networks (PSTN, ISDN, mobile, Internet) is an important aspect of any new network architecture. Moreover, it is probable that some people will not be willing to switch terminals or subscriptions to adopt new technology innovations, but there again some users may have more than one item of UE with totally different capabilities: WLAN-enabled UE for office usage, UMTS UE for outdoor usage, wire line UE for home. This will raise the issue of being able to reach people regardless of what kind of terminals they have or where they are located. To be a new, successful communication network technology and architecture the IMS has to be able to connect as many users as possible. Therefore, the IMS supports communication with PSTN, ISDN, mobile and Internet users.

The CS CN and the IMS differ regarding the service control model and service deployment. In the CS CN the visited service control is in use. This means that, when a user is roaming, an entity in a visited network provides services and controls the traffic for the user. This entity is called a Visited Mobile Switching Centre (VMSC). In contrast, home service control is in use in the IMS, meaning that Serving-Call Session Control Function (S-CSCF), which is always located in the home network, controls services. The importance of having a scalable service platform and the possibility to launch new services rapidly has meant that the old way of standardising complete sets of teleservices, applications and supplementary services no longer exists. This is the reasons the IMS provides a service framework that provides the necessary capabilities to support various multimedia applications within the IMS and on top of the IMS. Actually, the IMS is not a service itself; rather, it is a SIP-based architecture for enabling an advanced IP service and application on top of the PS network. The IMS provides the necessary means for invoking services. Examples of these kinds of applications are presence and conferencing. In the future, it is expected that the Open Mobile Alliance (OMA) will make maximum use of the IMS while developing various applications and services.

6.6 IMS Entities and Functionalities

This section discusses IMS entities and key functionalities. These entities can be roughly classified in six main categories:

- Session management and routing family (CSCFs).
- Databases (HSS, SLF).
- Inter-working functions (BGCF, MGCF, IMS-MGW, SGW).
- Services (AS, MRFC, MRFP).
- Support functions (THIG, SEG, PDF).
- Charging.

It is important to understand that IMS standards are set up such that the internal functionality of network entities is not specified in detail. Instead, standards describe reference points between entities and functionalities supported at the reference points. A good example is: How does the CSCF obtain user data from databases?

6.6.1 Call Session Control Functions (CSCFs)

There are three different kinds of Call Session Control Functions (CSCFs): Proxy-CSCF (P-CSCF), Serving-CSCF (S-CSCF) and Interrogating-CSCF (I-CSCF). Each CSCF has its own special tasks (these tasks are described in the following sections). All CSCFs play a role during registration and session establishment and form the SIP routing machinery. Moreover, all functions are able to send charging data to an offline charging function. There are some functions common to the P-CSCF and S-CSCF that they can perform. Both entities are able to release sessions on behalf of the user (e.g., when the S-CSCF detects a hanging session or the P-CSCF receives a notification that a media bearer is lost) and are able to check the content of Session Description Protocol (SDP) payload and to check whether it contains media types or codecs that are not allowed for a user. When the proposed SDP does not fit the operator's policy, the CSCF rejects the request and sends a SIP error message to the UE.

6.6.1.1 Proxy-Call Session Control Function (P-CSCF)

The P-CSCF is the first contact point for users within the IMS. This means that all SIP signalling traffic from the UE will be sent to the P-CSCF. Similarly, all terminating SIP signalling from the network is sent from the P-CSCF to the UE. There are four unique tasks assigned to the P-CSCF: SIP compression, IP security (IPSec) association, inter-action with the Policy Decision Function (PDF) and emergency session detection.

As SIP is a text-based signalling protocol, it contains a large number of headers and header parameters, including extensions and security-related information, which means that SIP message sizes are larger than those of binary encoded protocols. In the IMS architecture, setting up a SIP session is a tedious process involving codec and extension negotiations as well as QoS inter-working notifications. In general, this provides a flexible framework that allows sessions with differing requirements to be set up. However, the drawback is the large number of bytes and messages exchanged over the radio interface. The increased message size means that:

- Session set-up procedures when using SIP will take much more time to be completed than those using existing cell-specific signalling, which means that the end-user will experience a delay in session establishment that will be unexpected and likely unacceptable.
- Intra-call signalling will in some way adversely affect voice quality/system performance.

To speed up session establishment the 3GPP has mandated the support of SIP compression. The UE has a means of indicating to the P-CSCF that it wants to receive signalling messages compressed over the air interface.

The P-CSCF is responsible for maintaining Security Association (SA) and applying integrity and confidential protection of SIP signalling. This is achieved during SIP registration when the UE and P-CSCF negotiate IPSec SAs. After initial registration the P-CSCF is able to apply integrity and confidentiality protection to SIP signalling (Chapter 9 describes IPSec in more detail).

The P-CSCF is tasked with relaying session- and media-related information to the PDF when an operator wants to apply IP policy control. Based on the received information the PDF is able to derive authorised IP QoS information that will be passed to the GGSN when the GGSN needs to perform IP policy control prior to accepting a secondary PDP context activation. Moreover, via the PDF the IMS is able to deliver IMS-charging correlation information to the GPRS network, and, similarly, via the PDF the IMS is able to receive GPRS-charging correlation information from the GPRS network. This makes it possible to merge charging data records coming from the IMS and GPRS networks in the billing system.

IMS emergency sessions are not yet fully specified; therefore, it is a prerequisite that the IMS network detect emergency session attempts and guide a UMTS UE to use the CS network for emergency sessions. Detection is the task of the P-CSCF. This functionality will not vanish when IMS emergency sessions are supported because in certain roaming cases it is possible that the UE does not itself recognise that it has dialled an emergency number.

6.6.1.2 Interrogating-Call Session Control Function (I-CSCF)

The I-CSCF is a contact point within an operator's network for all connections destined for a subscriber of that network operator. There are four unique tasks assigned to the I-CSCF:

- Obtaining the name of the S-CSCF from the HSS.
- Assigning an S-CSCF based on received capabilities from the HSS. The assignment of the S-CSCF will take place when a user registers with the network or a user receives a SIP request while unregistered but still wants to receive services related to the unregistered state (e.g., voicemail).
- Routing incoming requests further to an assigned S-CSCF.
- Providing Topology Hiding Inter-network Gateway (THIG) functionality (THIG is further explained in Section 6.6.5).

6.6.1.3 Serving Call Session Control Function

The S-CSCF is the focal point of the IMS because it is responsible for handling the registration process, making routing decisions, maintaining session states and storing the service profile.

When a user sends a registration request it will be routed to the S-CSCF, which downloads authentication data from the HSS. Based on the authentication data, it then generates a challenge to the UE. After receiving the response and verifying it the S-CSCF accepts the registration and starts supervising the registration status. After this procedure the user is able to initiate and receive IMS services. Moreover, the S-CSCF downloads a service profile from the HSS as part of the registration process.

A service profile is a collection of user-specific information that is permanently stored in the HSS. The S-CSCF downloads the service profile associated with a particular public user identity (e.g., *joe.doe@ims.example.com*) when this particular public user identity is registered with the IMS. The S-CSCF uses information included in the service profile, which decides when and, in particular, which AS(s) is contacted when a user sends a SIP request or receives a request from somebody. Moreover, the service profile may contain further instructions about the kind of media policy the S-CSCF need to apply; for example, it may indicate that a user is only allowed to use audio and application media components but not video media components.

The S-CSCF is responsible for key routing decision as it receives all UE-originated and UE-terminated sessions and transactions. When the S-CSCF receives a UE-originating request via the P-CSCF it needs to decide whether ASs need to be contacted prior to sending the request further on. After possible AS(s) interaction the S-CSCF either continues a session in the IMS or breaks to other domains (CS or other IP networks). Moreover, if the UE uses an MSISDN number to address a called party, then the S-CSCF converts the MSISDN number (i.e., a tel URL) to SIP URI format prior to sending the request further because the IMS does not route requests based on MSISDN numbers. Similarly, the S-CSCF receives all requests that will terminate at the UE. Although the S-CSCF knows the IP address of the UE from the registration it routes all requests via the P-CSCF because the P-CSCF takes care of SIP compression

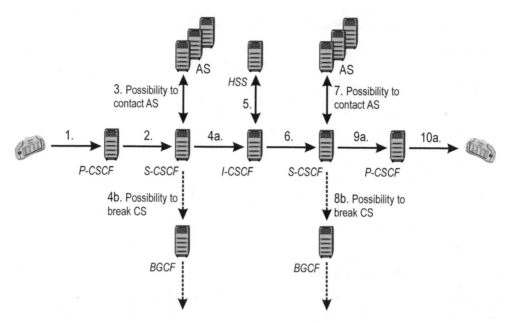

Figure 6.28 Serving-Call Session Control Function (S-CSCF) routing and basic IP Multimedia Subsystem (IMS) session set-up

and security functions. Prior to sending a request to the P-CSCF, the S-CSCF may route the request to an AS(s) (e.g., to check possible redirection instructions). Figure 6.28 illustrates the S-CSCF's role in routing decisions.

6.6.2 Databases

There are two main databases in the IMS architecture: Home Subscriber Server (HSS) and Subscription Locator Function (SLF).

The HSS is the main data storage for all subscriber- and service-related data of the IMS. The main data stored in the HSS include user identities, registration information, access parameters and service-triggering information [3GPP TS 23.002]. User identities consist of two types: private and public user identities (see Sections 6.2.1.1.7 and 6.2.1.1.8). Private user identity is a user identity that is assigned by the home network operator and is used for such purposes as registration and authorisation, while public user identity is an identity that other users can use for requesting communication with the end-user. IMS access parameters are used to set up sessions and include parameters like user authentication, roaming authorisation and allocated S-CSCF names. Service-triggering information enables SIP service execution. The HSS also provides user-specific requirements for S-CSCF capabilities. This information is used by the I-CSCF to select the most suitable S-CSCF for a user. In addition to functions related to IMS functionality, the HSS contains the subset of the HLR/AuC functionality required by the PS domain and the CS domain. Communication between different HSS functions is not standardised. There may be more than one HSS in a home network depending on the number of mobile subscribers, the capacity of the equipment and the organisation of the network.

The SLF is used as a resolution mechanism that enables the I-CSCF, the S-CSCF and the AS to find the address of the HSS that holds the subscriber data for a given user identity when multiple and separately addressable HSSs have been deployed by the network operator.

6.6.3 Interworking Functions

This section introduces the four interworking functions that are needed for exchanging signalling and media between the IMS and CS CN.

Section 6.6.1.3 explained that it is the S-CSCF that decides when to break to the CS CN. To do this the S-CSCF sends a SIP session request to the Breakout Gateway Control Function (BGCF), which chooses where a breakout to the CS domain should occur. The outcome of a selection process can be either a breakout in the same network in which the BGCF is located or another network. If the breakout happens in the same network, then the BGCF selects a Media Gateway Control Function (MGCF) to handle the session further. If the breakout takes place in another network, then the BGCF forwards the session to another BGCF in the selected network [3GPP TS 23.228]. The latter option allows routing of signalling and media over IP near to the called user.

When a SIP session request hits the MGCF it performs protocol conversion between SIP protocols and ISUP (or the Bearer Independent Call Control, BICC) and sends a converted request via the Signalling Gateway (SGW) to the CS CN. The SGW performs signalling conversion (both ways) at the transport level between the IP-based transport of signalling (i.e., between Sigtran SCTP/IP and SS7 MTP) and the Signalling System #7 (SS7)-based transport of signalling. The SGW does not interpret application layer (e.g., BICC, ISUP) messages. The MGCF also controls the IMS Media Gateway (IMS-MGW). The IMS-MGW provides the user plane link between CS CNs and the IMS. It terminates the bearer channels from the CS CN and media streams from the backbone network (e.g., RTP streams in an IP network or AAL2/ATM connections in an ATM backbone), executes the conversion between these terminations and performs transcoding and signal processing for the user plane when needed. In addition, the IMS-MGW is able to provide CS users with tones and announcements.

Similarly, all incoming call control signalling from IMS users is destined to the MGCF that performs the necessary protocol conversion and sends a SIP session request to the I-CSCF for session termination. At the same time, the MGCF interacts with the IMS-MGW and reserves the necessary IMS-MGW resources at the user plane.

Figure 6.29 visualises the inter-working concept. The leftmost part of the figure presents an IMS-originated session and the rightmost part of the figure shows a CS-originated call.

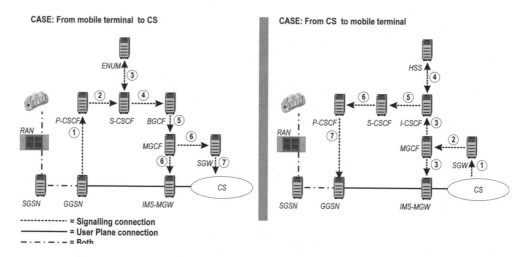

Figure 6.29 IP Multimedia Subsystem (IMS) and Circuit Switched (CS) inter-working

6.6.4 Service-related Functions

Three functions in this book are categorised as IMS service-related functions: Multimedia Resource Function Controller (MRFC), Multimedia Resource Function Processor (MRFP) and AS.

Keeping the layered design in mind, ASs are not pure IMS entities; rather, they are functions on top of IMS. However, ASs are described here as part of IMS functions because ASs are the entities that provide value-added multimedia services in the IMS, such as presence and PoC. An AS resides in the user's home network or in a third-party location. "Third party" here means a network or a stand-alone AS. The main functions of the AS are:

- The possibility to process and have an impact on an incoming SIP session received from the IMS.
- The capability to originate SIP requests.
- The capability to send accounting information to the charging functions.

The services offered are not limited purely to SIP-based services, since an operator is able to offer access to services based on the CAMEL Service Environment (CSE) and the Open Service Architecture (OSA) for its IMS subscribers [3GPP TS 23.228]. Therefore, "AS" is the term generically used to capture the behaviour of the SIP AS, OSA Service Capability Server (SCS) and CAMEL IMS Switching Function (IMS-SF).

Figure 6.30 shows how the different functions are connected. From the perspective of the S-CSCF SIP AS, the OSA service capability server and the IMS-SF exhibit the same reference point behaviour. An AS may be dedicated to a single service and a user may have more than one service; therefore, there may be one or more ASs per user. Additionally, there may be one or more ASs involved in a single session. For example, an operator could have one AS to control the terminating traffic to a user

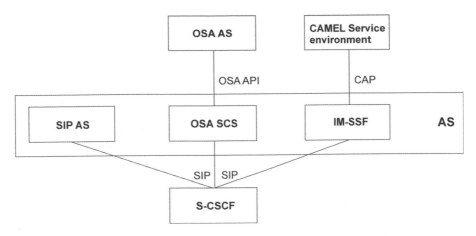

Figure 6.30   Relationship between different Application Server (AS) types

based on user preferences (e.g., redirecting all incoming multimedia sessions to an answer machine between 5 p.m. and 7 a.m.) and another AS to adapt the content of instant messages according to the capabilities of the UE (screen size, number of colours, etc.).

The MRFC and MRFP together provide mechanisms for bearer-related services, such as conferencing, announcements to a user or bearer transcoding in the IMS architecture. The MRFC is tasked to handle SIP communication to and from the S-CSCF and to control the MRFP. The MRFP in turn provides the user plane resources that are requested and instructed by the MRFC. The MRFP performs the following functions:

- Mixing of incoming media streams (e.g., for multiple parties).
- Media stream source (for multimedia announcements).
- Media stream processing (e.g., audio transcoding, media analysis) [3GPP TS23.228, TS 23.002].

Currently, the role of MRFC and MRFP in the IMS architecture is minor because the MRFC is co-located with an AS in the IMS conferencing work [3GPP TS 24.147] and the reference point between the MRFC and MRFP is not yet well defined.

6.6.5 Support Functions

Separation of the control plane and the user plane was maybe one of the most important issues of IMS design. Full independence of the layers is not feasible because, without interaction between the user plane and the control plane, operators are not able to control QoS, the source/destination of IMS media traffic and when the media starts and stops. Therefore, a mechanism to authorise and control the usage of the bearer traffic intended for IMS media traffic was created; this is based on the SDP parameters negotiated at the IMS session. This overall interaction between the GPRS and the

IMS is called a "Service Based Local Policy" (SBLP). It was later that the additional capability of exchanging charging correlation information was specified. The PDF is responsible for making policy decisions based on session- and media-related information obtained from the P-CSCF. It acts as a policy decision point for SBLP control.

Session establishment in the IMS involves end-to-end message exchange using SIP and SDP. During message exchange the UEs negotiate a set of media characteristics (e.g., common codec(s)). If an operator applies the SBLP, then the P-CSCF will forward the relevant SDP information to the PDF together with an indication of the originator. Correspondingly, the PDF allocates and returns an authorisation token which the P-CSCF will pass on to the UE. The PDF notes and authorises the IP flows of the chosen media components by mapping from SDP parameters to authorised IP QoS parameters for transfer to the GGSN via the Go interface. When the UE activates or modifies a PDP context for media it has to perform its own mapping from SDP parameters and application demands to some UMTS QoS parameters. PDP context activation or modification will also contain the received authorisation token and flow identifiers as the binding information. On receiving the PDP context activation or modification, the GGSN asks for authorisation information from the PDF. The PDF compares the received binding information with the stored authorisation information and returns an authorisation decision. If the binding information is validated as correct, then the PDF communicates the media authorisation details in the decision to the GGSN. The media authorisation details contain IP QoS parameters and packet classifiers related to the PDP context. The GGSN maps the authorised IP QoS parameters to authorised UTMS QoS parameters and, finally, the GGSN compares the UMTS QoS parameters with the authorised UMTS QoS parameters of the PDP context. If the UMTS QoS parameters from the PDP context request lie within the limits authorised by the PDF, then PDP context activation or modification will be accepted.

In addition to the bearer authorisation decision the PDF receives information about when an SBLP-governed PDP context is released, whether the UE has lost/recovered its radio bearer(s) and when an SBLP-governed PDP is using streaming or conversational traffic. Based on this information the PDF is able to inform the P-CSCF about the event that has occurred. This allows the P-CSCF to affect charging and it may even start to release an IMS session on behalf of the user. Moreover, the PDF is able to request the GGSN to deactivate a particular SBLP-governed PDP context.

The Security Gateway (SEG) has the function of protecting control plane traffic between security domains. A security domain refers to a network that is managed by a single administrative authority. Typically, this coincides with operator borders. The SEG is placed at the border of the security domain where it enforces the security policy of a security domain toward other SEGs in the destination security domain. In the IMS all traffic within the IMS is routed via SEGs, especially when the traffic is inter-domain, meaning that it originates in a different security domain from the one where it is received. When protecting inter-domain IMS traffic, both confidentiality as well as data integrity and authentication are mandated.

THIG functionality could be used to hide the configuration, capacity and topology of the network from outside an operator's network. If an operator wants to use a hiding functionality, then the operator must place a THIG function in the routing path when receiving requests or responses from other IMS networks. Similarly, the THIG must be placed in the routing path when sending requests or responses to other IMS networks. The THIG performs the encryption and decryption of all headers that reveal topology information about the operator's IMS network.

6.6.6 Charging Functions

The IMS architecture supports both online and offline charging capabilities. Online charging is a charging process in which IMS entities, such as an AS, interact with the online charging system. The online charging system in turn interacts in real time with the user's account and controls or monitors the charges related to service usage: for example, the AS queries the online charging system prior to allowing session establishment or it receives information about how long a user can participate in a conference. Offline charging is a charging process in which charging information is mainly collected after the session and the charging system does not affect in real time the service being used. In this model a user typically receives a bill on a monthly basis, which shows the chargeable items during a particular period. Due to the different nature of charging models different architecture solutions are required.

The central point in the offline charging architecture is the Charging Collection Function (CCF). The CCF receives accounting information from IMS entities (P-CSCF, S-CSCF, I-CSCF, BGCF, MGCF, AS, MRFC). It further processes the received data and then constructs and formats the actual Charging Data Record (CDR). The CDR is passed to the billing system, which takes care of providing the final CDR, taking into account information received from other sources as well (e.g., CGF). The billing system will create the actual bill. The bill could contain, for example, the number of sessions, destinations, duration and type of sessions (audio, text, video).

The S-CSCF, AS and MRFC are the IMS entities that are able to perform online charging. When the UE requests something from either the AS, the MRFC or the S-CSCF that requires charging authorisation, the entity contacts the Online Charging System (OCS) before delivering the service to the user: for example, the user could send a request to a news server asking for the latest betting odds or asking for a voice conference to be set up. The OCS supports two different authorisation models: immediate event charging and event charging with unit reservation. In the immediate event charging model the rating function of the OCS is used to find the appropriate tariff for an event. After resolving the tariff and the price, a suitable amount of money from the user's account is deducted and the OCS grants the request from the requesting entity (AS, MRFC or S-CSCF). When using this model the IMS entity should know that it could deliver the requested service to the user itself. For example, the AS could send a request and inform the OCS of the service (say, a game of chess) and the number of items (say, 2) to be delivered. Then the OCS uses the rating function to resolve the tariff (€0.3) and to calculate the price based on the number of delivered units (€0.6). Finally, €0.6 is deducted from the user's account and the OCS informs the AS that 2 units have been granted. In the event charging with unit reservation model the OCS uses the rating function to determine the price of the desired service according to service-specific information, if the cost was not given in the request. Then the OCS reserves a suitable amount of money from the user's account and returns the corresponding amount of resources to the requesting entity (AS, MRFC or S-CSCF). The amount of resources could be time or allowed data volume. When resources granted to the user have been consumed or the service has been successfully delivered or terminated, the IMS entity informs the OCS of the amount of resources consumed. Finally, the OCS deducts the amount used from the user's account [3GPP TS 32.200, TS 32.225, TS 32.260], but may require further interaction with the rating function. This model is suitable when the IMS entity (AS, MRFC or S-CSCF) is not able to determine beforehand whether the service could be delivered, or when the required amount of resources are not known prior to the use of a specific service (e.g., duration of the conference).

7

The UMTS Terminal

Lauri Laitinen

The Universal Mobile Telecommunication System (UMTS) terminal is the most visible network element of the UMTS system as far as the end-user is concerned. Therefore, it has been mostly considered as an element that provides the application interface and services to the end-user. Practically, however, this is just the tip of the iceberg and the terminal function is considerably wider. The aim of this chapter is to provide a short insight into the UMTS terminal architecture and those functionalities beyond this architecture, bearing in mind the factors that bring constraints and possibilities to the terminal architecture.

Note that, from the network standpoint, the overall protocol architecture and functions of the terminal are subject to standardisation. Despite this, their implementation and additional internal capabilities are strictly implementation-dependent and, hence, seen as key competitive factors in the mobile communication business. Therefore, the architectural perspective presented in this chapter mainly rests on the specifications and is just one way to approach this interesting subject.

7.1 Terminal Architecture

The mobile end-user's terminal-end equipment of the radio interface is officially called "User Equipment" (UE) in the UMTS. The UE is often briefly called a "terminal". From the network point of view, the UE is responsible for those communication functions that are needed at the other end of the radio interface, excluding any end-user applications. The mandatory functionality of a UMTS terminal is related mainly to the interaction between the terminal and the network. The following functions are considered mandatory for all UMTS terminals:

- An interface to an integrated circuit card for insertion of the Universal Integrated Circuit Card (UICC) containing Universal Subscriber Identity Module (USIM) and, optionally, IMS Identity Module (ISIM) application.
- Service provider and network registration and deregistration.
- Location update.

UMTS Networks Second Edition   H. Kaaranen, A. Ahtiainen, L. Laitinen, S. Naghian and V. Niemi
© 2005 John Wiley & Sons, Ltd   ISBN: 0-470-01103-3

- Originating and receiving of both connection-oriented and connectionless services.
- An unalterable equipment identification (IMEI).
- Basic identification of terminal capabilities.
- The terminal must be able to support emergency calls without a USIM.
- Support for the execution of algorithms required for authentication and encryption.
- Besides these mandatory functions, which are essential for network operation, the UMTS terminal should also support the following additional functionalities to facilitate future evolution:

  - An Application Programming Interface (API) capability.
  - A mechanism to download service-related information (parameters, scripts or even software), new protocols, other functions and even new APIs into the terminal.
  - Optional insertion of several UICC cards.

The UE is often presented as one single monolith device, mainly because the same vendor has delivered the physically indivisible equipment. But, in sophisticated mobile systems the UE is often seen—at least in standardisation—as a set of interconnected modules with an independent group of functions. These modules may also be sometimes physically implemented as separate parts. In any case, these functional groups or subgroups have their own counterparts at the network end.

For example, one of the main novelties in the Global System for Mobile Communication (GSM) system is the physical separation of the user-dependent Subscriber Identity Module (SIM) and the general telecommunication-system-dependent part by defining a standardised interface. This clever idea is also inherited in the UMTS. This idea enabled subscriptions and operators as well as the physical terminal to be independent of each other.

Another important separation is that between Radio Access Network (RAN) and Core Network (CN)-dependent parts. However, this is not a matter of standardisation, but instead is due to the practical arrangements followed by suppliers when implementing the terminals. This modularity clarifies the manufacturing of multimode/multinetwork terminals—at least internally—and facilitates the independence of radio access technology from CN technology. Figure 7.1 outlines the UE reference architecture as discussed here together with the related counterparts at the network end.

The UE consists of the Mobile Equipment (ME) and the UICC.

The UICC is the user-dependent part of the ME. It contains at least one or more USIMs and appropriate application software. The USIM is basically a logical concept that is physically implemented in the UICC. The UICC may also contain optional ISIM applications for IMS services. The operator, with whom users make the subscription, will in any case provide the information content of USIMs. The USIM is connected to service profiles rather than to a specific user profile. The ISIM is simply intended for subscriber and network authentication and key agreement for IMS services.

The ME is the user-independent part of the UE and consists of different modules.

The Terminal Equipment (TE) part of the ME is the equipment that provides end-user application functions, such as the call control client, IP and other data termination, IMS media coding and the session management client. The TE knows the possible

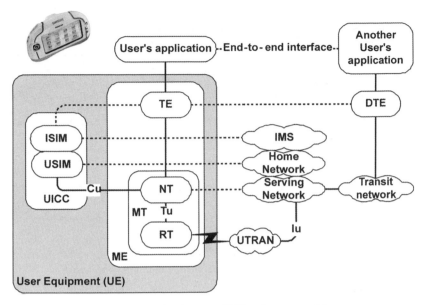

Figure 7.1   User Equipment (UE) reference architecture

standardised telecommunication services on behalf of user applications. It also terminates the telecommunication service platform.

The Mobile Termination (MT), on the other hand, is the part of the ME that terminates radio transmission to and from the network and adapts TE capabilities to those necessary for radio transmission. From the mobile system's point of view, the MT is basically the actual terminal device itself. The MT has the ability to change its location within the access network or move to a coverage area of another access network implementing the same access technology. The MT also terminates the services of UMTS network systems.

The Network Termination (NT) functional group of the MT is the CN-dependent part of the MT. The NT uses non-access stratum protocols for Mobility Management/GPRS Mobility Management (MM/GMM) and Communication Management/Session Management (MM/SM). Therefore, the NT may be seen as the terminal from the pure CN point of view. Non-access stratum protocols are presented in Chapter 10.

The Radio Termination (RT) functional group of the MT is related to the RAN only. It contains functions common to all services using the same RT-specific radio access technology, and uses such access stratum protocols as Medium Access Control (MAC), Radio Link Control (RLC) and Radio Resource Control (RRC) on top of the physical radio connection. Therefore, the RT is seen as a terminal from the UMTS Terrestrial Access Network (UTRAN) point of view. MAC, RLC and RRC are also discussed in Chapter 10.

In this book, terminal architecture is described by using the functional groups described above. A summary of their dependence on network end services and subsystems can be found in Table 7.1. Allocation of main functional entities to the above functional groups clarifies the nature of the functional groups in Figure 7.2.

Table 7.1   Summary of functional groups in UE

	Dependence	Termination at terminal end	Counterpart at network end
UE	User application or user-interface-independent part of user's telecom facilities	Terminates the application-independent telecom system between users	Corresponding fixed or mobile user equipment behind the transit networks
UICC	User-dependent and user-subscription-dependent part of UE	Finally terminates all user-dependent security and service control activities	Subscription management systems of telecom service providers
USIM	Users' UMTS-subscription-dependent part of UICC	Terminates user-dependent control functions in principle within his or her home network	Basically, users' home network registers (e.g., HLR, HSS) managed by the home operator
ISIM	Users' IMS-subscription-dependent part of UICC	Terminates user-dependent control functions within IMS	Users' IMS managed by an operator
ME	Users' subscription-independent, mobile-system-dependent, part of the UE	Terminates all control plane functions and UMTS bearer for user plane	Entire UMTS network
TE	Telecom service platform-dependent, basically mobile-system-independent, part of ME	Terminates the telecom services transported by UMTS bearers	Peer terminal equipment behind external networks
MT	UMTS-dependent, basically telecom-service-independent, part of ME	Terminates UMTS network system services	UMTS system managed by active access (visited) operator
NT	CN-dependent, RAN-independent, part of MT	Terminates UMTS CN services	UMTS CN
RT	Radio-access-technology-dependent, CN-independent, part of MT	Terminates UTRAN services	UTRAN

Strictly speaking, only the final "logical" counterparts in the CN are mentioned here. In some cases the actual protocols may terminate in another network element for optimisation reasons. For example, the MM control protocol between the USIM and the Home Location Register (HLR) often terminates in the Mobile Switching Centre/Visitor Location Register (MSC/VLR) or in the Serving GPRS Support Node (SGSN) in the visited network, because the actual information from the HLR is often duplicated there.

Although terminal and network end architecture differ from each other some interfaces that basically correspond can be identified at both ends. Naturally, the radio interface Uu that is used is exactly the same.

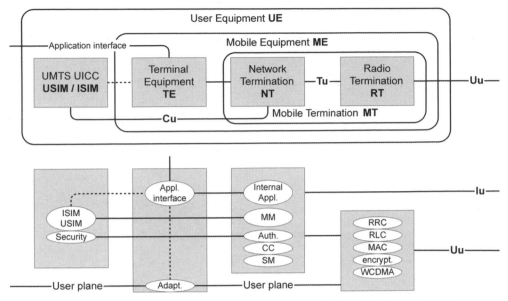

Figure 7.2   Allocation of main functional entities in User Equipment (UE)

The Tu reference point (see Figure 7.1) connects the UTRAN- and the CN-specific parts together at the terminal end as the Iu does at the network end. For performance reasons the Tu reference point is in practice proprietary and embedded inside UE hardware implementation. The corresponding Iu, in turn, is a standardised interface because there may be different vendors for UTRAN and CN devices.

The Cu interface corresponds to the D, C, Gr and Gc interfaces in the CN that connect switching (MSC and GMSC) or routing (SGSN and GGSN) serving network elements to register network elements (HLR, AuC, HSS) in the home network. These interfaces are standardised at both ends because, at the terminal end, there is an interface between operators and mobile equipment vendors in the Cu and, at the network end, there are interfaces between the home and visited operators.

7.2 Differentiation of Terminals

The UMTS terminal has very diverse requirements and these are not necessarily in line with each other as illustrated in Figure 7.3. This increases the complexity of the equipment and will have its effect on equipment price. However, if the price of the terminal is too high, this in turn increases the risk of users not taking up network implementations. These together with many other factors will cause differentiation in terminals. Differentiation development is already visible in the GSM market, where certain MS models are directed at consumers and others at business customers and so on. It seems that in 3G this differentiation will cause further segmentation in terminal markets.

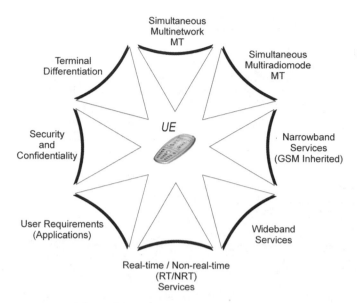

Figure 7.3   Factors affecting the UE

A remarkable feature of a 3G terminal is that it has available, even in its basic form, two different kinds of serving network domains: Packet Switched (PS) domain for packet services and Circuit Switched (CS) domain for circuit services. From the terminal point of view, it results in three different operating modes:

- In *PS/CS operation mode* the terminal is attached to both the PS and CS domain and the terminal is able to provide *simultaneously* both PS and CS services via both domains.
- In *PS operation mode* the terminal is attached only to the PS domain and may only provide services over the PS domain. However, this does not prevent circuit-like services being offered over the PS domain. The best known example of an original CS service that can also be implemented as a PS service is Voice over Internet Protocol (VoIP).
- In *CS operation mode* the terminal is attached only to the CS domain and may only provide services over the CS domain. However, this does not prevent packet-like services being offered over the CS domain. It is even possible that some real-time packet services, especially those with high Quality of Service (QoS) requirements, can be provided more practically over the CS domain using an allocated constant bit rate, although they may be packet-switched in external networks.

In PS/CS operation mode the terminal may have an opportunity to optimise location update procedures if the network supports the optional Gs interface between the MSC/VLR and SGSN for this purpose. In this case the terminal may choose according to its capabilities whether to use combined or separate update procedures. In the combined location update procedure the terminal only lets the SGSN know about a new location and the SGSN then lets the the MSC/VLR know about the location area update.

The separation of radio termination and network termination functions in mobile equipment allows the UE to be classified according to the MT's capability to use several access and CN technologies. There are some basic alternatives that are theoretically possible:

- A *single radiomode MT* can utilise just one type of radio interface for user traffic. In the UMTS, the preferred radio interface, at least initially, is Wideband Code Division Multiple Access-Frequency Division Duplex (WCDMA-FDD)—a typical example of a single radiomode MT.
- A *multi-radiomode MT* can use several radio terminations for user traffic. One interesting case in this category is a GSM/UMTS dual-mode terminal, the interoperation of which is well defined in 3G Partnership Project (3GPP) specifications. This kind of terminal enables 2G services to be used when outside the WCDMA radio network's coverage. Another implementation may be a UMTS terminal with UTRAN and GSM/EDGE Radio Access Network (GERAN) access capability.
- A *single network MT* can utilise just one type of CN. This is a UMTS terminal that is able to use at least one of the operation modes. PS, CS or PS/CS are examples of a single network MT.
- A *multi-network MT* is able to use several CNs. Besides the UMTS CN, the most typical terminal of this kind also supports GSM NSS.

As stated earlier, all of these alternatives offer the basis for terminal implementation, but some of the alternatives may not be commercially attractive due to complicated implementation and/or high costs. On the other hand, the UMTS will be introduced to a quite mature cellular market, where plenty of 2G services are already available. This in turn will create pressure for UMTS terminals to be able to provide the existing 2G services from the very outset as well.

The first UMTS networks will be implemented on top of GSM and this requires UMTS terminals to be able to roam in GSM networks. This is also required by the network operators since a terminal that can use both GSM and WCDMA radio access is more attractive and usable, benefiting as it would from both access network infrastructures. The CS voice service, in particular, causes this demand because subscribers expect their voice call not to be dropped due to possible limited coverage in the first UMTS network installations. In this respect the multi-radiomode MT is a good way of expanding network coverage.

Thus, the typical UMTS terminals will be from the beginning multi-radiomode and multi-network MTs.

Such a UMTS terminal will provide a platform for a very diverse set of services. On the other hand, the services the users require also vary a lot: some may be happy with a conventional voice service and uninterested in sophisticated streaming-type PS services. Others may see PS services as necessary. If a terminal has the ability to handle all kinds of services, its implementation will be very complex and expensive, and, of course, subscribers may not be happy to pay for additional features they do not consider to be relevant or necessary.

Using this as our base, we present a scenario that illustrates a possible way in which UMTS terminals may be segmented. Please note that this scenario does not represent the view of any manufacturer or any specification—its only purpose is to give an idea of how different segment and mobile market fragmentation could happen.

Based on subscribers and their needs—as well as considering the possibilities wide-band radio access offers—we can distinguish four main segments describing sub-scribers, their needs and relevant services:

- *Classic terminal*: this is equivalent to the present cellular phone. It is made to be cheap and, thus, contains limited facilities, such as CS voice access and limited data access with modestly low data rates, which are better than in the present GSM and General Packet Radio Service (GPRS). This terminal is able to handle both GSM and WCDMA radio access but not necessarily simultaneously. In other words, it implements a "selectable multi-network MT". This terminal can be considered to be a kind of "GSM extension", its value rests with its use in existing GSM networks (WCDMA access is used occasionally, mainly for PS connections).

- *Dual mode*: this terminal type can access both the GSM and WCDMA, and can automatically select the access method based on available coverage and requested service. For instance, voice calls are typically connected through the GSM, but data and multimedia services connect through the WCDMA. This terminal type is able to utilise the advantages of both accesses and is also able to perform inter-system handover in both directions. In the case of inter-system handover, the service in use is adapted to radio access whenever possible. Thus, this terminal type implements a "simultaneous multi-network MT". When UMTS networks are brought into wide commercial use, these terminals will most probably form the mass market.

- *Multimedia terminal*: this is like the previous terminal but more intelligent from the network point of view. Dual mode does not necessarily handle UTRAN radio bearers in the best possible manner, but the multimedia terminal is able to perform "optimal multiplexing" of the bearer(s) used for multimedia calls. This capacity-saving aspect will be a very important factor as UMTS networks develop. A multi-media terminal is a kind of combination of cellular phone and palm/laptop computer. It contains plenty of applications to handle multimedia connections and services. This terminal type is not necessarily for the mass market; rather, it is directed at business users, at least initially.

- *Special terminals*: these terminals do not necessarily have a "phone presence" like the previous ones. They serve special purposes and will be integrated with other equipment. This kind of terminal could be, for instance, located in a car and would work in cooperation with the vehicle's computer. If the car was stolen this kind of terminal could be "woken up". It could give detailed information (e.g., name of street) about the vehicle's location using GPS and the vehicle's own computer information. The three previous types more or less implement CS/PS operation mode. Nevertheless, it is likely that this type will only use PS operation mode where application areas can be very diverse. One such application might be intelligent houseware, such as a refrig-erator that can order more food when required. The special terminal described here could be the integrated communication equipment to establish the connection in such a case.

7.3 Terminal Capabilities

Because the life cycle of UMTS networks is expected to span several decades there will be many kinds of compatible terminals with different kinds of capabilities. An arbitrary terminal and an arbitrary network should therefore be able to somehow negotiate which basic possibilities or alternatives they can use with each other. The UMTS has adopted a GSM-like policy for negotiation: networks inform their UEs by broadcasting a lot of system information about network capabilities, while the UE knows its own capabilities and informs the UTRAN or CN.

The basic information set about the capabilities of a UE is called a *mobile station classmark*. During the evolution of mobile system technology the number of alternatives has been growing and, therefore, the classmark concept has been updated in a backward-compatible way. The size of the original, smallest classmark—classmark 1— used in the early GSM was 2 octets (1 octet is 8 bits). Classmark 2 was 5 octets and classmark 3 may have a maximum size of 14 octets. Classmarks 1 and 2 are also used in the GSM. In the UMTS classmark 2 can be characterised as a "CN classmark" and classmark 3 as an "RAN classmark". Which version of classmark is actually needed depends on the actual procedure. Typically, information about the basic capabilities of a UE held in mobile station classmark 3 includes:

- Available WCDMA modes (i.e., FDD or TDD).
- Dual-mode capabilities (i.e., support for different variants of GSM systems with supported frequency bands and other special features).
- Available encryption algorithms.
- The properties of measurement functions in the UE (i.e., availability of extended measurement capabilities and the time needed for the MT to switch from one radio channel to another to carry out neighbour cell measurement).
- Ability to use positioning methods and different kinds of positioning methods.
- Ability to use universal character set 2 (i.e., 16-bit character code standard, known also as "ISO/IEC10646" or "Unicode", instead of the default 7-bit GSM character set in SMSs.

Besides the classmarks there is another, similar kind of information element for describing the entirely radio-interface-specific properties of the UE. This information element is called the "mobile station radio access capability".

7.4 UMTS Subscription

Like the GSM, UMTS networks also separate the subscription from the ME. The subscription-specific information set is called a USIM (Figure 7.4). The USIM is also called a "SIM" because the services actually follow SIM card identification information in every respect. The corresponding information is originally stored in the HLR of the home network of the subscriber. Users of the packet data domain (PS) can also use an additional ISIM application in the UICC for IMS services. The corresponding

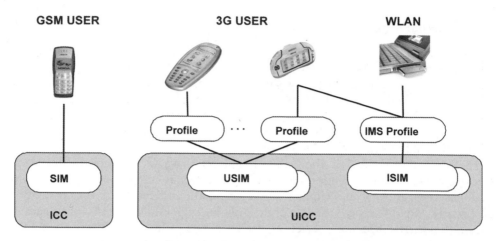

Figure 7.4 Subscriber Identity Module in GSM and 3G

information for ISIM applications is stored in the Home Subscriber Server (HSS) in the user's home network.

In the GSM the SIM is on a removable Integrated Circuit Card (ICC)—the physical place to store subscriber and, possibly, service information. In the UMTS the removable physical data storage is the UICC which contains the USIM holding the service information and identities. The USIM is accessed through profiles—"filters" that define how, for instance, stored information is shown to the user. These profiles are subject to change: the user is able to change profile settings and the network can make some changes to profile information as well.

One USIM may contain many profiles, each destined for certain purposes. Let us suppose that a user has two UMTS terminals: one is of the "classic type" (please see the terminal segmentation issues presented in Section 7.2) and the other is a "multimedia terminal". When a user inserts the USIM into either of his or her terminals the subscription is the same but the terminal shows information in a different way. A different profile is used for the same subscription depending on the TE used. For example, through a "multimedia terminal" the subscriber is able to gain access to information (e.g., picture archive) that is not available through the "classic terminal".

The clear difference between a GSM SIM and USIM is that a USIM is, by default, downloadable and its information is accessible and updatable through the radio path. A functionality making USIM information accessible to TE applications is the USIM Application Toolkit (USAT).

A USIM basically contains five types of data:

- *Administrative data*: these are data assigned by the USIM manufacturer and service provider/operator that cannot be altered, such as key values for security algorithms, IMSI and access class information.
- *Temporary network data*: these are mainly MM information, such as current location area ID, TMSI and calculated ciphering key value(s).
- *Service-related data*: these hold information about the availability or permissibility of

different services and their internal data. A service is available when the subscription has the capability and the permission to support the service. When a service not available message is received, this means that the service should not be used by the USIM user, even if the UE had the capability to support the service. For example, the USIM may contain (if supported by the operator): the local phone book of the user; mobile subscriber ISDN number; fixed dialling numbers; service dialling numbers; barred dialling numbers; outgoing and incoming call information; storage; status reports and service parameters for SMSs; advice about charging; a user- and operator-controlled PLMN selector with access technology; cooperative network list, etc.

- *Application data*: The USIM may store small applications needed for specific services. These applications may implemented as, say, Java applets, which are downloaded and stored in the USIM for later execution within the UE.
- *Personal data*: these cover the data the user stores in the SIM (e.g., SMSs and abbreviated dialling).

From these five classes, the first three are fixed in their size and format since they must appear in exactly the same way and cause similar actions in any TE. The fourth class, applications, is undefined. It can be considered as memory, but right now there are no exact ideas on how large this area should be. The fifth class has in principle a fixed format, but the size varies per operator and per subscription: some USIM cards are configured to reserve more memory for SMSs and abbreviated dialling storage than others.

The optional ISIM application for IMS subscription contains the user's security keys, the IP Multimedia Private user Identity (IMPI), IP Multimedia Public User identity (IMPU), home network domain name identifying the entry point at the network end, various administrative data and access rule reference information controlling the needs for verification. The data content of ISIMs is simpler than that of USIMs, because the ISIM does not contain radio or other access-technology-specific information. Basically, the ISIM needs the USIM to enable packet domain access to the home network. However, if some complementary access technology (e.g., WLAN) is used, the ISIM can carry out its role in the subscriber and network authentication and key agreement procedure alone (i.e., without any USIM).

7.5 User Interface

The 3GPP allows free hands to implement the user interface of the UMTS terminal (Figure 7.5). This arrangement aims to provide cost-effective and creative solutions for a terminal's user interfaces.

The user interface of UMTS terminals may or may not follow the "traditional" layout used in GSM terminals. Implementation of the user interface depends completely on the terminal manufacturer. It is, however, very likely that something similar to a normal numeric keypad will be presented in one form or another.

The requirements on the physical layout of input and output features are kept to a minimum in order to allow for differentiated types of UE and to ease the introduction

Figure 7.5 Conceptual UMTS terminals

of future developments. However, since the requirements on the physical input features are minimal, the control procedures may differ from one UE to another, depending on the solution of the manufacturers. The common denominator between these requirements is that the same logical actions have to be taken by the user. That is, the user has to provide the same information for call control and signalling no matter what the method is. This is also valid if an automatic device is used to carry out the same actions.

There are some applicable and mandatory functions the terminal must fulfil, such as "Accept", "Select", "Send", "Indication" and "End". These functions are essential in order to manage mobile-originated and mobile-terminated calls and supplementary services. These functions can be implemented in any suitable way (i.e., traditional numeric keypad, voice control, etc.). The main factor here is that the user must have access to all of these functions.

The "Accept" functionality is used to accept a mobile-terminated call. The "Select" functionality is used when entering information. The physical means of entering the characters 0–9, +, * and # (i.e., the "Select" function) may be a keypad, voice detection device, data terminal equipment or other, but there must be a means of entering this information.

The "Send" functionality takes care of sending the entered information (e.g., called subscriber number) to the network. The "Indication" functionality is used to give all kinds of call progress indications. The "End" functionality is used to terminate or disconnect a call. Either party involved in the call may cause the execution of the "End" functionality. In addition to this, the "End" functionality could be activated for system-level reasons, such as loss of coverage or invalidation of payment.

8

Services in the UMTS Environment

Heikki Kaaranen, Siamäk Naghian and Jan Kåll

In this chapter we highlight some end-user- and service-related issues. At first, we study the scope of service: what it is, where it comes from and what are the key factors related to a service. Then, we will move the focus to service realisation (i.e., the kind of technical requirements a service requires); this is known as "Quality of Service" (QoS). In Section 8.3 we will give some examples of services that are expected to be implemented using the Universal Mobile Telecommunication System (UMTS).

8.1 About Services in General

"Service" is a word that suffers from inflation in the telecommunication business area— businesses seem to use it in almost any context. This can be very confusing for anyone making a study of "service", made worse by the fact that all commercial bodies have a slightly different meaning and content for this term. In our context, the term "service" is used in its original meaning: the services that a network/operator/service provider offers to its customers.

In Chapter 2 we briefly presented the evolution from basic Global System for Mobile Communication (GSM) to UMTS multi-access. We saw in that presentation how end-user services can grab a greater slice of the development chain the further the evolution goes. This is where the "inflation" mentioned at the start of this section enters the stage: all commercial bodies need something to sell but the basic network technology is no longer an issue. There are some questions to be asked in this phase:

- What do users really want?
- How can we make money out of this?
- What are the most adequate design principles for a complex system?
- Do service-related facts in mobile networks differ from those in fixed networks?

UMTS Networks Second Edition   H. Kaaranen, A. Ahtiainen, L. Laitinen, S. Naghian and V. Niemi
© 2005 John Wiley & Sons, Ltd   ISBN: 0-470-01103-3

8.1.1 What Do Users Really Want?

In the telecommunication business, user expectations are not always the same as those the vendor has. A very typical example of this is when a vendor imagines that a new, technically sophisticated service will become a killer app. But, when it is launched it dawns on the vendor that very few people are willing to use it. On the other hand, there are cases where a service seems technically awkward but suddenly becomes a commercial success.

A good example of an unsuccessful service is the Wireless Application Protocol (WAP). WAP was originally based on a good, solid idea; it had exact specifications and a clear target to aim at. The idea was to open a browsing functionality for mobile users. All related protocols and procedures were specified so that every commercial body had the possibility to implement all the necessary parts for this service subsystem. The target was to get the users to fetch informative content (e.g., number enquiries) themselves whenever they want and wherever they want. What went wrong? WAP is now doomed to failure and is no longer much advertised.

The lack of success of WAP was the sum result of various different factors. As a technology, WAP was intended for Packet Switched (PS) information transfer. Nevertheless, it was first implemented on top of Circuit Switched (CS) connections (i.e., it was launched before GPRS was present in the networks). As a result of this, operators were forced to apply a CS data pricing policy, which got the users thinking that WAP browsing was expensive. Earlier versions of WAP also suffered from diverging terminal development. WAP, as a technology, gave users the opportunity to transfer and display pictures, to send content having both text and pictures and to handle changing content with scripting. All these functionalities are the very basic stuff of the Internet world and users have clear expectations of how the information should be displayed. Users were disappointed when mobile terminals were unable to show the content correctly. Terminal display was too robust, colours were missing and some vendors did not fully implement WAP's specifications. This led to a situation in which a similarly coded WAP page was displayed differently on different terminals.

In contrast to WAP, the surprising success story has been the SMS (Short Message Service). Nobody expected that a simple, text-based connectivity could be a killer app. But, that was what happened. What are the factors that led to the SMS success story?

Well, the first was simplicity. SMS is very simple to use, very simple to transfer within the network and does not require excessively complicated Value Added Service (VAS) equipment for processing. Users do like simple-to-use services, and, if they like it, they will use it more often. When service-related information is easy to transfer within the network, the operator does not have the need to make costly investments. Second, there was understandability. When a service is simple, users understand its functionality and their expectations are realistic. With SMS no one expects real-time delivery and everybody understands that it will take a couple of minutes to deliver the message, sometimes possibly longer. When users get accustomed to this, they make full use of the service at every opportunity.

The SMS success story has been studied very carefully many times. The main factors underlying a successful service, no matter what it is, can be found from these studies. In short, they can be expressed as follows:

- The service must be simple to use.
- The technical implementation of a service must be simple.
- The intra-network control functionality for the service must be as simple and light as possible.
- Service functionality as a whole must be easy to understand.
- The pricing policy of a service must follow the nature of the service.

8.1.2 How Can We Make Money out of This?

Technically, it is possible to do almost anything that will satisfy end-user needs, both realistic and imaginary things. A real-life factor that must be considered, however, is money. If service creation is too expensive, or its implementation is too costly, or its pricing does not meet end-user expectations, the service will not be a commercial success—as was explained earlier.

It is for these reasons that the telecommunication service market is heavily segmented and specialised: there are companies for content provisioning and others for content delivery. Some mobile operators partly undertake this business themselves, but, since the service field is a very complex issue as such, this trend of segmentation and specialisation has proven its robustness.

With this kind of business structure every commercial body is able to concentrate on its core business, thus generating more revenue. On the other hand, this chaining may cause difficulties with security and end-user identification confidentiality. For proper identification purposes the commercial bodies presented in Figure 8.1 have to transfer confidential data between themselves. In the worst case scenario there is a risk that the identification information of an end-user may end up in the wrong hands.

Of course, these security- and confidentiality-related problems have been recognised (they are discussed in more detail in Chapter 9).

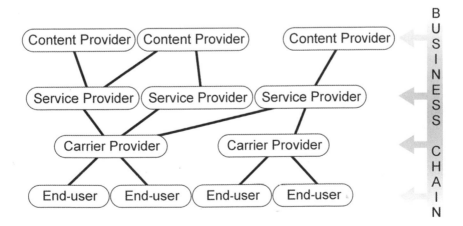

Figure 8.1   3G services and the business/value chain

8.1.3 What Are the Most Adequate Design Principles in a Complex System?

Telecommunication systems normally have a built-in trend within themselves toward containing increasingly more complexity as time goes by. There are many reasons for this development, but one is more important than all the others put together.

When interface openness was adopted as the principal design principle in telecommunication systems, it introduced a drawback to the design process. This drawback is officially known as "backward compatibility"; that is, a new system must support and adapt itself to older systems and environments as much as possible. A typical example of this is the case where an end-user uses, say, a 5-year-old mobile terminal in a modern network launched 3 months ago. There could be situations where the old terminal actually wastes network resources and does not perform optimally from the network point of view. In the worst cases, backward compatibility could be very costly to the operator, but for the end-user it will always be an expected, natural functionality.

The implementation of backward compatibility normally brings with it a lot of complexity, and, unfortunately, this complexity easily spreads to the end-user end as well. A good example of this complexity could be a service in which the end-user downloads a new ringing tone to his or her terminal.

Originally, all ringing tones were just monophonic and all terminals were able to handle them. Then, polyphonic ringing tones were brought into market, but only certain terminal types were able to handle them. This was followed by today's realistic ringing tones (which have a different file format). Again, there are certain terminal types able to utilise these. Normally, the newer the terminal the greater the repertoire. The newest terminal types are able to handle all of these three types of ringing tones. The older ones are only capable of handling monophonic ringing tones. Let us suppose we have a company providing this service and preparing all types of ringing tones. What would the history of this company be like? In the early days it had maybe one or two SMS addresses through which an end-user was able to request a new ringing tone. The service would have been simple and fast. Then the new polyphonic ringing tones came into the marketplace. The company would have been faced with making a choice: either it should add a couple of new SMS entries for new, polyphonic ringing tones or the SMS request format should be changed. Later still the same problem would have occurred with realistic ringing tones. The change in service complexity can be readily seen in the company's advertising. Earlier a couple of lines and short instructions were enough; nowadays, the company must list every single terminal type and explain which ringing tone type is eligible for a certain terminal. In other words, although the service itself is approximately the same, the varying content sets a number of obstacles for accessing terminals to overcome, thus adding more complexity to the system.

In conclusion, there are some important design principles that need to be taken into account:

- *Simplicity* (both the network and the service). This increases service usage and generates more revenue. Additionally, network management will be easier and more effective.

- A *uniform transport network* able to carry all kinds of services (i.e., there should not be separate equipment and domains for various traffic types). This increases the effectivity of transmission overall and makes the whole network more manageable.
- *Minimised control functions* within the network. This decreases network complexity and increases manageability.

8.1.4 Do Service-related Facts in Mobile Networks Differ from Those in Fixed Networks?

The answer is a resounding "no". The main challenge for mobile networks is to develop to the same level as fixed networks in all respects. So far, mobile networks have gained and even exceeded the fixed network service level in certain topics. The major bottleneck remaining is actually bandwidth.

As a result of the investment operators are obliged to make to their networks, a bit transferred through a mobile access is not as cheap as a bit transferred through a fixed access. This fact creates pricing pressures and again end-users will face charges they are not expecting. The fixed Internet access is relatively cheap—in most cases it involves a monthly charged flat rate. In mobile networks, pricing of basically the same service does not have a flat rate; instead, all transferred bits and packets are counted and, in some cases, the transferred content will have its own, content-dependent price. In other words, although the end-user and the fetched content are the same, the content is cheaper through one access type than through another.

From the end-user point of view this is not acceptable in most cases. The basic question behind this is the added value mobile networks bring to the end-user. The actual question to ask is: How much would an end-user be prepared to pay for mobility?

8.2 Quality of Service (QoS)

8.2.1 Traffic Classes and QoS Attributes

Sections 8.1.3 and 8.1.4 briefly discussed services in general. In this section we will look in more detail at the technical realisation of a service in the network. Every service sets some requirements for the transmission path carrying the service information. Some services set tighter requirements than others. A network typically carries many services and service requests from many users simultaneously. Each of these services has its own requirements. Since network resources are limited, the aim is to allocate just enough resources for every request—not too much, but not too little.

It is possible to make very complex QoS definitions, but by doing so all we would achieve would be a complex system as well. In other words, complex QoS classification leads inevitably to a very complex network, since a lot of QoS control must be added on top of the actual network delivering traffic.

The UMTS introduces a relatively simple QoS concept consisting of four traffic classes and some QoS attributes to define the traffic characteristics of the traffic classes.

Those four traffic classes are:

- QoS conversational.
- QoS streaming.
- QoS interactive.
- QoS background.

Currently, the best known scheme of conversational traffic class is telephony speech, but the Internet and multimedia will introduce more services requiring this traffic class. Good examples could be Voice over IP (VoIP) and videoconferencing services. Real-time conversation is always performed between real end-users (i.e., people). This is the only scheme in which the required traffic characteristics are strictly determined by human perception. In this traffic class, traffic is mostly characterized by a low transfer delay value and minimum delay variation between the Signalling Data Units (SDUs) delivered.

When an end-user is looking at/listening to real-time video/audio the streaming traffic class is applied. Traffic within this class always aims at a live/human destination and is a one-way transport by nature. In this traffic class, delay variation of an end-to-end data flow will be limited in order to keep delay variation at a manageable level. Normally, the receiving end of the streaming service is time-aligned and this time alignment is taken care of by an end-user's streaming application. Thus, the limits for delay variation are set by the end-user application. Since service communication is one-way transport, longer delays and greater delay variations are more acceptable than in the conversational traffic class.

Interactive traffic class describes the classical data communication scheme that, overall, is characterised by the request–response pattern of the end-user. At the message destination there is an entity that is expecting the message (response) within a certain time. Round trip delay time is therefore one of the key attributes. Another characteristic is that the content of packets should be transparently transferred with a low bit error rate.

Background traffic class is another classical data communication scheme that, overall, is characterised by the destination not expecting the data within a certain time. The scheme is thus more or less delivery-time-insensitive. Another characteristic is that the content of packets should be transparently transferred with a low bit error rate.

Each of these traffic classes carry actual end-user services with them. Table 8.1 shows some typical service examples and their traffic class. The behaviour of traffic classes is defined with QoS attributes. These adjustable parameters are:

- *Traffic class.* This parameter has one of the following values: conversational, streaming, interactive or background. By analysing this attribute the UMTS can make assumptions about the traffic source and optimise the transport for that traffic type.
- *Maximum bit rate* (kb/s). The maximum bit rate is the upper limit a user or application can accept or provide. All UMTS bearer service attributes can be fulfilled for traffic up to the maximum bit rate, depending on network conditions. The purpose of this attribute is to to limit the delivered bit rate to applications or external networks with such limitations and to allow the maximum wanted user bit rate defined for applications able to operate at different rates.

Table 8.1 Traffic classes and their characteristics

Traffic class	Conversational	Streaming	Interactive	Background
Fundamental characteristics	• Low delay • Small delay variation	• Moderate delay and variation (end-user-application-dependent)	• Round trip delay is a matter of importance • Moderate delay variation • Request–response pattern	• Destination (end-user application) does not expect a response within a certain time
Service examples	• Speech, VoIP, video conferencing	• Streaming video, streaming audio	• Web-browsing	• Email and file-downloading

- *Guaranteed bit rate* (kb/s). UMTS bearer service attributes (e.g., delay and reliability) are guaranteed for traffic up to the guaranteed bit rate. For traffic exceeding the guaranteed bit rate, UMTS bearer service attributes are not guaranteed. A guaranteed bit rate may be used to facilitate admission control based on available resources, and for resource allocation within the UMTS.

- *Delivery order* (Y/N). This attribute indicates whether the UMTS bearer should provide in-sequence SDU delivery or not. This attribute is derived from the user protocol (PDP-type) and specifies whether out-of-sequence SDUs are acceptable or not. Whether out-of-sequence SDUs are dropped or re-ordered depends on the specified reliability. Note that delivery order should be set to "N" for Packet Data Protocol (PDP)-type (IPv4 or IPv6).

- *Maximum SDU size* (octets). This attribute defines the maximum SDU size at which the network must satisfy the agreed (i.e., negotiated) QoS. Maximum SDU size is used for admission control and policing and/or optimising transport. Handling of packets larger than maximum SDU size is implementation-specific (i.e., they may be dropped or forwarded with a reduced QoS).

- *SDU format information* (bits). With this attribute it is possible to define a list of exact sizes of all possible SDUs. The reason for doing so is that Radio Access Network (RAN) needs SDU size information to be able to operate in transparent Radio Link Control (RLC) protocol mode, which is beneficial to spectral efficiency and delay when RLC re-transmission is not used. Thus, if the application can specify SDU sizes, the bearer will be less expensive.

- *SDU error ratio*. This attribute indicates the fraction of SDUs lost or detected as erroneous. SDU error ratio is only defined for conforming traffic. By reserving resources, SDU error ratio performance is independent of the loading conditions, whereas without reserved resources, such as in interactive and background traffic classes, SDU error ratio is used as a target value. This attribute is used to configure protocols, algorithms and error detection schemes (primarily, within RAN).

- *Residual bit error ratio*. This attribute indicates the undetected bit error ratio of the SDUs delivered. If no error detection is requested, residual bit error ratio indicates the bit error ratio in the delivered SDUs. This attribute is used to configure radio interface protocols, algorithms and error detection coding.

- *Delivery of erroneous SDUs* (Y/N/−). This attribute indicates whether SDUs detected as erroneous should be delivered or discarded. "Y" implies that error detection is employed and that erroneous SDUs are delivered together with an error indication; "N" implies that error detection is employed and that erroneous SDUs are discarded; "−" implies that SDUs are delivered without considering error detection.
- *Transfer delay* (ms). This attribute indicates maximum delay for the 95th percentile of the distribution of delay for all delivered SDUs during the lifetime of a bearer service, where delay for an SDU is defined as the time from a request to transfer of an SDU from one Service Access Point (SAP) to its delivery at another. This attribute relates to the delay tolerated by the application. In conjunction with the SDU error ratio attribute, care needs to be taken in deriving the value for the 95th percentile when an application wants, say, 99.9% of all transmitted packets delivered within a certain time. This attribute allows the RAN to set transport formats and Hybrid Automatic Repeat Request (HARQ) parameters.
- *Traffic-handling priority*. This attribute specifies the relative importance of handling all SDUs belonging to the UMTS bearer rather than the SDUs of other bearers. Within the interactive traffic class, in particular, there is a definite need to differentiate between bearer qualities. This is handled by using the traffic-handling priority attribute, which allows the UMTS to schedule traffic accordingly. By definition, "traffic-handling priority" is an alternative to absolute guarantees and, thus, these two attribute types cannot be used together for a single bearer.
- *Allocation/retention priority*. This attribute specifies the relative importance, compared with other UMTS bearers, of allocation and retention of the UMTS bearer. The allocation/retention priority attribute is a subscription attribute that is not negotiated by the mobile terminal. This attribute is used for differentiating between bearers when performing allocation and retention of a bearer. In situations where network resources are scarce, the relevant network elements can use this allocation/retention priority to give preferential treatment to bearers with a high priority over bearers with a low priority when performing admission control.
- *Source statistics descriptor* ("speech"/"unknown"). This attribute specifies the characteristics of the source of the SDUs submitted. Conversational speech follows a well-known statistical behaviour or has a Discontinuous Transmission (DTX) factor. Once informed that the SDUs for a UMTS bearer are generated by a speech source, the RAN, Serving GPRS Support Node (SGSN) and Gateway GPRS Support Node (GGSN) as well as the User Equipment (UE) may, based on experience, calculate the statistical multiplex gain to be used for admission control on the relevant interfaces.
- *Signalling indication* (Y/N). This attribute indicates the signalling nature of the SDUs submitted, is additional to the other QoS attributes and does not over-ride them. It is only defined for the interactive traffic class. If signalling indication is set to "Y", the UE should set the traffic handling priority to "1". Signalling traffic can have different characteristics from other interactive traffic (e.g., higher priority, lower delay and increased peak). This attribute permits RAN operation to be accordingly enhanced. An example of signalling indication is its use in IMS signalling traffic. Signalling indication is sent by the UE in the QoS information element of a message.

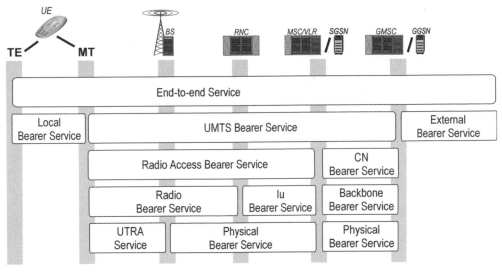

Legend:
TE = Terminal Equipment
MT = Mobile Termination

Figure 8.2 Bearer/QoS architecture in UMTS

The technical realisation of a QoS attribute value combination that forms a certain traffic class with certain characteristics is called a "bearer". A bearer is a collection of allocated network resources forming a "bit pipe" which fulfils QoS requirements and is thus the response to an end-user service request.

Figure 8.2 indicates the way in which bearers make use of these parameters and their combinations. The requirements of the actual service can be seen at the UMTS bearer level. Within the terminal the local bearer keeps the same definitions. If a connection is going beyond the network, service requirements are met by external bearer definitions. Note that external bearer definitions do not have the same scope as UMTS specifications. This may cause problems for services since QoS cannot be necessarily guaranteed outside the UMTS network.

Since the UMTS network consists of various parts, each having its own characteristics, there must be various bearers covering these system parts as well. Bearers covering only a certain part of a system and being closer to the physical connection always have more stringent QoS requirements. For example, at the UMTS bearer level, transfer delay is defined in milliseconds, whereas this is defined at the microsecond level within the Universal Terrestrial Radio Access (UTRA) Service. This is necessary because, otherwise, the original QoS requirement may not be fulfilled.

Recalling the Admission Control (AC) and packet-scheduling mechanisms presented in Chapter 5, the RNC plays a very central role as far as QoS is concerned. In a QoS context the procedure is referred as bearer management (Figure 8.3), meaning the overall procedure involved in how these parameters are actually handled in the system.

The maximum limits of a bearer are set in the subscription and managed by the Core Network (CN). Since the radio interface is the most likely bottleneck, the real, allocated

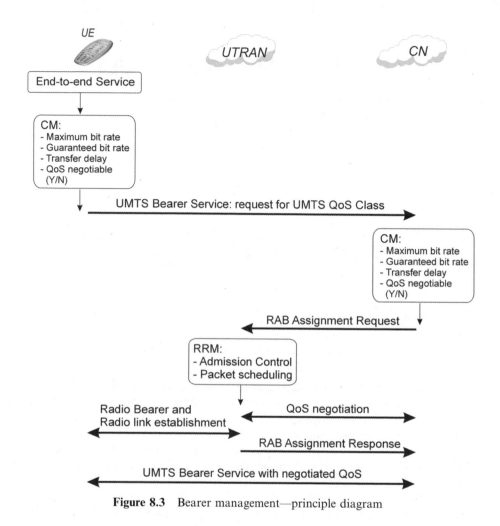

Figure 8.3   Bearer management—principle diagram

bearer is defined by the Radio Network Controller (RNC) based on existing radio interface load and interference.

8.2.2 About QoS Mechanisms

From the release of 3G Partnership Project (3GPP) R5 the aim has been to convert the 3G network to completely work over the Internet Protocol (IP). IP has proven itself as a data transfer concept over the last few decades. It is especially suitable for networks where connection quality can be described as "best effort". The most famous network functioning according to this principle is the normal, public Internet.

When the world moves towards mobility and tele- and data-communication convergence is realised, this traditional IP best effort approach will face new challenges. The origin of IP is in data-communication where "best effort" has been satisfactory in the majority of transactions. In mobile networks, end-users have become used to employing connection-oriented services with well-defined quantity and quality characteristics. This has created pressure for mechanisms guaranteeing QoS on top of standard IP to be devised.

A very straightforward method of bringing QoS to a system is to make an unlimited amount of bandwidth available to end-users. This method is called "overprovisioning". Overprovisioning is unfortunately a very expensive way to guarantee QoS and, thus, a lot of specification work has been done to generate alternative methods for guaranteeing QoS to the many users with numerous needs in an environment where bandwidth is a limited resource and the number of users continuously changes.

In the following subsections we briefly present some of these mechanisms and their basic principles at a general level. Note that the list is not exhaustive in any sense, but the items handled are not only the most common alternatives discussed in the context of QoS, but have also been study items in the 3GPP.

8.2.3 ReSerVation Protocol (RSVP)

RSVP is maybe the most complex mechanism for QoS provision. However, it does offer possibilities to simulate CS connection. Actually, RSVP is used to launch integrated services. An integrated service is an entity consisting of two alternatives:

- *Guaranteed*: this is an integrated service state emulating a dedicated virtual circuit. It provides stable bounds on end-to-end connection delays based on agreed traffic parameters.
- *Controlled load*: this integrated services state is equal to the definition "best effort service under unloaded conditions". In other words, it is better than "best effort", but this superiority cannot be guaranteed in all conditions.

As shown in Figure 8.4, the sender (service) mainly describes outgoing traffic in terms of bandwidth limits (minimum and maximum), delay and jitter. This information is sent by RSVP in the PATH message to the receiver address. Each RSVP-enabled router along the route towards the receiver receives the PATH message along with its parameters and establishes a path or, to be exact, a path state that always includes the previous address from which the PATH message has come. Finally, the PATH message reaches the receiver (service).

In order to make a resource reservation, the receiver (service) sends an RESV message towards the sender (service). This message contains the original QoS attributes of the PATH message and, in addition, a so-called "request specification" which defines the kind of integrated service requested for use. The RESV message also contains filtering information defining packets for which the reservation is made (transport protocol and port number). All of these definitions together form a so-called "flow descriptor", which is used by every RSVP-enabled router along this route to identify resource reservations.

The reception of an RESV message activates the RSVP-enabled routes to start AC and other required procedures to authenticate the request and allocate required resources. If it is not possible, for instance, to allocate the resources, the router sends

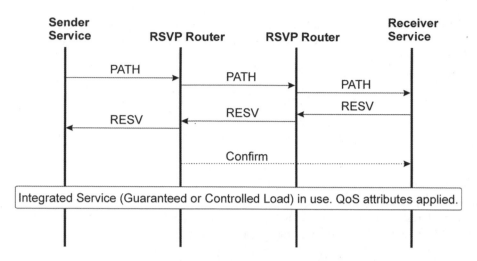

Figure 8.4   The main principle of RSVP

an error message back to the previous address. If all AC-related procedures are success-ful, the RSVP-enabled router sends the RESV message towards the next router.

When the last RSVP-enabled router receives the RESV message, it sends a confirma-tion message back to the receiver (service) indicating that the traffic path is now ready for traffic and has the desired and requested QoS characteristics. The last router mentioned here is either the RSVP-enabled router closest to the sender (service) or a common reservation point in the case of multicast flows.

The route between sender (service) and receiver (service) may contain both RSVP-enabled and non-RSVP routers. In this case the non-RSVP routers are transparent (i.e., PATH and RESV messages pass through them without handling). Note that the non-RSVP routers in this kind of environment constitute a risk, since no one guarantees how they make resource reservation related to the connection.

8.2.4 Differentiated Services (DiffServ)

As the name suggests, DiffServ is a relatively simple method that classifies services and then allows these classes to be treated differently within the network. It is possible to have many traffic classes with DiffServ, but there are only two very important service levels, or traffic classes. Traffic classes have predefined behaviour profiles and limits, called *codepoints* or *DiffServ values*:

- *Expedited forwarding*. This class minimises delay and jitter and thus provides the highest level of QoS. Any traffic unsuitable for class definitions is simply discarded. Expedited forwarding implements a single DiffServ codepoint.
- *Assured forwarding*. This traffic class can be broken down into four subclasses and three drop precedences. Thus, assured forwarding contains altogether $4 \times 3 = 12$ DiffServ codepoints.

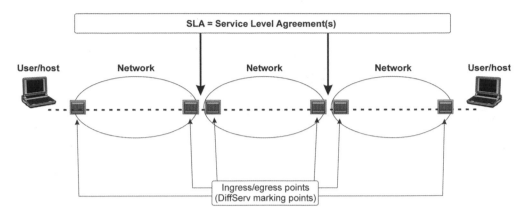

Figure 8.5 DiffServ principle diagram

DiffServ is basically a network-internal QoS protocol invisible to end-users. If used, it will be applied to end-user connections at the edge/border of the network. In terms of DiffServ, this network border entry is called a *network ingress point*. Respectively, the other edge of a network where the DiffServ definitions are torn down is called a *network egress point*. It is possible for end-user applications, or hosts, to perform DiffServ marking directly. This method may bring advantages, especially if one is considering how to arrange end-to-end QoS (e2e–QoS).

By default, DiffServ assumes that a Service Level Agreement (SLA) exists between networks. The SLA, in turn, defines the technical parameters describing the quality of the connection between the networks. This technical parameter set is called (in terms of DiffServ) a *policy criterion*. Traffic is policed according to the policy criteria in the network ingress/egress points.

If traffic is out-of-policy, its treatment follows SLA definitions; for example, this kind of traffic could simply be dropped or could be delivered at an extra cost. Traffic policy definitions and the SLA could be based on time, day, source and destination addresses, application ID values, etc.

Figure 8.5 illustrates the principle underlying DiffServ and where it is applied if several networks have SLAs with each other. If DiffServ marking is carried out at the user/host end, the SLA has an effect on that ingress/egress point as well. At these points the original SLA is typically applied if the user/host end is, say, an Internet Service Provider (ISP). If the user/host is, say, a mobile subscriber, the SLA is actually called a "subscription". Subscription does not necessarily indicate whether DiffServ is used or not, but with a legally binding subscription the operator guarantees certain level of service for a customer and this relationship is very similar to an SLA.

8.2.5 Multi Protocol Label Switching (MPLS)

MPLS shares many common characteristics with DiffServ. For instance, it marks the traffic in the network ingress/egress points as well. The scope provided by MPLS differs

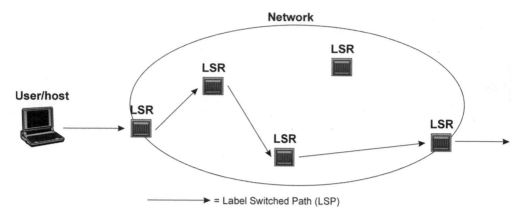

Figure 8.6 MPLS principle diagram

from that of DiffServ: DiffServ aims to classify traffic over the network, whereas MPLS aims to classify traffic only over the next router hop.

MPLS is not controlled by any application (i.e., it does not have an Application Programmable Interface or API) and does not have any end-user/host component. MPLS is only located in servers.

One characteristic worthy of note is that MPLS is protocol-independent. In other words, it can be used with any network protocol. IP is naturally the first option, but ATM and frame relay may also be considered. It can even be used directly on top of the data link layer, when the environment is called an "MPLS framework".

Figure 8.6 describes the principle underlying MPLS in a network.

MPLS-enabled routers are called "Label Switching Routers" (LSRs). The first LSR in the MPLS network receives packets and makes forwarding decisions based on the destination address of the packet or any other information present in the packet header. The information used for the forwarding decision is determined by local policy. The first LSR attaches an appropriate label to the packet and forwards the packet to the next LSR.

Upon receiving the packet the next LSR investigates the attached label. The label acts as a pointer to a table that contains both further indexed information about the next hop and a new label for this packet. The LSR attaches this new label to the packet and sends the packet on its way.

This same procedure is repeated for as long as necessary—there can be any number of LSRs in the chain through which the packet passes and every MPLS label in every LSR changes for routing purposes. This chain is called the "Label Switched Path" (LSP). If all the LSRs in the network have proper configuration information and valid MPLS routing tables, the LSP fulfils the QoS requirements from the network ingress point to the network egress point.

What all of this and Figure 8.6 suggests is that MPLS, as such, is a straightforward solution for implementation of QoS in a network. With MPLS the tasks dedicated to routers could be simpler than when MPLS is not used. MPLS, however, will not work properly alone. As explained, every LSR has its own indexed table for next hops and labels for these hops. The previous LSR looks at this table to determine the hop. In order to guarantee QoS, these tables must be accurate. If a network contains many LSRs the number of tables increases accordingly. In addition, if one LSR change occurs in a major MPLS network, there are going to be many tables to update. For this purpose MPLS needs another protocol/mechanism, one whose main task is to keep LSR tables up-to-date. There are some options for this purpose: one could be the Label Distribution Protocol (LDP), another could be a combination of RSVP and MPLS.

8.3 About Service Subsystems

So far, we have introduced the philosophical background to services, to technical quality requirements and to their implementation possibilities. In this section we will give a short, but not exhaustive, overview of possible services UMTS networks could offer to end-users.

When 3GPP R5 is introduced, one of its aims will be to streamline the transport network, thus leaving more "space" for service implementation. In other words, if a service could trust that the transport network is uniform, service structure and functionality would not impose any limitations on the various types of transport that networks may bring to the marketplace.

One of the leading principles underlying UMTS networks is that the main task of the network is to provide transport resources for service purposes. These transport resources, in turn, should fulfil the quality requirements the service sets (as explained earlier in this chapter). The equipment required for service-handling is located on top of the transport network and forms its own entities, called "service subsystems".

Parts of these service subsystems are inherited from earlier network structures and other parts belong to the newer network entities introduced in the context of 3GPP R5. In this approach, the service subsystems are:

- Services inherited from the GSM.
- UMTS SIM Application Toolkit (USAT).
- Browsing facilities.
- Location Services (LCSs).
- IMS service mechanisms:
  - Messaging.
  - Presence.

8.3.1 Services Inherited from the GSM

Since the UMTS has its background in the GSM and can be interpreted as a massive evolutionary step in mobile networking, it basically supports all the services defined and supported by the original, standard GSM. This applies especially to 3GPP R99 implementation, which is actually a very straightforward GSM-based implementation offering wider bandwidth and higher end-user data rates.

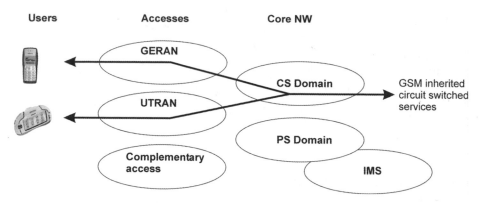

Figure 8.7   GSM-inherited Circuit Switched (CS) service connectivity in 3GPP R5

Ever since 3GPP R4 things have not been as clear as they used to be in 3GPP R99. 3GPP R4 introduced capacity and call control scalability within the CN by bringing Media Gateway (MGW) and Mobile Switching Centre (MSC) server elements to the network architecture. With 3GPP R5 the transport network aims to be uniform within the CN and service interfaces have to be changed somewhat. Since backward compatibility is an issue in the telecommunication world, new ways of integrating services with 3GPP R5 must be considered very carefully.

As a result of 3GPP R5 architecture specification (TS 23.002 v. 5.12.0) and backward compatibility, R5 implementation of the network continues to support both CS and PS traffic. However, the emphasis has transferred to PS traffic and the PS domain contains new interfaces for the IP Multimedia System (IMS) and, thus, creates a platform for future services.

Figure 8.7 shows the high-level connectivity of GSM-inherited CS services compared with the 3GPP R5 architecture at a domain level. As indicated, CS services use the CS domain for connectivity and the CN is accessed either through the GSM/EDGE Radio Access Network (GERAN) or UMTS Terrestrial Access Network (UTRAN). GSM terminals naturally represent the main terminal segment utilising GSM-inherited services. These access the UMTS network through the GERAN and the network needs to guarantee a service level equal to that of the GSM for these.

The other branch of the black line in Figure 8.7 going through the UTRAN indicates CS traffic that may come from actual UMTS terminals. This could be in use in networks where 3GPP R5 is only partially implemented and a part of the network still follows 3GPP R99 functioning principles. In 3GPP R99, QoS conversational class services use CS connections.

A very typical service representative of Figure 8.7 is "Voice" or, more officially, Teleservice TS11 as "Speech".

The PS-inherited service called "2G GPRS" should also be supported in the UMTS environment. 2G General Packet Radio Service (GPRS) is used by GSM/GPRS terminals, and by employing this service the end-user is able to establish QoS "best effort" connections without any exact quality guarantees (Figure 8.8).

2G GPRS is used for messaging and browsing applications (e.g., SMSs and MMSs).

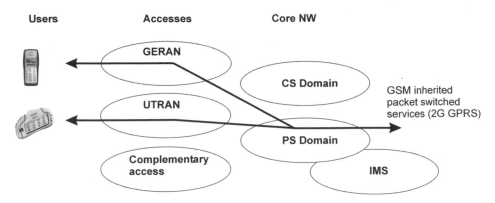

Figure 8.8   GSM-inherited Packet Switched (PS) service connectivity in 3GPP R5

2G GPRS, as such, is also in principle able to utilise IMS services, but 2G terminals would need some additional applications before it could access the IMS.

8.3.2 *UMTS SIM Application Toolkit (USAT)*

Physically, a UMTS SIM card (USIM) is similar to a GSM Subscriber Identity Module (SIM), but it has more advanced features. USIM contains more memory space, more processing power and is downloadable. This creates room for new kinds of services not necessarily present in GSM networks. Note that USIM is backward-compatible with a GSM SIM, although compatibility is not guaranteed under all conditions.

The USAT is a separate platform containing facilities needed for USIM manipulation (Figure 8.9).

With USAT the operator can provide services that modify the menu structures in the terminal. In this way the operator can arrange easy "one-click-type" browsing access, which in turn allows the provision of more services from the same operator or from a specific service provider. This mobile browsing is briefly discussed in Section 8.3.3.

The USAT is a very powerful tool in that the operator may completely change all of the subscription parameters, such as the IMSI number. This may be required in the

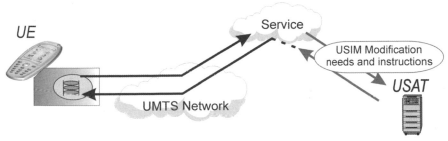

Figure 8.9   USAT—principle diagram

context of some services; however, it is not recommended because of the high risks that are involved.

8.3.3 Browsing Facilities

As its definition implies a mobile browser is an application in a mobile device whose main function is to show content on its display and, based on the request generated by the consumer interaction with the displayed content or the programmed features of the content itself, fetch the new content from the location specified by the request. Sounds difficult? Not really, since most of us have been using browsing applications for many years through the traditional Internet. In this section we take a brief glance at what happens and will happen with browsing when mobility is added onto it.

From the end-user point of view, a browser offers one way to access services either located in the mobile network or through the mobile network. Basically, the end-user is required to understand some basic concepts, which fortunately are mostly the same as those on the traditional Internet side. These concepts are a browser (application), a page, an address and a link. Since these concepts are self-explanatory and familiar, we are not going to define them further. Instead, it will be helpful to keep in mind some principles that should always be applied to browsing and content irrespective of access type. These principles are:

- The service concept must be simple to understand. A service must be a simple collection of a number of interactive pages that work together to form the end-user service.
- Addition of new services must be simple. It is indicative of failure when every new service requires the end-user to install new applications. Instead, new services should be available through a new address.
- Services must use standardised interface elements. Since services vary a lot from each other, the interface between services should contain constant characteristics that can always be located in an expected way (e.g., where to start an application, what the link looks like, where to close an application, etc.).
- Visual richness. Instead of pure voice or text, browsing services should also contain visual elements. This characteristic is strictly tied to terminal development.

In order to handle these characteristics technically, a mobile browsing application must be relatively complex and able to handle the various media types the servers may provide. Figure 8.10 roughly shows the principle behind a mobile browsing application.

Pages shown to the end-user are coded with a mark-up language the browser must be able to present correctly. Nowadays, this mark-up language is Extended Hypertext Mark-up Language (XHTML), which contains Wireless Mark-up Language (WML) and HTML definitions in it. WAP technology using WML had some serious limitations affecting the end-user browsing experience and this highlighted the need to have a more complete solution for a mobile browser mark-up language.

Every page described using XHTML can have all manner of content: pure text, pictures, voice and so on. Text elements are handled and described by XHTML, but

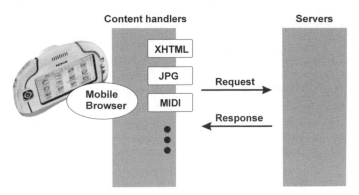

Figure 8.10   Mobile browsing—rough principle diagram

pictures, their resolution and other graphics-related issues are handled by separate, more suitable handlers to guarantee the best possible viewing experience. The same principle applies to voice elements as well. As Figure 8.10 suggests, there are many content handlers and their number will most likely increase. A good example of the use of these is graphics: nowadays, the Internet supports several tens of available picture formats, all of which are necessary for pictures to be shown to an end-user. The most common graphic types in this sense are: BMP (bitmap picture, various sub-versions of BMP exist), JPG (for photos, various versions exist), GIF (Web-tailored format for photos, various versions exist), TIFF (heavy picture format with high resolution possibilities) and PNG (portable network graphics supported by standard browsers).

From the network point of view mobile browsing is a basic service facility, which does not set too many requirements for the transport network itself, because mainly it is a terminal issue. The network's task in mobile browsing is to be available and able to open data connection(s) according to an end-user subscription's QoS definitions for browsing purposes (Figure 8.11).

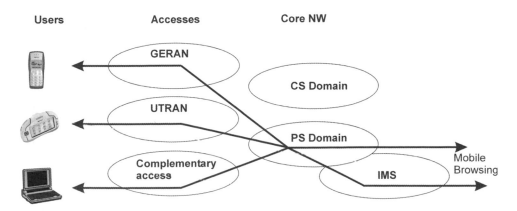

Figure 8.11   Mobile Browsing—connectivity

Mobile browsing can also be interpreted as a platform providing possibilities for different kinds of extensions, thus giving further possibilities for service development. These extensions can be shortlisted as:

- *Location information handling*: LCSs are introduced in Section 8.3.4. With mobile browsing it is possible to offer, for example, location-aware responses to end-user requests.
- *Mobile commerce*: this is one of the expected principal business areas of the future. Mobile browsing will act as a platform and user interface for e-commerce applications.
- *Device management*: since terminal equipment contains software applications that may need some maintenance and upgrades from time to time, mobile browsing could be a handy interface to provide upgrade mechanisms and terminal configuration support to end-users. Note that such a development could impact on security matters.
- *Service discovery and installation*: with mobile browsing an operator could suggest/ advertise new services for end-users and provide fast service activation mechanisms.
- *New XML content types*: when services utilising multimedia eventually emerge, there will be a need to provide the end-user with new content handlers that are able to present new content types.
- *Offline browsing*: a terminal need not always be attached to a network, and there are some services that could be used offline. Mobile browsing will provide the possibility of using such services in this way. This is based on the concept of XHTML downloadable service packages, where a service or a relevant part of it is downloaded to the terminal for local offline use.
- *Voice browsing* offers up the chance to replace standard "link-clicking" by voice commands. Another possibility is that the received content will also be hearable rather than just being displayed.
- *Semantic Web* is a vision about the future Internet, where content is no longer tied to presentation media. With the semantic Web the network will be able to adjust the requested information in such a way that it is suitable for various terminals automatically. There is a clear advantage here for a mobile user: the required content would always be returned in the best possible format, thus giving a better browsing experience.

8.3.4 Location Communication Services (LCS)

Mobile positioning is becoming an important feature of cellular systems. There are two main paths along which LCSs could evolve: Accurate location information may be services using approximate position information based on the coverage of the radio cell and services using accurate location information obtained with the help of GPS (Global Positioning System) or using advanced time-based calculation methods in the radio network and assistance data delivered to the mobile phone.

Examples of commercial location-based applications are fleet management, traffic information management, transportation, nearest services, emergency services, naviga-

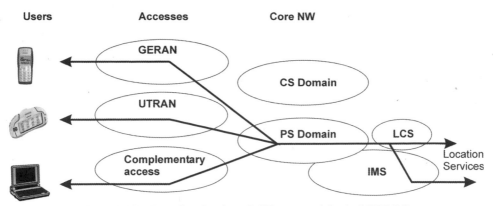

Figure 8.12   Location Services (LCS) connectivity in 3GPP R5

tion services, gaming, etc. Another example is when the network operator applies location-based load information when planning the network and optimising the performance and capacity of the mobile system.

Cellular technology allows the end-user to move around and use various communication services regardless of his or her position. However, this freedom of movement also makes it difficult to know the exact location of the user.

In this subsection we first describe the basic principles of the different positioning methods that have so far been developed for mobile networks. Then, we describe the positioning methods that have been specified for the UMTS. We further give an insight into the UMTS positioning architecture as well as the positioning functions allocated to different network elements. Location information can also be communicated directly between the mobile phone and the location server—this type of solution is described in Section 8.3.4.5.

In order to maintain consistency with 3GPP specification terminology, we use the term "position" as a general concept in the radio network, but stick to "location" when referring to the service aspects of location information. 3GPP and its specifications refer to the overall concept as "Location Services" (LCSs).

Figure 8.12 shows the access networks, core network domains and LCS connectivity in 3GPP R5 networks.

8.3.4.1 Overview of Positioning Methods

Various methods have been investigated and researched in order to determine the geographic position of a mobile phone in a cellular network. The most important ones are:

- Cell-coverage-based positioning.
- Round Trip Time (RTT)-based positioning.
- Time Difference Of Arrival (TDOA) positioning.
- Enhanced Observed Time Difference (E-OTD).
- Global Positioning System (GPS).

- Time Of Arrival (TOA) positioning.
- Angle Of Arrival (AOA) positioning.
- Reference Node Based Positioning (RNBP).
- *Galileo*'s positioning system.

The basic principles that explain how these methods could be used for UE positioning are described in the following subsections, including a short analysis of the accuracy that can be achieved using these methods. In 3GPP Release 5, the 3GPP chose to standardise the cell-coverage-based method, TDOA (3G), E-OTD (2G) and GPS-based methods; so, these positioning methods are described in greater detail.

8.3.4.1.1 Cell Coverage Based Positioning
The position of the terminal can be estimated by discovering the radio cell where the terminal is camping or has recently been camping. The Radio Network Controller (RNC in 3G), Base Station Controller (BSC in 2G) or a separate so-called Serving Mobile Location Centre (SMLC) estimates the approximate geographical coordinates that correspond to the indicated cell and sends the result to the location server in the network.

Note that in 3G specifications a cell's actual identity is seen as radio information and may therefore not be transported to the CN. Instead, the cell ID is mapped to a "service area identification", which may be sent to the CN.

As cell coverage/size varies according to radio channel characteristics and radio network planning strategy, the cell coverage method on its own may not meet the quality of positioning needed for certain location-based services. Therefore, this method is often combined with the RTT positioning method (also known as the Timing Advance, or TA, positioning method) to determine terminal coordinates more accurately.

8.3.4.1.2 Round Trip Time (RTT) Based Positioning
The RTT of a radio signal may also be used to increase the accuracy of position estimates. Calculation of the terminal position may use the RTT measurements of signal branches from several Base Stations (BSs) (Figure 8.13). RTT is the propagation delay time of a signal travelling from the terminal to the BS and back. Therefore, the distance between the terminal and the BS can be obtained based on the time t and the velocity c of the radio wave. This gives us:

$$D = \frac{RTT}{2} c + \varepsilon$$

where D depicts the distance from the mobile terminal to the BS; c is the speed of the cellular radio wave, which is equal to the speed of light (3×10^8 m/s) in a free space medium; RTT is the round trip time; and ε is measurement error.

Note that distance D is in fact the radius of a circle around the BS. Therefore, for more accurate estimates, the location of the mobile terminal can be figured out by using additional RTT measurements from neighbouring BSs, if available. Then, the position estimate will be within the intersection of the three neighbouring circles with the BSs

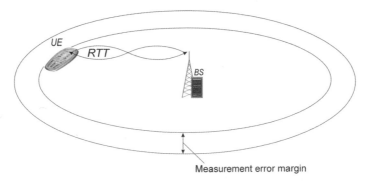

Figure 8.13   Round Trip Time (RTT) positioning

located in the centre of those circles. The resulting position estimate accuracy is determined by the error margins of all RTT measurements.

8.3.4.1.3 Time Difference Of Arrival (TDOA) Positioning

The basic principle of the TDOA method is different from that of the TOA method (see Section 8.3.4.1.5). In the TDOA method the terminal observes the TDOA of the radio signals from neighbouring BSs. The unknown terminal position is estimated by processing TDOA measurements between the terminal and at least three BSs with known coordinates. The TDOA measured by the terminal consists of two components:

$$TDOA = RTD + GTD$$

where the Geometric Time Difference (GTD) comes from the geometry (e.g., propagation delay differences between the handset and the two BSs). GTD is the actual quantity containing information concerning the handset position, since it defines a hyperbola between the two BSs. In addition, in a system like WCDMA, where BSs are not synchronised, the Relative Time Difference (RTD) has to be known. RTD is the transmission difference between the signals of the neighbouring BSs.

Figure 8.14 depicts the principles of the TDOA positioning method. The position of the terminal is calculated based on the TDOA between the serving BS and the neighbouring BSs, which define a hyperbola, the foci of which coincide with coordinates of the corresponding BS transmitter antenna. The position of the terminal can be estimated by utilising the least squares of the distances of the terminal to the hyperbolas, if there are more than two TDOA values available.

Therefore, the basic idea in the TDOA method is to determine the relative position of the terminal by examining the difference in time at which the signal arrives at the target (terminal or BS), rather than the absolute arrival time. Therefore, if this time difference between neighbouring BSs is available and there is a Line Of Sight (LOS) between the terminal and the BSs, the mobile terminal is located on the hyperbola:

$$D_1 - D_3 = c\Delta t = c * GTD = \sqrt{(X_i - x_m)^2 + (Y_i - y_m)^2 + (Z_i - z_m)^2}$$

where D_i is the distance from the mobile handset to neighbouring BSs; c is the speed of light; Δt is the TDOA of the signal from neighbouring BSs; X_i, Y_i and Z_i are the

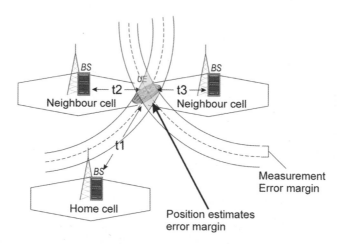

Figure 8.14   Time Difference Of Arrival (TDOA) positioning

coordinates of neighbouring BSs; and (x_m, y_m, z_m) are the coordinates of the mobile terminal. As shown in Figure 8.14, by measuring two TDOAs of three different BSs the position of the terminal can be estimated in the intersection(s) of these hyperbolas. However, to obtain a unique and more accurate estimate there should be at least three time difference measurements comparing three or more BSs.

8.3.4.1.4 Enhanced Observed Time Difference (E-OTD)
The basic principle underlying E-OTD is similar to that of the TDOA method. Like TDOA, the E-OTD method estimates the mobile position by utilising measured time differences between neighbouring BSs. Therefore, this method is occasionally considered identical to the TDOA method. However, in E-OTD more precise measurement procedures are employed to improve positioning accuracy. The measurement method used in E-OTD is called "Enhanced-OTD". In this respect, OTD refers purely to a measured time value and, therefore, should not be confused with E-OTD, which refers to the positioning method itself. There is a need for more accurate measurement methods (particularly when the network is unsynchronised); this applies both to the GSM and Wideband Code Division Multiple Access (WCDMA).

OTD measurements are made at both Location Measurement Units (LMUs) and Mobile Stations (MSs). The BSC, or SMLC, determines a location estimate by comparing the two sets of OTD measurements. As OTD measurement reporting causes additional signalling, the LMU's OTD measurements are only reported to the SMLC at certain intervals. How frequently OTD results are reported depends on both the availability of BS resources and the level of accuracy required for the requested service.

8.3.4.1.5 Time Of Arrival (TOA) Positioning
In the TOA method, position calculation is based on the propagation delay of the radio signal from the mobile device to the BS. When there are at least three TOA measurements available, the position of the terminal can be ascertained by applying a triangulation technique, minimising the least square distances between the terminal and

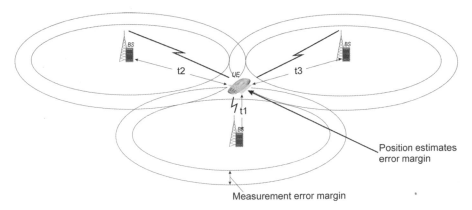

Figure 8.15   Time Of Arrival (TOA) positioning

corresponding TOA circles. Figure 8.15 shows the principle used to estimate the terminal position. This is done by measuring the TOA of three neighbouring BSs.

In the TOA method the distance from the terminal to the BS is proportional to the propagation time t_i. Thus, if there is an LOS between the handset and BS, the distance between them is given by:

$$D_i = ct_i$$

where c is the speed of an electromagnetic wave and t_i is the delay time for the signal to travel from a BS to the terminal or vice versa. However, in practice there is always some error regarding TOA measurements due to Non Line Of Sight (NLOS), signal fading, reflection and shadowing and, consequently, coverage variations at the edge of the cells. Therefore, a measurement error margin should be taken into account when estimating the accuracy of TOA positioning.

As can be seen from Figure 8.15, in order to obtain a unique position estimate of the mobile terminal the TOA of at least three BSs is needed. Then, the position of the handset can be estimated by calculating the distance between the terminal and several BSs (TOA circles) in a least square sense. Thus:

$$D_i = \sqrt{(X_i - x_m)^2 + (Y_i - y_m)^2 + (Z_i - z_m)^2 + \varepsilon}$$

where (X_i, Y_i, Z_i) are the coordinates of the neighbouring BSs involved in the positioning process and (x_m, y_m, z_m) are the coordinates of the terminal to be positioned.

The TOA method would require very accurate BS synchronisation, which may bring challenges to unsynchronised cellular systems. The position calculation entity should also be capable of discerning the time difference between the transmitted and received signal. The most important asset of the TOA is that it does not necessarily require mobile device upgrading, though the cost of network upgrading may become a problem area for its wide acceptance in the mobile network.

8.3.4.1.6 Angle Of Arrival Positioning (AOA)
The position of the terminal can also be determined by calculating the intersection of the two lines representing pilot signal branches, each formed at an angle from the BSs

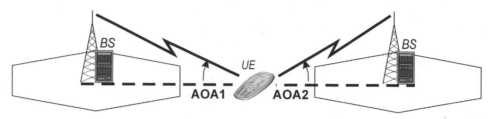

Figure 8.16   Angle Of Arrival (AOA) positioning

to the mobile terminal. A single measured angle forms pairs of lines and provides the target handset position. Therefore, as shown in Figure 8.16, if there is a LOS between the mobile terminal and two BSs and the AOA measurements of two neighbouring BSs are available, the mobile terminal will be located at the intersection of the lines defined by the AOAs.

Similarly to the TOA and TDOA methods, AOA method accuracy can be improved by using more than two measurements. Nevertheless, reflection and diffraction cause serious errors, especially when AOA is used in an urban environment. Normally, the mobile terminal is unable to determine AOA by itself; so, it should be measured at the BSs.

AOA is prone to signal fading; so, AOA in combination with, say, frequency re-use needs careful network design and makes network capacity planning very demanding. The AOA method is also quite complicated and expensive to implement, because several large antennae are needed at BS receiver sites. On the other hand, the many existing antennae could be utilised for this purpose.

8.3.4.1.7 Reference Node Based Positioning (RNBP)

Also occasionally called "local positioning" or "self-positioning", the RNBP method is based on the principle that a reference node is chosen to provide auxiliary positioning measurements of the mobile device. The reference node may be movable or fixed and may have relay or repeater functionality. The reference node may be a positioning service device, a GPS receiver or any device with a known position that can be used as a reference point when determining the position of terminals. In certain circumstances, position information may be delivered to the mobile terminal from another device close by with a known geographical location when a short-range radio is used, such as Bluetooth, Wireless LAN or Ultra Wideband (UWB). The mobile terminal can hence calculate its own position by measuring, say, the received signal strength or traverse time from the reference nodes. As the number of mobile devices equipped with short-range radios is rapidly increasing, the feasibility and accuracy of RNBP may improve considerably. In particular, it is envisaged that RNBP in the future will become a common method for indoor applications.

In addition to the basic signal traverse time and signal strength, the RNBP can, at least in principle, be combined with other positioning methods, such as TDOA, TOA or AOA and so forth. The main point is that RNBP may provide enhanced positioning, by utilising additional reference devices in the network. RNBP could be used to improve network-based positioning methods (e.g., by inserting so-called "seed nodes" in the

RAN. This approach has some cost drawbacks, since RNBPs would have to be placed at a BS or somewhere similar, and each RNBP would have to have a separate antenna. However, the cost drawback might not be very significant in that the existing radio network infrastructure already contains sites for nodes and repeaters. In any case, careful network planning would be needed to avoid increased interference levels as a result of repeaters.

8.3.4.1.8 Global Positioning System (GPS)

GPS is a satellite-based positioning system, which in many ways offers the best radio navigation aid currently available. It is operated by the US Government and is used both for military and public purposes.

GPS can be combined with cellular applications in several ways. The first step is to implement a GPS receiver in the terminal, which offers the same benefits as a stand-alone GPS receiver. The accuracy and speed of the GPS receiver can be radically improved by sending GPS assistance data to the terminal. This is what makes GPS very useful in cellular networks, because GPS assistance data can be generated by a few GPS reference receivers and delivered to the mobile terminals over cellular connections.

The principle behind GPS is based on the TOA method, but, due to accurate timing, the implementation of GPS can be complicated. The GPS satellite transmits a spread spectrum signal with very precise timing to the Earth on the L-band. A precise clock at the receiver measures the signal's time delay between the satellites and the receiver. This permits calculation of the distance from the receiver to each satellite. When a receiver observes three satellites, its position can be estimated by utilising the triangulation approach. In practice, the clock of the receiver need not be so accurate, if the signals from a fourth satellite are used to correct receiver clock errors. The radio signal from a GPS satellite could be seen to form a sphere around the satellite that can be drawn based on the signal travelling time. Having defined three spheres from three satellites, the receiver's position is calculated as the intersection of these spheres, providing co-ordinates for latitude, longitude and altitude. Figure 8.17 illustrates GPS-based position estimation.

GPS positioning is one of the most accurate positioning systems currently in existence, especially when using Differential GPS (DGPS) assistance. However, GPS receivers normally should have an LOS to some four satellites, and this sometimes causes problems, especially indoors. Moreover, the GPS receiver has a dedicated satellite antenna that is typically relatively large and may interfere with cellular usage. The GPS receiver increases manufacturing costs and the power consumption of the mobile terminal. In order to overcome or at least ease such problems, GPS positioning and the delivery of GPS assistance data have been standardised both for the GSM and UMTS.

8.3.4.1.9 Galileo satellite navigation system

Galileo is a joint initiative by the EU and the European Space Agency.

Galileo will comprise a constellation of 30 satellites orbiting at an altitude of nearly 24,000 kilometres and will be operational from 2008. It will be compatible and interoperational with GPS. This means improved access to navigation satellites and a possibility to improve accuracy or response times. *Galileo* is planned to offer greater accuracy than GPS.

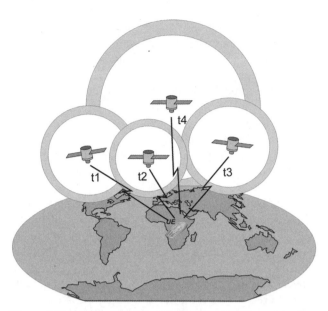

Figure 8.17   Global Positioning System (GPS) positioning

The operating principle of *Galileo* is similar to GPS: the satellites are fitted with a clock measuring time very accurately and this precise time is indicated in the signal they transmit. The mobile terminal receives signals from several satellites and measures the time taken by the signal to arrive. Using the measurement results and the indicated time the mobile can calculate its own position based on the distances to the measured satellites. More information on *Galileo* is available at *http://europa.eu.int/comm/dgs/energy_transport/galileo/index_en.htm*.

8.3.4.2 Positioning Accuracy

As previously described, the main principles of any positioning method, excluding the cell-information-based method, are founded on radio signal propagation characteristics, which are not easy to predict. The accuracy of cellular positioning methods depends on the cellular environment, the quality of the radio connection (e.g., how close it is to the LOS), signal measurements, receiver features and calculation method.

The accuracy of the cell-ID-based method depends mainly on the cell structure and cell size of the radio network. In a network structure with macro-cells, position accuracy ranges from a few kilometres to tens of kilometres. In a pico- and micro-cell environment the degree of accuracy will vary from a few tens to hundreds of metres.

The accuracy of the TOA method depends on signal fading, shadowing and the number of signal branch measurements available to different BSs. It is essential to have an LOS to several BSs to obtain accurate position estimates of the mobile terminal. Unfortunately, this requirement contradicts normal cellular network planning principles; so, optimum accuracy cannot be achieved everywhere in a cellular network.

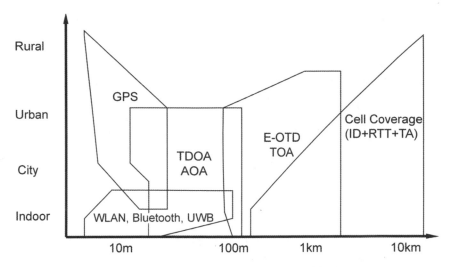

Figure 8.18   Accuracy comparison of different positioning systems

As in the TOA method, the accuracy of the TDOA method is subject to the inherent characteristics of the radio signal in the cellular network environment and the number of available measurements, especially the Non-LOS cases. When TDOA is used, however, some errors are common to all BSs and cancel each other out in the positioning process because only time differences are measured. Therefore, TDOA-based methods normally offer better results than TOA-based methods if preconditions remain similar.

All in all, there are no reasonable absolute estimates to generalise the accuracy of different positioning methods. Nevertheless, the degree of accuracy of different methods can be roughly outlined as shown in Figure 8.18. There is a cost increase when using more accurate methods; as a result it is most cost-efficient to select a method that offers "good enough" accuracy for a certain purpose. Different types of LCSs have different accuracy requirements—the reason several methods may be used within one network.

8.3.4.3 Positioning Methods Specified for UMTS and GSM

All the positioning methods described above were investigated and discussed in 3GPP from a standardisation point of view. As a result, three methods were selected for UMTS UTRAN networks:

- Cell-ID-based positioning, the cell ID corresponds to the so-called Service Area ID, SAI in the UMTS.
- Observed Time Difference Of Arrival (OTDOA) positioning, which—as the name suggests—is based on the TDOA principle described above.
- Assisted GPS positioning.

For GERAN networks the 3GPP selected the following methods for standardisation:

- Cell-ID-based positioning.
- Enhanced Observed Time Difference (E-OTD) positioning.
- Assisted GPS positioning.

These methods can, in principle, be network-based, mobile-based, network-based mobile-assisted or mobile-based network-assisted. The difference between these variants is whether the position calculation is carried out in the network or in the terminal. Other variants are defined depending on where the assistance data are generated, so the method may be termed network-assisted or mobile-assisted, respectively. The following subsections give an insight into the positioning methods specified for the UMTS and GSM.

8.3.4.3.1 Cell ID Based Positioning in UMTS and GSM

In the cell-ID-based method, the Serving RNC (SRNC) maps the cell ID into the corresponding SAI. Additional operations may be needed if UE is in soft-handover state or in Radio Resource Control (RRC) state, where the cell is not defined. In GERAN networks the cell ID is handled by the BSC and is used as such for LCSs.

In soft-handover state the UE may have different signal branches connected to different cells, reporting different cell IDs. In this case the cell-coverage-based position estimate can be improved by combining all available information in the SRNC.

Normally, cell ID selection is based on parameters defining the quality of received signal branches. That is, the SRNC selects the cell ID that has the strongest signal branch as a reference when determining the UE position estimate. Alternatively, the first cell ID that was generated during connection establishment may be chosen as the cell for position estimation. In addition to these common mechanisms, other criteria have been specified for determining the cell ID during handover.

The cell ID can be determined according to the Primary Common Pilot Channel (CPICH in WCDMA-FDD) in the active set of BSs received by the terminal. In addition, a CPICH that is not included in the active set may become better than the primary CPICH that belongs to the active set. The excluded CPICH can then be used to determine which cell ID to utilise.

The cell-ID-based method should support UE positioning regardless of the RRC state of the connection (i.e., URA PCH, cell PCH, cell DCH, cell FACH, cell reselection, inter-system modes and idle mode).

It is not possible to obtain the cell ID if the UE is not in an active state (i.e., if there is no connection between the UE and at least one of the cells). For example, in the UMTS the cell ID can be provided only when an RRC connection exists between the UE and at least one BS. Therefore, if the UE is in an idle state, the UE is forced into a state where the cell identifier can be provided. For example, in UTRAN Registration Area Paging Channel (URA PCH) state the cell ID may not be available, so the UE is then forced into a channel state having the cell ID (say, cell FACH), to determine the cell identifier. The network can also prevent the UE from entering the URA update state to receive cell ID updates when the UE selects a new cell. When the UE is in idle mode or URA PCH mode, it needs to be paged for positioning purposes by either the CN or UTRAN.

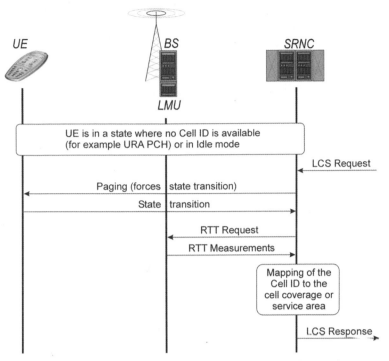

Figure 8.19   Cell-coverage-based positioning; general procedure

In the UMTS, UTRAN factors should be hidden from the CN. As a result of this the cell identifier is mapped to the SAI, which is sent from the UTRAN to the CN over the Iu interface. The service area could in principle include more than one cell, but currently there is a one-to-one relationship between the cell ID and SAI in the UTRAN.

8.3.4.3.2 Observed Time Difference Of Arrival-Idle Period Down Link (OTDOA-IPDL) in UMTS

The second UMTS positioning method Observed Time Difference Of Arrival-Idle Period Down Link (OTDOA-IPDL) consists of two parts. First, there is OTDOA, which is theoretically similar to TDOA described previously. However, in this concept it is emphasised that the UE measures the observed time differences between neighbouring BSs in order to estimate the position of the UE.

Measuring the OTDOA and RTD is not very straightforward in WCDMA-FDD (Frequency Division Duplex) for the following two reasons:

- In some cases there might not be enough downlink pilot signals available for measurement by the UE. This may occur when the UE is located so close to the serving BS that its receiver is blocked by strong BS transmission. This is referred to as the *hearability effect*.

- BSs are normally unsynchronised in WCDMA-FDD, so the synchronisation difference between BSs—the RTD—should be known or measured before the position of the terminal can be calculated.

To overcome these problems two possible solutions were investigated, first in the Japanese standardisation bodies and later in the 3GPP, to enable support for UE positioning in WCDMA-FDD networks.

The first solution would be to temporarily increase the transmission power of neighbouring BSs to enable the UE to measure them. However, because WCDMA-FDD, like every CDMA-based radio technology is interference-limited, the power-up approach was not feasible at all, because it would substantially increase interference levels in the network.

The other solution would be to have the serving BS cease its transmission for short periods of time to reduce the hearability problem. This method is called "Idle Period Down Link" (IPDL). During an idle period of the serving BS, the UE can measure the signals of neighbouring BSs. During such idle periods, RTDs can also be measured more accurately. The measurement units used for this purpose can be placed in, or close to, BS sites, when IPDL is used.

There are drawbacks to using IPDL, because normally the connection should be kept as good as ever possible and regular idle periods in BS transmission may cause disturbances. Another difficulty in WCDMA is that cells have little overlap, so mobiles may not be able to receive and measure a sufficient number of adjacent cells, even when IPDL is used.

8.3.4.3.3 Assisted GPS

The 3GPP has standardised assisted GPS methods for use in both GSM and UMTS networks. There are two basic types of assisted GPS: UE-based and UE-assisted. The difference lies in where position calculation is actually carried out. Note that the UE might also receive GPS assistance data from the network in the UE-assisted GPS method (Figure 8.20).

In the UE-based assisted GPS solution the UE contains a complete GPS receiver and position calculation is carried out by the UE. The GPS assistance data sent by the network to UE include:

- Measurement assistance data (i.e., GPS reference time, visible satellites list, satellite signal Doppler and code phase search window). These data are valid for 2 to 4 hours.
- Assistance data for position calculation (i.e., reference time, reference position, satellite ephemeris and clock corrections). These data are valid for 4 hours.

When differential GPS is utilised, differential correction data are also sent to the terminal. This type of assistance data is only valid for about 30 s, but is relevant for quite a large geographical area. It is therefore possible to generate assistance data using only one centrally located reference receiver.

In the network-based UE-assisted GPS solution, the UE has a reduced GPS receiver. Using this receiver the UE carries out pseudo-range measurements and transmits the measurement results to a calculation unit in the network, which carries out the rest of

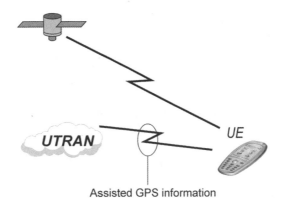

Assisted GPS information

Figure 8.20   The basic principle of assisted GPS in UMTS

the GPS operation. In this solution the network only sends limited assistance data to the UE, but the UE, on the other hand, has to send all measurement results in the uplink direction. The reference time must also be more accurate in UE-assisted mode than in UE-based mode.

8.3.4.4 System Architecture for Location Services (LCSs)

The Location Services (LCS) functionality in the UMTS is distributed among existing network elements in basically the same way as in the GSM. One main new network element, the *Gateway Mobile Location Centre* (GMLC), is added to the overall system architecture to support LCSs both in the GSM and UMTS.

Figure 8.21 illustrates the overall architecture of LCSs in the UMTS. LCS equipment and functionality is located in both the CN and UTRAN. UTRAN entities measure and collect positioning data and determine the location of the mobile using the methods described above. In the CN the GMLC acts as a connection point through which positioning data can be delivered to other service capabilities or client applications.

The UMTS LCS system architecture is very similar to the LCS architecture in the GSM, but several optional network solutions were omitted in UMTS LCSs in order to decrease the complexity of the solution. WCDMA-FDD-based UTRAN brings both constraints and advantages for cellular positioning applications when compared with the GSM.

In the UMTS the SRNC has a key role in controlling radio access, including radio resource management. Radio link measurements, including soft-handover measurements, are terminated and controlled in the SRNC. Since some 30–40% of all calls may be involved in soft handover, the SRNC in many cases will have access to measurement results that can be used for positioning purposes. It was therefore deemed reasonable to utilise this WCDMA-FDD advantage when considering the LCS architecture in the UMTS. As a result, the functions corresponding to the SMLC in the GSM were allocated to the SRNC in the UMTS.

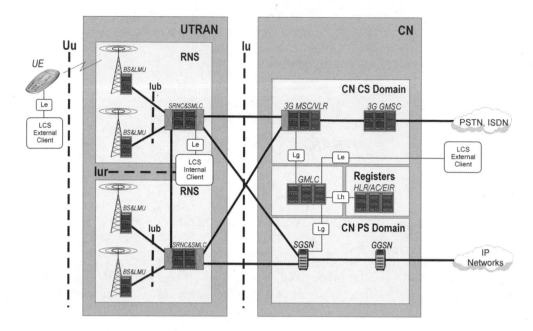

Figure 8.21   General Location Services (LCS) architecture in UMTS

For assisted GPS, however, a stand-alone SMLC has been specified that may be used either for mobile-based or mobile-assisted GPS.

The solution where the SRNC contains the main SMLC functionalities simplifies the signalling specifications of the Iu, Iur and Uu interfaces when compared with the many alternative solutions in the GSM. In the UMTS the leading principle is to separate radio-related aspects from CN aspects; this is also valid for LCSs.

In the GSM, the SMLC can either be a dedicated physical network element or integrated in the BSC. In the early phases of LCS standardisation the SMLC could reside either in the access network or in the CN.

The SMLC in the GSM and the SRNC in the UTRAN choose the positioning method, control positioning measurements and calculate the position of the target terminal using measurement results obtained from the target terminal itself and from dedicated LMUs in the BSs (or those connected to the BS).

LCS specifications in the 3GPP indicate a more formal view of the LCS functionality in different network elements and the corresponding interfaces between the network elements. The 3GPP LCS reference model in the GERAN is shown in Figure 8.22 and the LCS reference model in the UTRAN is shown in Figure 8.23. The interfaces and network entities in the Figures are specified in the corresponding 3GPP LCS specifications, and the LIF-MLP (Logical InterFace-Mobile Location Protocol) is specified in the Open Mobile Alliance (OMA). The interface between the requestor shown in the Figures and the LCS client, however, is not standardised. The requestor is the originator of the location request and may be another terminal or a location-based application server. The user of the target terminal often also acts as the requestor of his or her own location.

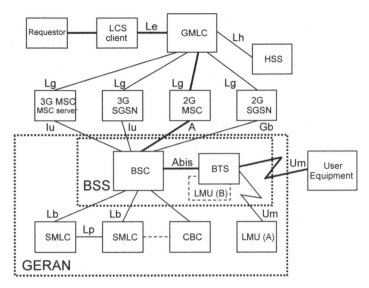

Figure 8.22 Location Services (LCS) reference model in GERAN

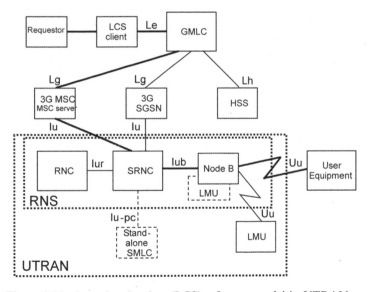

Figure 8.23 Location Services (LCS) reference model in UTRAN

8.3.4.4.1 LCS Functions in the Core Network (CN)

LCS functionality in the CN is allocated to the MSC/VLR (Visitor Location Register), MSC server, SGSN, Home Location Register/Home Subscriber Server (HLR/HSS) and GMLC. The MSC/VLR and SGSN may initiate paging for LCS purposes and handle UE authentication as well as call-related and non-call-related position requests and charging.

The user can control whether and how the position of his or her terminal is disclosed to different LCS clients. This is defined by LCS privacy classes in the subscription profile, which is kept in the HLR or HSS. For VASs the user may want to be notified each time his or her position is being requested. The user may then choose to accept or decline a position request from an indicated LCS client. The network will send this privacy invocation request only to such terminals that are able to handle the request and can show it to the user. The terminal informs the network about its capability in this respect using so-called "UE classmarks". The UE sends classmark 2 (i.e., the CN classmark) to the MSC/SGSN indicating whether the UE supports privacy verification or not.

The UE also sends classmark 3 (i.e., the RAN classmark) to the RNC (and to the BSC and MSC in the case of the GSM) indicating the supported LCS methods. This information is not needed in the CN.

The GMLC, MSC and SGSN handle all charging and billing related to LCS. This includes charging and billing of both clients and subscribers, including home subscribers and roaming subscribers from other Public Land Mobile Networks (PLMNs) as well. The MSC and SGSN also authorise the provision of positioning services for a particular UE in the same way as is done for other cellular services. The MSC/VLR handles LCS in the CN CS domain and the SGSN supports LCS in the CN PS domain.

The HLR contains subscription data and privacy profiles for subscribers using LCS as well as information about the VLR and/or SGSN to which the mobile is currently registered. The GMLC requests routing information (i.e., to which VLR and/or SGSN is the mobile currently registered) from the HLR via the Lh interface using Mobile Application Part (MAP) signalling.

The GMLC acts as a gateway between external LCS clients and the mobile network and forwards the LCS client's positioning request to the MSC/VLR or SGSN as indicated by the HLR. The GMLC authenticates the LCS client with the help of the HLR (i.e., the GMLC verifies that the LCS client is allowed to be informed of the position of a given subscriber). The GMLC determines whether the position estimate received from the network satisfies the QoS requirement set by the LCS client. The GMLC also provides flow control of positioning requests between simultaneous or periodical requests and may convert position information received from the network to local coordinates. The 3GPP has specified the way in which the GMLC should handle the complex service combinations of so-called "deferred and periodic location requests". The LCS client may have requested a location report whenever a given mobile becomes active in the network, and at the same time there may a periodic location request pending for the same subscriber.

8.3.4.4.2 LCS Functions in UTRAN and GERAN
The SRNC in the UTRAN (or the SMLC in the GSM) selects the appropriate positioning method by which the indicated QoS level for the position request is likely to be met. The SRNC also controls how the positioning method is carried out in the UTRAN and UE.

When the OTDOA-IPDL method is used the SRNC controls and defines the idle periods in different BSs to minimise the impact on UTRAN performance. This is done either according to predefined patterns or on demand. The RNC also coordinates the UTRAN resources involved in the positioning of the terminal.

In network-based positioning methods the SRNC calculates the position estimate and indicates the achieved accuracy. It also controls a number of LMUs at BSs for the purpose of obtaining radio measurements needed to position or help to position a UE. Signalling between the SRNC and LMU is transferred via the Iub interface and in certain soft-handover cases also via the Iur interface.

In the UMTS the LMU is normally integrated in the BS and a stand-alone LMU will probably not be specified. The main functions of an LMU are to measure the RTD between different BSs, the Absolute Time Differences (ATD) from a reference clock or any other kind of radio-interface-timing measurement of the signals transmitted by BSs. Some measurement results returned by the LMU to the SRNC can be used for more than one positioning method. All position and assistance measurements obtained by an LMU are sent only to that SRNC associated with the corresponding BS. The SRNC controls the timing and any periodicity of LMU measurements in the BS, either directly or on a predefined basis.

8.3.4.4.3 LCS Functions in the UE

Depending on the positioning method, the UE may be involved in positioning procedures in various ways. For instance, in a network-based positioning method the UE does not calculate its position, whilst in a mobile-based positioning method it does. The UE may also be equipped with a GPS receiver, which enables it to determine its position independently. A GPS mobile may request GPS assistance data from the network in order to speed up the positioning process and to improve accuracy or GPS receiver sensitivity. In network-based assisted GPS methods the position of the UE having a GPS receiver is determined in the network as a result of the GPS measurement results reported by the UE.

A mobile terminal can request the network to determine its position (network-assisted positioning) or a mobile may itself determine its own position using its own measurement results and assistance data from the network.

The use of IPDL requires that the UE be capable of measuring and storing radio signal timing during idle periods as well as correlating different Broadcast Channel (BCH) codes with different idle period patterns. The UE needs to determine the arrival time of the first detectable path, both for the serving BS and other BSs. The UE then reports the measurement results to the SRNC or SMLC.

8.3.4.4.4 LCS Client and GMLC

The LCS client may reside in a server that provides services and applications using the location information of the mobile terminal. The LCS client requests location information from the cellular network (PLMN) for one or several target terminal(s) via the GMLC. The request contains a set of parameters that defines, among other things, the requested accuracy and response time.

Many of the main interfaces of the UMTS (i.e., Uu, Iub, Iur and Iu) support LCSs. In addition, there are specific LCS interfaces defined for the GMLC as shown in Figures 8.22 and 8.23:

- The interface between the LCS client and GMLC is called "Le". The LCS client sends the LCS request on this interface and the GMLC sends the corresponding results back to the LCS client.
- Lh is the interface between the GMLC and the HLR (HSS). The GMLC uses this interface to obtain routing information from the HLR both for the CS and PS domain.
- The GMLC is connected to the MSC/VLR or SGSN over the Lg interface. The GMLC sends the location request for an indicated target terminal to the MSC/VLR or SGSN currently serving the UE using the Lg interface, and the MSC/VLR and SGSN return the position result to the GMLC.

Both the Lh and Lg interfaces are based on the MAP protocol.

8.3.4.5 Location Information in the User Plane

The LCS functions described above all take place in the so-called "control plane" of the cellular network, which means that several network nodes are involved in determining the location of the terminal and that there is standardised signalling between network nodes. All involved network elements must be able to communicate with each other, and one possible consequence is that the positioning capability of the network may be limited to the capability of the least advanced network node. There may also be inter-working problems in networks where the equipment is delivered from several vendors or in networks where the equipment belongs to different generations, some of which may need an upgrade to achieve full LCS capability.

One way to overcome some of the problems described above is to establish a PS data connection between the terminal and a server in the network. The connection may be requested either by the server or the terminal, and the cellular network sets up the connection in the same way as any other common data connection. The data connection carries user data and is therefore said to be in the "user plane". Using this data connection the server might deliver GPS assistance data to the terminal and an application in the terminal might deliver location information to the server.

The user plane location solution is being specified in the OMA and 3GPP2, where the aim is to achieve similar capabilities to those reached in the control plane solution. The user plane solution also needs to comply with regulatory privacy requirements and must support charging of GPS assistance data. Due to regulatory requirements in some countries the user plane solution would also have to support emergency calls, since location information is crucial in emergency situations.

8.3.4.6 User Privacy in Location Services (LCSs)

Location information is quite sensitive from the privacy point of view and people would not feel comfortable about letting their location be known by anyone at any time. There are also strict regulations regarding how location information can be handled (e.g., storage of location information is not allowed). The importance of privacy has been recognised by the standardisation bodies and several mechanisms have been described to enforce privacy rules on location information. According to 3GPP specifications the user of the target mobile should be able to define the rules that apply to the way in which location information is handled in different cases. Privacy rules may apply for a specific LCS client or for an identified requestor. In case the user has not defined any rules for an LCS client, the default treatment is not to allow positioning. The user may also wish to be notified about an incoming location request and given the chance to deny or allow the request. In addition, the user may set a codeword that can be given to friends or trusted applications. When such a codeword is defined, only requestors or applications that can provide the correct codeword will be given location information. Location services like navigation need accurate location information to be useful, while other services, such as those that offer local information, do not need such accuracy. According to the specifications the user may therefore also set the so-called "service-type" privacy rules that allow certain types of service but prevent other services from getting location information.

8.3.4.7 Example Services that Use Location Information

Location information and location-based services are often mentioned as examples of advanced services in cellular networks. The mobile phone user is free to move around and make or receive calls or activate services almost anywhere. It is evident that the current benefits of mobile services can be enhanced if the information is somehow made applicable or improved according to the current location of the mobile. There are also services that are dedicated or only applicable when location information is available. In this chapter some services that take advantage of location information are described, starting with the most obvious one—navigation.

8.3.4.7.1 Navigation

Navigation services are quite useful for most mobile users in certain situations. When you enter a town or step down from a bus or train in an area that you do not know very well, you would welcome a map of the area. You may need to find a particular address, but you do not have a clue in what part of the city that address is in relation to your current location (Figure 8.24). Another obvious need is to discover the shortest route to the address in question. LCSs are directly applicable in these cases. First, the navigation service needs to determine your current location and, second, you need to indicate your target address (i.e., the place you need to find). This perhaps sounds simple, but there are several difficulties that need to be resolved for the service to work well. First of all, you and the service provider need to find a common way of describing locations, such as the house name, the street address or geographical coordinates. Since coordinates are specific and well-defined they are ideal for navigation services. However, people do not normally refer to locations in coordinates, so there is a need to translate coordinates into a street address or the house name. A house name or address can also be indicated as a certain square on a map using a map grid, an example is shown in Figure 8.24. Navigation services are normally used with graphical maps displayed on your phone. Route guidance can also be given as text messages or spoken, but even in such cases the map is useful as a back-up.

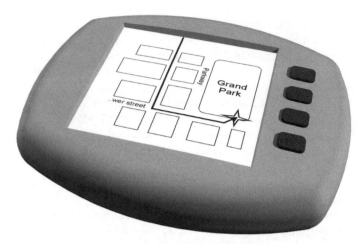

Figure 8.24   An example of a "show the way" LCS service

8.3.4.7.2 Find Something Close By

The above example shows how useful it would be to get local information when arriving in a city. Location-related information could be pushed to the user's terminal when the user enters a given location, or the user could request local information whenever needed. This type of service is close to marketing and could be offered free of charge or with minimum cost to the user. On the other hand, the user may well be prepared to pay for valuable information, so there will probably be a cost/benefit balance established for this type of service. The information offered should be kept up-to-date, however, because telephone directories and Yellow Pages continue to be strong competitors in this area, and they are often made available online.

8.3.4.7.3 Transport Management

Transport management is an obvious user of location information, because the information can be used to increase the efficiency of transports and, thereby, lower costs. When real-time location information is available to the transport vehicles in a fleet, it is easier to plan and maybe combine transports. Depending on such circumstances as change in traffic conditions, changing transport plans, etc., the transport control centre can optimise the operation schedule and inform the drivers. Location information is useful to track packages being transported and to check their progress. It is possible for tracking to take place in real time, but that can be costly and often unnecessary. During the course of transport the accuracy of location information can vary from a few kilometres in a long-haul operation to a few metres at the end points. However, provision of accurate location may be vital when least expected (e.g., when tracking stolen vehicles or in emergency situations).

Among the many transport-related services are taxi allocation, taxi navigation and surveillance. Using the location information of available taxis enables a central office to allocate a taxi that is closest to the client. This shortens the waiting time for the client, and improves the efficiency of the taxi fleet as well. The client could be given an estimate of arrival time of the taxi and the taxi driver can use the location system for navigation as described above.

8.3.4.7.4 Games
Network gaming is a hobby that fascinates many people, and those most addicted spend a lot of time playing games. Many network games need to make full use of all the graphical and calculation capabilities of a fully equipped PC for the experience to be satisfying, whilst some other games are played simply because the game idea is so compelling. Mobile devices obviously do not have the capabilities of PCs, but, on the other hand, *they are mobile*. If the game device is also a mobile phone there may be circumstances when the user will keep it active at all times in order to take part in certain games: as the mobile phone is moving around with its user, it is now possible to combine games in the phone with the location information of the phone.

There are a number of such games available and one example is treasure-hunting wherein the game player would receive a text message that a new treasure has been hidden at a given location. The first person arriving close enough to the treasure will collect the treasure. The value of the treasure could increase over time until found. Another game is "hide and seek": attempting to find other people using mobile phones and their location information.

8.3.4.7.5 Network optimisation
In addition to the ordinary location-based services perceived by the end-user, location assistance data can be applied to optimise system performance by feeding it into network planning, radio resource management mechanisms, mobility management mechanisms, network maintenance and performance management and so forth. This will allow network designers and operators to optimise their network based on the location information of mobile devices. For network planning, the basic information of subscribers will be valuable when dimensioning the network, developing the charging policy and automating the network management system (e.g., in conjunction with call drops, blocks and failure removing). Also, location information can be used as input to control the BS transmit power and beam-forming mechanism to provide adaptive coverage based on mobile device behaviour. These types of applications may not require highly accurate location information and can be readily reused once location information has been extracted for other services.

Handover control is a typical example of a network-optimisation-related feature that may use location information. Assume that the mobile is within a micro-cell where there is also a macro-cell overlying the micro-cell cluster (Figure 8.25). Suppose that a fast-moving mobile user is crossing the micro-cell. The fast-moving mobile has to reselect the micro-cell very frequently, putting an undesirable signalling load on the network. Based on location information, predictive actions can be taken and the fast-moving mobile can be handed over to the macro-cell, reducing the extra signalling load. In contrast, a slow-moving mobile can be handed over from the macro-cell to the micro-cell to decrease its power consumption and the interference it generates.

Location information will bring many new opportunities for operators to optimise their networks and network management systems. The value location-based measurements bring to network optimisation can be dramatically increased when applied in

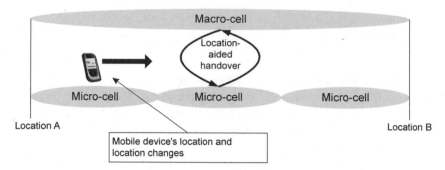

Figure 8.25 An example of the use of location-aided handover

association with network control functions, increasing the intelligence of the radio network. In this respect, user privacy will likely be the greatest difficulty that needs to be resolved before these applications become feasible.

8.3.5 IMS Service Mechanism—Messaging

Currently, networks can offer many forms of messaging. Since messaging, put simply, is just a means of sending a message from one entity to another, messages can not only have many forms they can also contain many types of media. Even delivery can be arranged in many ways. In addition, the IMS (see Section 6.5) contains dedicated messaging facilities.

There are three forms of IMS messaging:

- Immediate messaging.
- Session-based messaging.
- Deferred delivery messaging.

Each of these IMS messaging types has its own characteristics that actually make it act like a separate service (Figure 8.26). As far as the end-user is concerned, however, all these messaging types look very similar, since the differences between them are mainly technical and occur within the network. All these IMS messaging types directly utilise the Session Initiation Protocol (SIP) and the IMS architecture.

In immediate messaging the sender uses SIP's message method to send a message to the receiver (Figure 8.27). The message method is used in so-called "page mode", where the communication involves just sending a message and getting a response. The nature of this traffic is random. In other words, this is not a session consisting of a number of message pairs. The message sent could just consist of pure text but it may also contain other media types like sounds and images.

Session-based messaging is already in use in the traditional Internet where it is called "Internet Relay Chat" (IRC). In this form of messaging the traffic between related parties is expected and each party attends the session by means of an invitation procedure. In IRC, messages exchanged within the session are textual, but in the

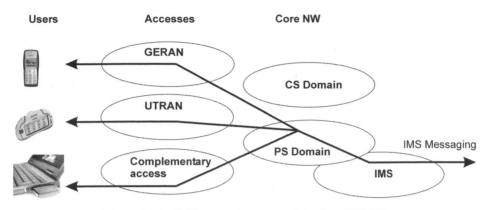

Figure 8.26   IMS messaging connectivity in 3GPP R5

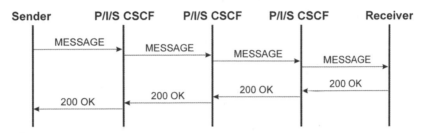

Figure 8.27   Example of an immediate messaging flow

IMS environment the aim is to enrich the content by enabling other media components to be a part of session-based messaging. This opens up very interesting possibilities for new services. For example, video calls providing parallel content through text channels would be an interesting service for hearing-impaired people.

Deferred delivery messaging is already implemented when things are looked at from the end-user point of view: this is provided by the Multimedia Messaging Service (MMS). Competent implementation of deferred delivery messaging in the IMS requires some specification changes and definitions to be done and these are handled in 3GPP R6.

8.3.6 IMS Service Mechanism—Presence

Another IMS-related service mechanism we would like to briefly introduce here is called "presence". Presence means basically two things: it makes one's personal status available to other (defined) people and shows, respectively, other people's statuses to a given requestor.

Presence information may consist of:

- Personal availability.
- Terminal availability.

- Communication preferences.
- Terminal capabilities.
- Current activity.
- Location.

A presence service will not necessarily show all of this information; every user has the possibility of modifying his or her presence information.

This kind of service is already available in the traditional Internet. Some service providers offer the possibility of downloading an application that partially provides the presence information for a given, selected, predefined list of people. In the Internet the presence service is used in the context of messaging (i.e., enabling the user to check whether the desired receiver is online or not).

As shown in Figure 8.28, in the mobile world the scope is wider. Instead of pure messaging the presence service could provide information about other service possibilities as well (e.g., about the ability to receive voice calls, availability for gaming, etc.). Thus, mobile presence applications will be extended versions of those used in the Internet. A very likely development of presence service usage may be a dynamic phonebook containing embedded presence information about the phonebook entries. If this information is available to the contact initiator he or she could immediately see whether it is feasible to establish communication or not.

Security is, of cource, an important area of concern with presence-related services. Under no circumstances should anyone's presence information be made available to everybody. However, there are mechanisms to permit this:

- Publish presence information. The person who owns the presence profile that holds the presence information is called a "presentity". Presentities define how they want to publish their presence information and who is able to see it and at which level.
- Watch presence information. The person who requests information of a person who has published his or her own presence information is called a "watcher". Watchers are only able to see what they are allowed to see; the target person/terminal of the presence query always defines these limits.

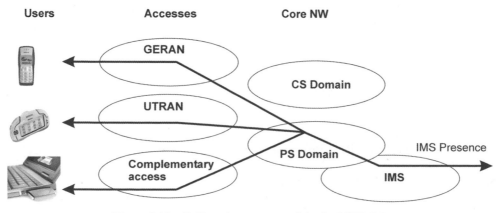

Figure 8.28   IMS presence connectivity in 3GPP R5

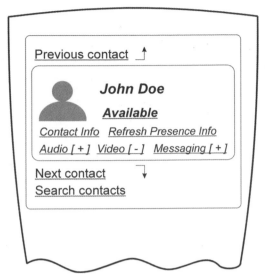

Figure 8.29 An entry in a dynamic phonebook may look like this

- Subscribe presence information is a SIP message flow through which the watcher requests presence information.

A very natural application based on the presence service is something called a "dynamic phone book" (Figure 8.29). With this application the terminal equipment is able to show the presence profile information of each contact the user has added to the terminal equipment's phone book. The added value over existing phone books is that the dynamic phone book informs the user whether the desired contact is available and, if he or she is available, the acceptable media components available.

8.4 Conclusions

This chapter provided a short overview of service-related matters in the UMTS environment.

As was shown, it is possible to create almost any service in the UMTS environment, but still some basic questions remain: What do users really want and where is the business going? Answers to these questions are vital to the telecommunication business. Moreover, uncertainty about these answers has affected the mobile communication business in an adverse way.

As far as system design principles and service differentiation are concerned, 3GPP R5 and ongoing solutions provide new, effective methods for service creation and provision. These changes are not necessarily visible to end-users in all cases, but new implementation methods will give more tools and alternatives to operators. The end result of all these new methods could well be an increase in the effectiveness of the business.

9

Security in the UMTS Environment

Valtteri Niemi and Heikki Kaaranen

Security has always been closely associated with mobile networks due to the very nature of wireless communications, which are easily accessible not only by the intended mobile phone users but also by any potential eavesdropper. However, security is a much wider issue that needs to be considered by all the players in the mobile communication business. In this chapter we will consider security both from the carrier/service providers' and from the end-users' point of view. At the end of this chapter lawful interception is also discussed, which is yet another aspect to mobile communications that has become part of modern society.

Analogue mobile networks (1G) did not have a lot of security-related features, and, in those times, subscribers did not necessarily understand the meaning of secure connections. In contrast, today's radio path (which as such is the most unsecured part of the system) can be secured very effectively but this arrangement has its pros and cons. If radio connection is of the point-to-point type (between two users) a dedicated security arrangement can be used. It is in this way that military radio connections are most often secured. In public cellular networks the radio path is essentially point-to-multipoint (i.e., many terminals may receive the same transmitted signal) and, thus, any dedicated security system on the radio path is unusable as such.

In the basic Global System for Mobile Communication (GSM), security is concentrated on radio path security (i.e., the access network part). In Universal Mobile Telecommunication System (UMTS) networks, security is a larger topic. Access network connection must naturally be secured, but, in addition to this, security must be taken into account in many other respects as well. The various commercial/business models lead to situations where sometimes very sensitive information is transferred between different parties and networks. Obviously, there exist severe security risks in this situation. In addition, local and international authorities have set out their directives in this respect. The UMTS integrates the telecommunication and data communication worlds and, once again, this creates threats to security. In the Internet Protocol (IP) world, security has been an issue for many years and numerous security threats have been identified and defence mechanisms developed.

UMTS Networks Second Edition   H. Kaaranen, A. Ahtiainen, L. Laitinen, S. Naghian and V. Niemi
© 2005 John Wiley & Sons, Ltd   ISBN: 0-470-01103-3

The chapter is organised as follows: in Section 9.1 we study the access security of the UMTS. This includes both secure user access to UMTS networks and security of the connection at the access network level. The security of UMTS access is based on the GSM access security model, but several enhancements have been made.

In Section 9.2 we list a few of the security features of release 1999, some of which are inherited from GSM.

Section 9.3 deals with security at the network element layer. Here the main issue is how to secure connections inside a UMTS network and between networks that are controlled by different mobile operators. As the UMTS Core Network (CN) (in its 3GPP R99 version) is based on the GSM CN, there are no major security enhancements on this front when directly compared with the 2G case. The situation has changed a lot in subsequent releases of UMTS as CN architecture evolves. Section 9.3 overviews the mechanisms that have been introduced.

Section 9.4 provides an overview of the security issues and mechanisms at the upper layers of service and content providers. These are largely independent of the structure of the UMTS network itself but, nevertheless, they play a major role in the overall security of the system. This section also discusses the access security of the IP Multimedia Subsystem (IMS) inside the UMTS network. Finally, in Section 9.5 we show how the regulatory requirement for lawful interception capability is satisfied in the UMTS.

A more thorough description of UMTS security can be found in Niemi and Nyberg (2003).

9.1 Access Security in UMTS

Radio access technology will change from Time Division Multiple Access (TDMA) to Wideband Code Division Multiple Access (WCDMA) when 3G mobile networks are introduced. Despite this shift, requirements for access *security* will not change. In the UMTS it will still be a requirement that end-users of the system are *authenticated* (i.e., the identity of each subscriber is verified)—nobody wants to pay for calls that are made by an impostor.

The *confidentiality* of voice calls is protected in the Radio Access Network (RAN), as is the confidentiality of transmitted user data. This means that the user has control over the choice of parties with whom he or she wants to communicate. Users also want to *know* that confidentiality protection is really applied: *visibility* of applied security mechanisms is needed. *Privacy* of the user's whereabouts is generally appreciated. Most of the time an average citizen does not care whether anybody can trace where he or she is. However, if persistent tracking of users were to occur, they would end up pretty uptight. On the other hand, exact information about the location of people would be very useful to burglars! The privacy of user data is a critical issue when data are transferred through the network. Privacy and confidentiality are largely synonymous in this presentation.

Of course, the *availability* of UMTS access is important for a subscriber who is paying for it. Network operators consider the *reliability* of network functionality to be important: they need control inside the network to function effectively. This is guaranteed by *integrity checking* of all radio network signalling, in which all control messages are checked to ensure they have been created by authorised elements of the network. In general, integrity checking protects against any manipulation of a message (e.g., insertion, deletion or substitution).

The most important ingredient in providing security for network operators and subscribers is *cryptography*. This consists of various techniques with their roots in the science and art of *secret writing*. It is sometimes useful to make a communication deliberately incomprehensible (i.e., by using ciphering or, synonymously, encryption). This is the most effective way to protect communications against malicious intent.

9.1.1 Legacy from 2G

The transition from analogue 1G systems to digital 2G systems made it possible, among many other things, to utilise advanced cryptographic methods. The most important security features in the GSM system are:

- Authentication of the user.
- Encryption of communication in the radio interface.
- Use of temporary identities.

As GSM and other 2G systems became increasingly successful, the usefulness of these basic security features also became more and more evident. Naturally, it has been a leading principle in the specification work of UMTS security to carry these features over to the new system.

The success of GSM also emphasises its security limitations. A popular technology is tempting for fraudsters. The properties of the GSM that have been most criticised on the security front are the following:

- Active attacks towards the network are possible in principle: this refers to somebody who has the required equipment to masquerade as a legitimate network element and/ or legitimate user terminal (see Figure 9.1 for a scenario).
- Sensitive control data (e.g., *keys* used for radio interface ciphering) are sent between different networks without ciphering.
- Some parts of the security architecture are kept secret (e.g., cryptographic algorithms). This does not create trust in them in the long run because they are not publicly available (where they could be analysed by new methods) and, on the other hand, global secrets tend to be revealed sooner or later.

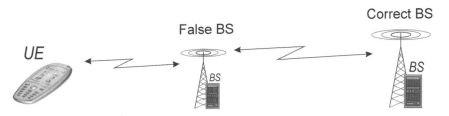

Figure 9.1   Active attack

- Keys used for radio interface ciphering eventually become vulnerable to massive *brute force* attacks where somebody tries all the possible keys until one matches.

These limitations were left in the GSM system on purpose. The threat imposed by them was considered of little consequence when compared with the added cost of trying to circumvent them. However, as technology has advanced, attackers have gained access to better tools. This is why the outcome of a similar comparison between cost and security led to a different conclusion in the 3G case.

In the UMTS, countermeasures for perceived weaknesses in the GSM have been developed. This is another leading principle that has guided the design of 3G security architecture. The most important security features for access security of the UMTS are:

- Mutual authentication of the user and the network.
- Use of temporary identities.
- RAN encryption.
- Protection of signalling integrity inside the UMTS Terrestrial Access Network (UTRAN).

Note that publicly available cryptographic algorithms are used for encryption and integrity protection. Algorithms for mutual authentication are operator-specific. Each of these features is described in Sections 9.1.2, 9.1.4, 9.1.5 and 9.1.6. Details can be found in 3G Partnership Project (3GPP) specification TS 33.102.

9.1.2 Mutual Authentication

There are three entities involved in the authentication mechanism of the UMTS system:

- Home network.
- Serving Network (SN).
- Terminal, more specifically Universal Subscriber Identity Module (USIM), typically in a smart card.

The basic idea is that the SN checks the subscriber's identity (as in the GSM) by a so-called "challenge-and-response" technique while the terminal checks the SN has been authorised by the home network to do so. The latter part is a new feature in the UMTS (compared with the GSM) and through it the terminal can check that it is connected to a legitimate network.

The mutual authentication protocol itself does not prevent the scenario in Figure 9.1, but (in combination with other security mechanisms) it does guarantee that the active attacker cannot get any real benefit out of the situation. The only possible gain for the attacker is to be able to disturb the connection. Nevertheless, clearly no protocol methods exist that can circumvent this type of attack completely (e.g., an attacker can implement a malicious action of this kind by radio-jamming).

The cornerstone of the authentication mechanism is a *master key K* that is shared between the USIM of the user and the home network database. This is a permanent

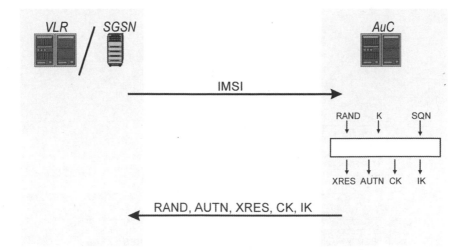

Figure 9.2 Authentication data request and authentication data response

secret with a length of 128 bits. The key K is never made visible between the two locations (e.g., the user has no knowledge of his or her master key).

At the same time as mutual authentication, keys for encryption and integrity checking are derived. These are temporary keys with the same length (128 bits). New keys are derived from the permanent key K during every authentication event. It is a basic principle in cryptography to limit the use of a permanent key to a minimum, and instead derive temporary keys from it for protection of bulk data.

Let us now describe the Authentication and Key Agreement (AKA) mechanism at a general level. The authentication procedure can be started after the user is identified in the SN. Identification occurs when the identity of the user (i.e., permanent identity IMSI or temporary identity TMSI) has been transmitted to the Visitor Location Register (VLR) or Serving GPRS Support Node (SGSN). Then, the VLR or SGSN sends an *authentication data request* to the Authentication Centre (AuC) in the home network.

The AuC contains the master keys of users, and, based on knowledge of the International Mobile Subscriber Identity (IMSI), the AuC is able to generate *authentication vectors* for the user. The generation process involves the execution of several cryptographic algorithms, which are described in more detail in Section 9.1.3. The vectors generated are sent back to the VLR/SGSN in the *authentication data response*. This process is depicted in Figure 9.2. These control messages are carried on the Mobile Application Part (MAP) protocol.

In the SN, one authentication vector is needed for each authentication instance (i.e., for each run of the authentication procedure). This means that (potentially long distance) signalling between the SN and AuC is not needed for every authentication event and that in principle it can be done independently of user actions after the initial registration. Indeed, the VLR/SGSN may fetch new authentication vectors from the AuC well before the number of stored vectors runs out.

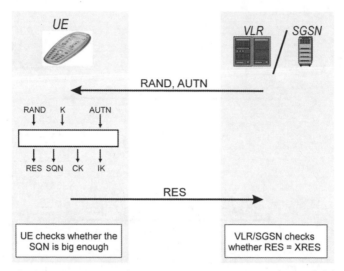

Figure 9.3 User authentication request and user authentication response

The SN (VLR or SGSN) sends a *user authentication* request to the terminal. This message contains two parameters from the authentication vector, called "RAND" and "AUTN". These parameters are transferred to any USIM that exists inside a tamper-resistant environment (i.e., in a UMTS Integrated Circuit Card or UICC). The USIM contains the master key K, and using it with the parameters RAND and AUTN as inputs the USIM carries out a computation that resembles the generation of authentication vectors in the AuC. This process also involves the execution of several algorithms, as is the case for corresponding AuC computation. As a result of the computation, the USIM is able to verify whether the parameter AUTN was indeed generated in the AuC and, if it was, the computed parameter RES is sent back to the VLR/SGSN in the *user authentication response*. Now the VLR/SGSN is able to compare the user response RES with the expected response XRES which is part of the authentication vector. If they match, authentication ends positively. This part of the process is depicted in Figure 9.3.

The keys for RAN encryption and integrity protection (namely, CK and IK) are created as a by-product of the authentication process. These temporary keys are included in the authentication vector and, thus, are transferred to the VLR/SGSN, where they are later further transferred to the Radio Network Controller (RNC) in the RAN when encryption and integrity protection are started. At the other end, the USIM is also able to compute the Cipher Key (CK) and Integrity Key (IK) after it has obtained the RAND (and verified it through the AUTN). Temporary keys are subsequently transferred from the USIM to the mobile equipment where the encryption and integrity protection algorithms are implemented.

9.1.3 Cryptography for Authentication

Next we take a closer look at the generation of authentication vectors in the AuC. A more detailed illustration of the process is given in Figure 9.4. The process begins by

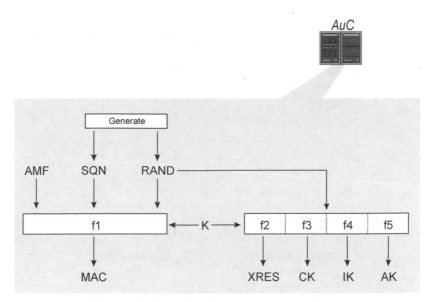

Figure 9.4 Authentication vector generation

picking up a correct Sequence Number (SQN). Roughly speaking, what is required is that the sequence numbers are picked up in increasing order. The purpose of the sequence number is to prove later to the user that the generated authentication vector is *fresh* (i.e., that it has not been used before). In parallel with the choice of sequence number, a random bit string RAND (128 bits long) is generated. This is a demanding task in itself, but in this presentation we just assume that a cryptographic pseudorandom generator is in use that produces large amounts of unpredictable output bits when a good physical random source is available as a seed for it.

The key concept in authentication vector computation is a *one-way function*. This is a mathematical function that is relatively easy to compute but practically impossible to invert. In other words, given the input parameters there exists a fast algorithm to compute the output parameters, but, on the other hand, if the output is known there exist no efficient algorithms to deduce any input that would produce the output. Of course, one simple algorithm to find the correct input is to try all possible choices until one gives the wanted output. Clearly, this *exhaustive search* algorithm becomes extremely inefficient as the length of input increases.

In total, five one-way functions are used to compute the authentication vector. These functions are denoted f1, f2, f3, f4 and f5. The first differs from the other four in the number of input parameters, taking four input parameters: master key K, random number RAND, sequence number (SQN) and, finally, an administrative *Authentication Management Field* (AMF). The other functions (f2 to f5) only take K and RAND as inputs. The requirement of a one-way property is common to all functions (f1–f5) and all of them can be built around the same *core* function. However, it is essential that they differ from each other in the following fundamental way: that no information about the outputs of the other functions can be deduced from the output of any one function. The

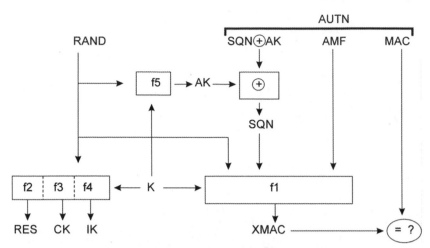

Figure 9.5   Authentication handling in USIM

output of f1 is Message Authentication Code (MAC, 64 bits), while those outputs of f2, f3, f4 and f5 are, respectively, XRES (32–128 bits), CK (128 bits), IK (128 bits) and AK (64 bits). The authentication vector consists of the parameters RAND, XRES, CK, IK and AUTN. The last one is obtained by concatenating three different parameters: SQN added bit-by-bit to AK, AMF and MAC.

Let us now take a closer look at the handling of authentication at the USIM end. This is illustrated in Figure 9.5. The same functions (f1–f5) are involved at this end as well, but in a slightly different order. Function f5 has to be computed before function f1 since f5 is used to conceal SQN. This concealment is needed to prevent eavesdroppers from getting information about the user identity through SQN. The output of function f1 is marked XMAC on the user side. This is compared with the MAC received from the network as part of the parameter AUTN. If there is a match it implies RAND and AUTN have been created by some entity that knows K (i.e., by the AuC of the user's home network).

Nevertheless, there is a possibility that some attacker who has recorded an earlier authentication event replays RAND and AUTN. As mentioned above, SQN protects against this threat. The USIM should simply check that it has not seen the same SQN before. The easiest way to check this is to require that SQNs appear in increasing order. It is also possible that the USIM allows some SQNs to arrive out-of-order should it maintain a shortlist of the greatest SQNs received so far. As a result of the transfer of authentication vectors from the AuC and the actual use of these vectors for authentication is done somewhat independently, there are several reasons for the possibility of authentication vectors being used in a different order from that in which they were originally generated. The most obvious reason for such a case is a consequence of the fact that Mobility Management (MM) functions for Circuit Switched (CS) and Packet Switched (PS) domains are independent of each other. This implies that authentication vectors are brought to the VLR and SGSN independently of each other and are also used independently.

The choice between algorithms f1–f5 is in principle operator-specific. This is because

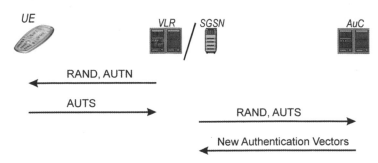

Figure 9.6   Re-synchronisation procedure

they are only used in the AuC and the USIM and the same home operator controls them both. An example set of algorithms (called "MILENAGE") exists in the 3GPP specification TS 35.206. SQN management is also operator-specific in principle. There are two basic strategies for creating SQNs: each user may have an individual SQN or SQN generation may be based on a global counter (e.g., universal time). A combination of these two strategies is also possible (e.g., the most significant part of the SQN is user-specific but the least significant part is based on a global counter).

The mutual authentication mechanism is based on two parameters stored both in the AuC and the USIM: a static master key K and a dynamic sequence number SQN. It is vital that these parameters are maintained in a synchronised manner at both ends. For a static K this is easy, but it is possible that the dynamic information about SQNs runs out of synchronisation for whatever reason. As a consequence, authentication would fail. A specific *re-synchronisation procedure* is used in this case (see Figure 9.6). By using the master key K as the basis for secure communication, the USIM informs the AuC of its current SQN.

A parameter AUTS is delivered during re-synchronisation. It contains two parts: the SQN of the USIM concealed by AK and a message authentication code MAC-S computed by another one-way function f1* from the input parameters SQN, K, RAND and AMF. The last two parameters are obtained from the failed authentication event. The one-way function f1* has to be different from f1 because otherwise recorded AUTN parameters could in principle be accepted as valid AUTS parameters during re-synchronisation and an attacker could at least disturb the authentication mechanism.

9.1.4 Temporary Identities

The permanent identity of the user in the UMTS is the IMSI, as it is for the GSM. However, the identification of the user in the UTRAN is in almost all cases done by temporary identities: Temporary Mobile Subscriber Identity (TMSI) in the CS domain or Packet TMSI (P-TMSI) in the PS domain. This implies that confidentiality of user identity is almost always protected against passive eavesdroppers. Initial registration is an exceptional case where a temporary identity cannot be used since the network does not yet know the permanent identity of the user. Subsequently, it is in principle possible to use temporary identities.

The mechanism works as follows. Assume the user has already been identified in the SN by the IMSI. Then, the SN (VLR or SGSN) allocates a temporary identity (TMSI or P-TMSI) for the user and maintains the association between the permanent identity and the temporary identity. The latter is only significant locally and each VLR or SGSN simply takes care that it does not allocate the same TMSI/P-TMSI to two different users simultaneously. The allocated temporary identity is transferred to the user once encryption is turned on. This identity is then used in both uplink and downlink signalling until a new TMSI or P-TMSI is allocated by the network.

The allocation of a new temporary identity is acknowledged by the terminal and, then, the old temporary identity is removed from the VLR (or SGSN). If allocation acknowledgement is not received by the VLR/SGSN it should keep both the old and new (P-)TMSIs and accept either of them in uplink signalling. In downlink signalling the IMSI must be used because the network does not know which temporary identity is currently stored in the terminal. In this case, the VLR/SGSN tells the terminal to delete any stored TMSI/P-TMSI and a new re-allocation follows.

Nevertheless, one problem remains: How does the SN obtain the IMSI in the first place? Since the temporary identity only has a meaning locally, the identity of the local area has to be appended to it to obtain an unambiguous identity for the user. So, a Location Area Identity (LAI) is appended to the TMSI and a Routing Area Identity (RAI) is appended to the P-TMSI.

If User Equipment (UE) arrives in a new area, then the association between the IMSI and (P-)TMSI can be retrieved from the old location or routing area if its address is known to the new area (based on the LAI or RAI). If the address is not known or a connection to the old area cannot be established, then the IMSI must be requested from the UE.

There are some specific places where lots of IMSIs may be transmitted over the radio interface (e.g., airports as people switch on their mobile phones after the flight). Despite this problem, the tracking of people is usually easier in such places (e.g., people can be identified as they exit the aircraft). All in all, the user identity confidentiality mechanism in the UMTS does not give 100% protection, but it does offer a relatively good protection level. Note that protection against an active attacker is not very good since the attacker may pretend to be a new SN to which the user has to reveal his or her permanent identity. The mutual authentication mechanism does not help here since the user has to be identified before he or she can be authenticated.

9.1.5 UTRAN Encryption

Once the user and the network have authenticated each other, they can begin secure communication. As described earlier, a CK is shared between the CN and the terminal after a successful authentication event. Before encryption can begin, the communicating parties must also agree on the encryption algorithm. Fortunately, when the UMTS is implemented according to 3GPP R99, only one algorithm is defined. Encryption/decryption takes place in the terminal and in the RNC at the network end. This means that the CK has to be transferred from the CN to the RAN. This is done in a specific RAN Application Part (RANAP) message called a *security-mode command*. After the

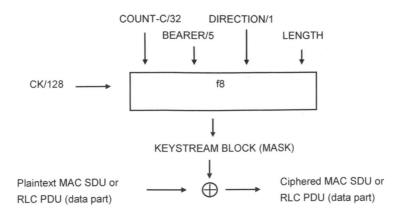

Figure 9.7 Stream cipher in UMTS

RNC has obtained CK it can switch encryption on by sending a Radio Resource Control (RRC) security-mode command to the terminal.

The UMTS encryption mechanism is based on a *stream cipher* concept as described in Figure 9.7. This means that plaintext data are added bit-by-bit to random-looking *mask* data, which are generated based on the CK and a few other parameters. This type of encryption has the advantage that the mask data can be generated even before the actual plaintext is known. In this case, final encryption is a very fast bit operation. Decryption at the receiving end is done in exactly the same way since adding the mask bits twice has the same result as adding zeros.

As mask data do not depend in any way on the plaintext, there has to be another input parameter that changes every time a new mask is generated. Otherwise, two different plaintexts, say P_1 and P_2, would be protected by the same mask. Then, the following unwanted phenomenon would happen: if we add P_1 to P_2 bit-by-bit, and we do the same to their encrypted counterparts, then the resulting bit string would be exactly the same in both cases. This is, again, due to the fact that two identical masks cancel each other during bit-by-bit addition. Therefore, the bit-by-bit sum of P_1 and P_2 would become known to any attacker who eavesdrops on the corresponding encrypted messages on the radio interface. Typically, if two bit strings of meaningful data are added to each other bit-by-bit, both of them could be given away completely by the resulting bit string. Hence, this would mean a breakdown in encryption for the two messages P_1 and P_2.

Encryption occurs in either the *Medium Access Control* (MAC) layer or in the *Radio Link Control* layer (RLC). In both cases, there is a counter that changes for each Protocol Data Unit (PDU). In MAC this is a *Connection Frame Number* (CFN) and in RLC it is a specific *RLC Sequence Number* (RLC-SQN). If these counters are used as input for mask generation the problem explained in the previous paragraph would still occur since these counters wrap around quickly. This is why a longer counter called a *Hyperframe Number* (HFN) is introduced. This is incremented whenever the short counter (CFN in MAC and RLC-SQN in RLC) wraps around. The combination of HFN and the shorter counter is called "COUNT-C". It is used as ever-changing input for mask generation in the encryption mechanism.

In principle, the longer counter HFN could also eventually wrap around. Fortunately, it is reset to zero whenever a new key is generated during the AKA procedure. Authentication events are frequent enough to rule out the possibility of HFN wrap-around.

The radio bearer identity BEARER is also needed as input to the encryption algorithm since the counters for different radio bearers are maintained independently of each other. If the input BEARER is not in use, then this would again lead to a situation where the same set of input parameters would be fed into the algorithm and the same mask would be produced more than once. Consequently, the problem outlined above would occur and the messages (this time in different radio bearers) encrypted with the same mask would be exposed to the attacker.

The core of the encryption mechanism is the mask generation algorithm that is denoted as function f8. The specification is publicly available as 3GPP TS 35.201 where it is based on a novel *block cipher* called KASUMI (for which there is another 3GPP specification TS 35.202). This block cipher transforms 64-bit input to 64-bit output. Transformation is controlled by the 128-bit CK. If the CK is not known there are no efficient algorithms to compute the output from the input or vice versa. In principle, transformation can only be done if either:

- All possible keys are tried until the correct one is found.
- An enormous table of all 2^{64} input–output pairs is somehow collected.

In practice, both approaches are impossible.

It is possible for authentication not to be carried out at the beginning of a connection. In this case the previous CK is used for encryption. The key is stored in the USIM in-between connections. In addition, the most significant part of the greatest HFN used so far is stored in the USIM. For the next connection, the stored value is incremented by one and used as the starting value for the most significant part of the HFN.

A ciphering indicator is used in terminals to show the user whether encryption is applied or not, thus providing some visibility to security mechanisms. Note that although the use of ciphering is highly recommended it is still optional for the UMTS network.

9.1.6 Integrity Protection of Radio Resource Control (RRC) Signalling

The purpose of integrity protection is to authenticate individual control messages. This is important since a separate authentication procedure only assures the identities of the communicating parties at the time of authentication. Figure 9.1 can be used to illustrate the issue: a *man-in-the-middle* (i.e., a false BS) acts as a simple relay and delivers all messages in their correct form until the authentication procedure is completed. Afterwards, the man-in-the-middle may begin to manipulate messages freely. However, if messages are protected individually, deliberate manipulation of messages can be observed and false messages can be discarded.

Integrity protection is implemented at the RRC layer. Thus, it is used between the terminal and RNC, just like encryption. The IK is generated during the AKA

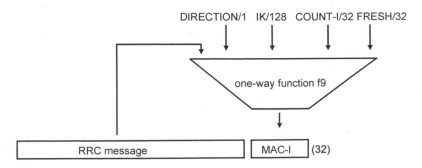

Figure 9.8 Message Authentication Code (MAC)

procedure, again similar to the CK. Additionally, the IK is transferred to the RNC together with CK in a security-mode command.

The integrity protection mechanism is based on the concept of a MAC. This is a one-way function that is controlled by the secret key IK. The function is denoted by f9 and its output is MAC-I: a 32-bit random-looking bit string. The MAC-I is appended to each RRC message and is also generated and checked at the receiving end. Any change in input parameters are known to influence the MAC-I in unpredictable ways. Function f9 is depicted in Figure 9.8. Its inputs are IK, the RRC message itself, a counter COUNT-I, direction bit (uplink/downlink) and a random number FRESH. The parameter COUNT-I resembles the corresponding counter for encryption. Its most significant part is an HFN and the four least significant bits consist of the RRC-SQN. Altogether, COUNT-I protects against replay of earlier control messages: it guarantees that the set of values for input parameters is different for each run of integrity protection function f9.

The parameter FRESH is chosen by the RNC and transmitted to the UE. It is needed to protect the network against a maliciously chosen start value for COUNT-I. Recall that the most significant part of an HFN is stored in the USIM in-between connections. An attacker could masquerade as the USIM and send a false value to the network, forcing the starting value of an HFN to be too small. If the authentication procedure is not run, then the old IK is used. If the parameter FRESH was not present this would create a chance for the attacker to replay RRC signalling messages from earlier connections with recorded MAC-I values. By choosing FRESH randomly, the RNC is protected against these kinds of replay attacks, which are based on recording of earlier connections. As already explained, the ever-increasing counter COUNT-I protects against replay attacks based on recording during the connection when FRESH is active. Note that the radio bearer identity is not used as an input parameter for the integrity algorithm, although it is an input parameter for the encryption algorithm. Because there are also several parallel radio bearers for the control plane, this seems to leave room for a possible replay of control messages that were recorded within the same RRC connection but on a different radio bearer. However, this is not the case, since the radio bearer identity is always appended to the message when the MAC is calculated (although it is not transmitted with the message). Therefore, the radio bearer identity has an effect on the MAC-I value and we also have protection against replay attacks based on recordings on different radio bearers.

It is clear that there are a few RRC control messages whose integrity cannot be protected by the mechanism. Indeed, messages sent before the IK is in place cannot be protected. A typical example is the *RRC connection request* message.

The algorithm for integrity protection is based on the same core function as encryption. Indeed, the KASUMI block cipher is used in a special mode to create a MAC function.

The integrity protection mechanism used in the UTRAN is not applied to the user plane for performance reasons. On the other hand, there is a specific (integrity-protected) control plane procedure that is used for *periodic local authentication*. As a result of this procedure, the amount of data sent during the RRC connection is checked. Hence, the *volume* of transmitted user data is integrity-protected.

9.1.7 Summary of Access Security

In this section we present a schematic overview of the most important access security mechanisms and their relationships with each other. For the sake of clarity, many parameters are not shown in Figure 9.9. For instance, HFN and FRESH are important parameters that are transmitted between different elements and, yet, they are omitted from the figure.

9.2 Additional Security Features in 3GPP R99

There exist additional security features in the 3GPP R99 specification set. Some of these are directly inherited from the GSM as such, while others have been added for the first time in 3GPP R99. In the following subsections we take a brief look at individual features.

9.2.1 Ciphering indicator

There is a specific *ciphering indicator* in the ME that is used to show the user whether encryption is applied or not, thus providing some visibility of the security mechanisms to the user. Note that although the use of ciphering is highly recommended it is still optional for the UMTS network. Details of the indicator are left as implementation-specific and the best way to inform the user is highly dependent on the characteristics of the terminal itself (e.g., different display types may utilise different types of indicators).

In general, it is important that the security level is not dependent on whether the user is directly making active checks. Nevertheless, for some specific actions, users may appreciate visibility of those security features that are active.

9.2.2 Identification of the UE

In the GSM, Mobile Equipment (ME) can be identified by its *International Mobile Equipment Identity* (IMEI). This identity is not directly associated with the user because a SIM card can be moved from one terminal to another. There are,

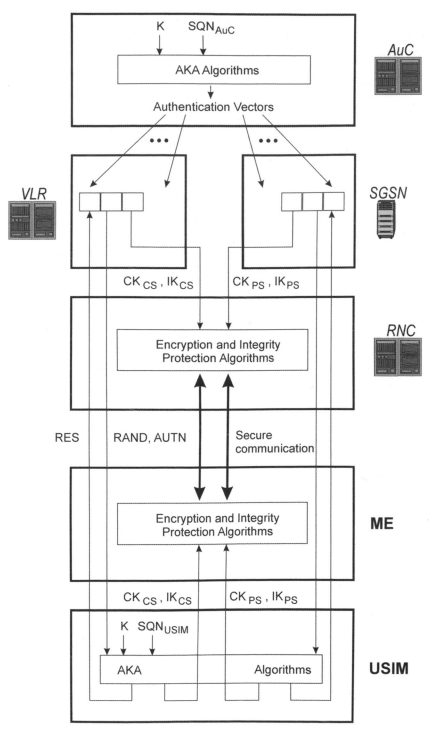

Figure 9.9 UMTS access security summary

however, important features in the network that can only be based on the value of the IMEI (e.g., it is possible to make *emergency calls* with a terminal without a SIM card). The only identification method in this case is to require the terminal to provide its IMEI.

This feature has been carried over to the UMTS system as well. There are no mechanisms in either the GSM or UMTS to actually *authenticate* the provided IMEI. So, protection methods for IMEI have to be based solely at the terminal end: it must be made difficult to modify the terminal in such a way that it provides an incorrect IMEI when requested by the network.

9.2.3 Security for Location Services (LCSs)

User location information is clearly sensitive. People are not comfortable with the idea of their whereabouts being tracked at any time. Security mechanisms have been defined to protect against leakage of location information to unauthorised parties. The *privacy profile* concept plays a central role here: users must have control over who knows their whereabouts.

9.2.4 User-to-USIM Authentication

This is another feature that has been carried over from the GSM to UMTS. It is based on a PIN known only to the user and the USIM. The user has to give the correct (4–8-digit-long) PIN to the USIM before further access to the USIM is granted.

9.2.5 Security in Universal Subscriber Identity Module (USIM) Application Toolkit

Similarly to what happens in the GSM, it is possible to build applications that are executed in the USIM. This feature is called a *(U)SIM Application Toolkit*. Part of the concept concerns the possibility for the home operator to send messages directly to the USIM. A further part specifies the kind of protection that can be provided for message transfer. Many details about protection mechanisms are implementation-specific.

9.3 Security Aspects at the System and Network Level

In this section we briefly discuss potential security threats at the network level and how to protect against them. The aim is to provide confidentiality and integrity protection for communication between different network elements. These elements can belong either to the same network or to two different networks. Especially in the latter case, fully standardised security solutions are needed to ensure interoperability.

In the 3G business chain there are four parties involved: "subscriber", "carrier provider", "service provider" and "content provider" (Figure 9.10). When considering a service the subscriber uses and is charged for, we can see that every party in this chain

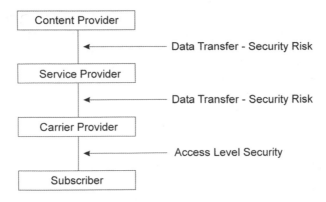

Figure 9.10   Security risks in the business chain

is involved: the carrier provider provides the platform through which the connection is established; the service provider arranges USIM (identification information) and actual service; and the content provider takes care of the content a service generally requires. The carrier, service and content provider could physically be the same company, but this need not necessarily be the case. In practice, all kinds of combinations concerning these three parties exist in the GSM world. Clearly, the subscriber is charged for the service and its use, while the other parties in the business chain share the income. The system controlling the monetary flow between the carrier, service and control provider is a potential security risk, since very sensitive information is transferred between these parties.

9.3.1 Typical Security Attacks

Many threats to communication between UMTS network elements are similar to the those at the application layer. Clearly, there are already major differences between applications, but the attack examples presented in this subsection have to be taken into account in all cases.

There are plenty of ways to carry out security attacks, confined only by imagination and effective protection. The following list presents some examples. But, note that the list is definitely not exhaustive.

- Social engineering.
- Electronic eavesdropping (sniffing).
- Spoofing.
- Session hijacking.
- Denial of Service (DoS).

Social engineering is not usually identified as a security risk by the normal user/subscriber, though it plays an essential role in many attacks. As far as subscribers are concerned, social engineering might mean ways of gaining access to their terminals (i.e., by ascertaining the PIN number in some way). Subscribers can protect themselves against social engineering by keeping a PIN enquiry in the terminal active and memorising their PIN numbers. Social engineering is also surprisingly common at the network end. It is not infrequent for people working with network elements in the operator's premises to receive weird calls where the caller explains he needs a user ID and password to some equipment there and the person responsible for that equipment is on holiday or otherwise unreachable. Often, these calls are nothing more than social engineering where secure information could end up in the wrong hands. Actually, hackers typically try this procedure as a first step. As a result, social engineering can be used to give hackers access to both vital or not-so-vital network elements. In the IP world social engineering is relatively common, but in the telecommunications world social engineering is not so widely used: the equipment is not subject to public access and the personnel maintaining them are very aware of their responsibilities.

Electronic eavesdropping, also known as sniffing, is another commonly used attack method as already explained in Sections 9.1.1 and 9.1.5. Sniffing is very difficult to detect and physically prevent. With sniffing the hacker aims to collect user ID and password information. Unfortunately, sniffing programs are publicly available on the Internet for anyone to download. A sniffing program itself is only a tool, and in the right hands it is used for network monitoring and possible fault detection. In the wrong hands it is a powerful tool for a hacker to silently monitor large numbers of Internet connections.

The information gathered by sniffing can be utilised in the next step: a hacking method called "spoofing". Spoofing allows a hacker to use someone else's IP address and *receive* packets from other users. In other words, the hacker takes the place of the correct receiver in the connection. Armed with this information, surely a hacker could open a connection using someone else's IP address. Fortunately, this is more difficult. However, receipt of "one-way traffic" is often all the hacker wants to achieve. Nowadays, with people spending a lot of time working remotely at home, this could be one way to gain access to company information (i.e., when employers and employees exchange data over the Internet).

The next step our hacker could take is called "session hijacking", where he attempts to take over an existing connection. As mentioned earlier, even a strong authentication mechanism at the beginning of a connection does not protect against it being hijacked later. We need integrity protection for the whole session.

In a DoS attack the hacker does not aim to collect information, rather he is aiming to cause harm and inconvenience to other users and service provider(s). In a typical DoS attack the hacker generates "disturbing" traffic, which, in the worst case, jams the target server in such a way that it is no longer able to provide a service. The idea behind this is to block up the server's service request queue with requests and then ignore all acknowledgements the server sends back. Consequently, the server occupies resources for an incoming connection that never occurs. When the timers for the connection expire, the resources are freed to serve another connection attempt. When the buffer containing connection attempts is continuously filled with new requests, the server becomes jammed with these requests and, hence, is unable to provide a "real" service.

Unfortunately, there are more sophisticated DoS attacks and, amazingly, plenty of tools available to hackers for DoS attacks on the Internet. All in all, protection against DoS is very difficult.

An advanced DoS attack may be combined with the other methods described above. For example, DoS can be initiated from "stolen" IP addresses and, when used in a distributed way, there can even be hundreds of computers taking part in the DoS attack. DoS is a very dangerous and powerful attack and can easily result in serious financial losses.

The preceding text briefly described the most common security attacks one may experience. These threats cannot be ignored since the tools to implement them are provided free-of-charge on the Internet. So, what can be done to eliminate such threats? Well, security should be thought of as a chain, where the entire system (i.e., communication security, data security and signalling security) is only as strong as the weakest link. In other words, everything must be secure; this covers algorithms, protocols, links, end-to-end paths, applications, etc.

9.3.2 Overview of 3GPP Network Domain Security

As mentioned earlier, one of the weaknesses in the GSM security architecture stems from the fact that authentication data are transmitted unprotected between different networks. For instance, CKs are used to protect the traffic on the radio interface. However, these keys are themselves transmitted "in clear" between networks. The reason for this state of affairs lies in the closed nature of Signalling System #7 (SS7) networks: only a relatively small number of large institutions have access to them. In UMTS Release 99 the CN structure is still very much like that in the GSM. This is the reason no major enhancements were made in the security of traffic between CNs.

In subsequent releases of the UMTS the situation has changed: the CN structure has evolved and IP has become a more important protocol on the network layer. Although this does not mean that signalling between different CNs would be carried over truly open connections, there is certainly a shift towards easier access to CN traffic. There are many more players involved and there exists a community of hackers who are skilful in IP matters.

The basic tool used for the protection of network domain traffic is the IP Security (IPSec) protocol suite. It provides confidentiality and integrity of communication in the IP layer. Communicating parties can also authenticate each other using IPsec. The critical issue is key management: how to generate, exchange and distribute the various keys needed in algorithms that are used to provide confidentiality and integrity protection. We present a brief introduction to IPSec in Section 9.3.3.

In addition to the protection of IP-based networks, the security of purely SS7-based networks is enhanced in UMTS Release 4. In particular, a specific security mechanism has been developed for the MAP protocol. This is called MAPSec and provides confidentiality and integrity protection.

9.3.3 IP Security (IPSec)

IPSec is standardised by the Internet Engineering Task Force (IETF). It consists of a dozen Requests for Comment (RFCs 2401–2412) and is a mandatory part of IPv6. In

IPv4 IPSec can be used as an optional "add-on" mechanism to provide security in the IP layer. The main IPSec components are the following:

- Authentication Header (AH).
- Encapsulation Security Payload (ESP).
- Internet Key Exchange (IKE).

The purpose of IPSec is to protect IP packets: this is done by ESP and/or AH. In short, ESP provides both confidentiality and integrity protection while AH only provides the latter. There are more fine-grained differences between the two, but ESP and AH are largely overlapping mechanisms. One of the reasons for including this kind of redundancy in IPSec standards is export control: there have been severe restrictions on the export of confidentiality protection mechanisms in most countries while integrity protection mechanisms have typically been free from restrictions. Currently, such export restrictions have been eased and, as a consequence, the importance of AH compared with ESP is decreasing.

Both ESP and AH need keys. But, more generally, the concept of a "Security Association" (SA) is essential in IPSec. In addition to encryption and authentication keys, an SA contains information about the used algorithm, lifetime of the keys and the SA itself. It also contains an SQN to protect against replay attacks, etc.

SAs must be negotiated before ESP or AH can be used, one for each direction of communication. This is done in a secure way by means of the IKE protocol. There are several IKE modes, but the idea is the same: the communicating parties are able to generate "working keys" and SAs, which are used to protect subsequent communication. IKE is based on the ingenious idea of public key cryptography where secret keys for secure communication can be exchanged over an insecure channel. However, authentication of the parties who run IKE cannot be done without some long-term keys. These are typically based on either the manual exchange of a shared secret or by means of a Public Key Infrastructure (PKI) and certificates. Both solutions are clearly non-trivial to bring about: the first requires a lot of configuration effort while the latter implies dedicated infrastructure elements with special functionality.

The negotiation of SAs by IKE is independent of the purpose for which these SAs are used. This is why IKE can also be used for negotiation of keys and SAs to be used in, say, MAPSec.

We conclude this section by describing ESP in a little more detail. There are two ESP modes, *transport* mode and *tunnel* mode. Transport mode functions basically as follows. Everything in an IP packet except the IP header is encrypted. Then, a new ESP header is added between the IP header and the encrypted part, which contains among other things the ID of the SA in use. In addition, encryption typically adds some bits to the end of the packet. Finally, a Message Authentication Code (MAC) is calculated over everything except the IP header and is then appended to the end of the packet. At the receiving end, integrity is checked first. This is done by removing the IP header from the beginning of the packet and the MAC from the end of the packet, then running the MAC function (using the algorithm and key found, based on the information in the ESP header) over the remainder of the packet and comparing the result with the MAC in the packet. If the outcome of the integrity check is positive, then the ESP header is

Transport Mode:

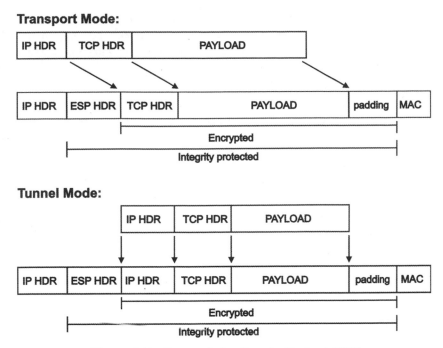

Figure 9.11 Encapsulation Security Payload (ESP)

removed and what remains is decrypted (again based on information in the ESP header). See Figure 9.11 for an illustration.

Tunnel mode differs from transport mode in the following way. A new IP header is added to the beginning of the packet and, then, the same operations as in transport mode are carried out for the new packet. This means that the IP header of the original packet is protected, as illustrated in Figure 9.11.

Transport mode is the basic use case of ESP between two end points. However, when applied in 3GPP networks there are two problems: the communicating network elements have to:

- Know the IP address of each other.
- Implement all the IPSec functionality.

The typical use case of tunnel mode is related to the concept of a Virtual Private Network (VPN). IPSec is used between two middle nodes (security gateways) and end-to-end protection is provided implicitly as the whole end-to-end packet is inside the payload of the packet that is protected between the gateways. Clearly, in this scenario each element has to trust the corresponding gateway. In addition, the leg from the end point element to the gateway has to be protected by other means (e.g., implemented in a physically protected environment). The preferred protection method for UMTS CN control messages is to use ESP in tunnel mode between the security gateways. Details of the the network domain security architecture that is based on security gateways and the profiling of IPSec can be found in 3GPP TS 33.210. For Release 6, the 3GPP has specified how PKI mechanisms can be used in the 3GPP environment to support network domain security mechanisms (see TS 33.310).

9.3.4 MAPSec

We now briefly discuss some of the basic properties of MAPSec. The aim is to protect the confidentiality and integrity of MAP operations. In protection mode 2 of MAPSec, both confidentiality and integrity are protected, while in protection mode 1 integrity alone is protected. In protection mode 0 there is no protection.

The basic functionality of MAPSec can be described as follows. A plaintext MAP message is encrypted and the result is put into a "container" inside another MAP message. At the same time a cryptographic checksum (i.e., a MAC) covering the original message is included in the new MAP message. To be able to use encryption and MACs, keys are needed. MAPSec has borrowed the concept of an SA from IPSec. The SA not only contains cryptographic keys but also other relevant information (e.g., key lifetimes and algorithm identifiers). MAPSec SAs resemble IPSec SAs, but the two are not identical.

The key management to support MAPSec is planned to be provided by similar techniques to those used to support IPSec. However, these mechanisms are not yet included in 3GPP Release 6.

9.4 Protection of Applications and Services

Network security is implemented according to the Open System Interconnection (OSI) model (i.e., each OSI layer has its own means of security protection—Figure 9.12). Security protection can be broken down into two main branches: link-by-link security and end-to-end security.

In link-by-link protection the idea is that a connection (path) is formed by communication links. Everything going through a particular link is protected (e.g., by encryption), and if this is done for every link along a path, the whole connection is secured. The drawback here is that if traffic is misrouted the contents of the link can be exposed, and this is not necessarily visible to the end-user.

If security is applied at higher OSI layers, we usually deal with end-to-end security. In this case, data remain protected (e.g., encrypted) until they reach their destination. Transmitted data are well secured, but now it is possible to carry out *traffic analysis*: an eavesdropper is able to discover, say, who sent it, who received it, when the transmission was done and how frequently such a transmission is made. If a total network security solution is implemented, one must combine both link-to-link and end-to-end security mechanisms. Encryption of each link makes any analysis of the routing information impossible, while end-to-end encryption reduces the threat of unencrypted data at the various nodes in the network.

In certain cases, such as the cellular network, link-layer authentication is required to control access to the cellular network. In the 3G network, access-level security procedures take care of link-level encryption and integrity protection within the access

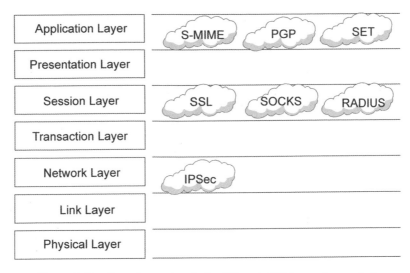

Figure 9.12   Security protocols in different OSI layers (examples)

network. Within the CN and in the connections to other networks various methods can be used, including those discussed in Section 9.3. In the 3GPP system, the IMS is an important special case of application-layer security. Before discussing generic application-layer security mechanisms, we take a closer look at IMS security in Section 9.4.1.

9.4.1 IP Multimedia CN Subsystem (IMS) Security

A major part of 3GPP Release 5 is devoted to specification of the IMS security. This is a complete application-layer system that is built on top of the UMTS PS domain, but it is designed in such a way that it is independent of the underlying access technology. Therefore, security for the IMS cannot be provided solely by UMTS Release 99 security features. Section 6.4 contained a general description of the IMS. In this section we take a more detailed look at IMS security features (see 3GPP TS 33.203 for full details).

All security features of Release 5 IMS are described in Figure 9.13.

When a User Agent (UA) wants to access the IMS, it first creates a Packet Data Protocol (PDP) context with the PS domain. In this process, UMTS access security features are utilised: mutual authentication between the UE and the PS domain, integrity protection and encryption between the UE and the RNC. It is through the Gateway GPRS Support Node (GGSN) that the UA is able to contact IMS nodes and does so by means of SIP signalling. The first contact in the IMS is the Proxy Call Session Control Function (P-CSCF). Through it the UA is able to register with the home IMS. At the same time, the UA and the home IMS authenticate each other, based on a permanent, shared, master secret. They also agree on temporary keys to be used for further protection of SIP messages.

SIP traffic between the visited IMS and the home IMS is protected by network

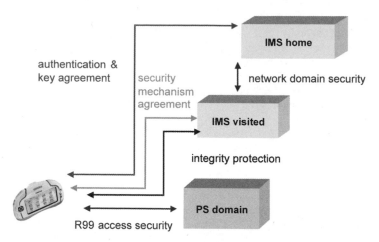

Figure 9.13   IMS security features according to 3GPP R5

domain security mechanisms. The SAs used for this purpose are not specific to the UA in question.

Next, the UA and the P-CSCF negotiate *in a secure manner* all the parameters of the security mechanisms to be used to protect further SIP signalling (e.g., cryptographic algorithms are negotiated). Finally, integrity protection of all first-hop SIP signalling between the UA and the P-CSCF is started, based on the temporary keys agreed during the authentication phase.

The purpose of all these features is largely analogous to CS and PS domain access security: the network is protected against fraudulent attempts to get access and communication between the IMS network and the UA is protected against (both active and passive) attackers.

To summarise, the IMS security architecture relies on three main components:

1. A permanent security context between the UE and the Home Subscriber Server (HSS), achieved by using a security module called the "IMS Subscriber Identity Module" (ISIM) that dwells in a tamper-resistant environment at the user's end, and at the network end the HSS has a secure database. Both the ISIM and HSS contain the IMS subscriber's identity and a corresponding master key, which is the basis on which the AKA procedure can be carried out.
2. A temporary security context exists between the terminal equipment at the user's end and a P-CSCF at the network end, consisting of IPSec ESP SA(s) and a link between these SAs and IMS subscribers. The security context is used to authenticate each SIP message individually in the first hop (from the UE to the P-CSCF or vice versa).
3. All control plane (both SIP and other protocols) traffic between different network nodes is protected using mechanisms described in Section 9.3.3. The related security contexts are not specific to the user.

In the following subsections we present more detailed descriptions of each IMS security mechanism. There are two SIP procedures that play a central role both in SIP itself and in IMS security: REGISTER for registration and INVITE for session establishment. We describe the security solution for the IMS around these two basic functions.

9.4.1.1 Authentication and Key Agreement (AKA)

Use of the IMS is subscription-based. The user enters into an agreement with the IMS operator, resulting in the user having an IMS Private Identity (IMPI) stored in both the ISIM and HSS. There is also a cryptographic 128-bit master key stored within the IMPI.

Before a subscriber can begin to use services provided by the IMS, his or her registration must be active. This is obtained by sending a REGISTER message to a P-CSCF, which forwards the REGISTER request to an Interrogating CSCF (I-CSCF), which in turn contacts the HSS to allocate a suitable Serving CSCF (S-CSCF) for the user. All this communication and all subsequent communication between network elements is protected by network domain security methods using SAs that are not specific to the subscriber.

After an S-CSCF is selected the REGISTER message is forwarded to it, and, then, the S-CSCF fetches Authentication Vectors (AVs) from the HSS. These AVs have the same format as these used for CS and PS domain authentication. Afterwards, the S-CSCF picks up the first AV and sends three or four parameters (excluding XRES and possibly CK) to the P-CSCF via the I-CSCF. The P-CSCF extracts the IK but forwards RAND and AUTN to the UE. The SIP message used to carry all this information is 401 UNAUTHORISED and, hence, from the SIP perspective the first registration attempt has failed.

Anyway, the ISIM in the UE is now able to check the validity of AUTN, and (if the result of the check is positive) RES and IK are also computed. The parameter RES is included in a new REGISTER request that is already integrity-protected by the IK. Integrity protection is done using IPSec's ESP protocol.

The new REGISTER goes first to the P-CSCF from where it is forwarded (via the I-CSCF) to the S-CSCF who compares RES against XRES. If these two match, an OK message is sent all the way back to the UE.

The AKA procedure is now complete and the end result is:

- The UE and P-CSCF share IPSec ESP SAs which can be used to protect all further communication between them.
- The S-CSCF and HSS have both changed the status of the subscriber from "un-registered" to "registered".

The S-CSCF always initiates an AKA at the time of initial registration. Authentication may be skipped for re-registrations, depending on the choice of S-CSCF. It is also possible for the S-CSCF to force a re-registration of the UE at any time and, therefore, the S-CSCF may authenticate the UE whenever it wants.

Pure SIP is an IETF protocol on top of which the 3GPP IMS needs its own extensions. One such extension is the use of the 3GPP AKA for a mutual AKA. A specific

RFC (3310) has been devoted to this issue, which is an extension of *HTTP Digest* (RFC 2617).

9.4.1.2 Security Mode Set-up

When a communication channel is protected by security mechanisms, the critical point is the very start of protection:

- Which mechanisms are activated?
- At what point in time does the protection start for each direction?
- Which parameters (e.g., keys or SAs) are activated?

Preventing the execution of the mechanism in the first place is an obvious attack against any security mechanism: even the strongest mechanisms are useless if you do not turn them on. This attack, and other similar attacks, may be carried out by a *man-in-the-middle*.

A specific RFC (3329: Security Mechanism Agreement for the Session Initiation Protocol) was created for the purpose of securing negotiation of security mechanisms and parameters for SIP.

The basic idea is, first, to exchange *security capability lists* between the client and the server in an unprotected manner, and then check the validity of the mechanism choice when protection is later turned on. The message flow is depicted in Figure 9.14.

9.4.1.3 First-hop Protection Using IPSec ESP

Mutual authentication is not enough to guarantee all charging is made against the correct subscription; this is the reason that individual SIP signalling messages are also authenticated. This is especially required for INVITE messages that are used to set up sessions.

Integrity protection (or message authentication) is done by the IPSec ESP protocol that was described in Section 9.3.3. Use of the IP-layer protection mechanism involves some additional requirements. The identity against which messages are authenticated is the IP address in the case of ESP. On the other hand, charging related to IMS signalling

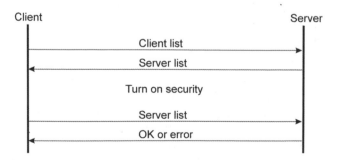

Figure 9.14   Security agreement message flow

is based on IMS identities, which are only visible at the SIP layer. Therefore, SIP-layer identities, especially the IMPI, have somehow to be tied to IP addresses. This problem is solved in the P-CSCF by checking message by message that the IP address used for integrity protection in the SIP layer is allowed for the IMPI in question. The link between the IP address and IMPI is originally created during the AKA (i.e., at the same time as the ESP SA is created).

9.4.2 Examples of Application-layer Security Mechanisms

At the application layer, popular security mechanisms include Secured Multipurpose Internet Mail Extension (S-MIME) and Pretty Good Privacy (PGP). The former is a security protocol that adds digital signatures and encryption to Internet MIME messages. It was originally developed by RSA Data Security Inc. and is based on the triple Data Encryption Standard (triple-DES) and X.509 digital certificates. S-MIME uses an RSA public key encryption method and the Diffie–Hellman system for key management. Secure Hash Algorithm #1 (SHA-1) is adopted for data integrity protection purposes.

PGP was originally freeware for email security designed by Philip Zimmermann. It brings privacy and authentication to email by using encryption and digital signatures. One of the most interesting aspects of PGP is its distributed approach to key management. There is no Certification Authority (CA) for keys; instead, every user generates and distributes his or her own public key. Users sign each other's public keys, creating an interconnected community of PGP users. The benefit of this approach is that there is no CA for everyone to trust. Each user keeps a collection of signed public keys in a file called a "public key ring". The weakest link of the PGP is key revocation. If someone's private key is compromised (e.g., stolen), a key revocation certificate has to be sent out. Unfortunately, there is no guarantee that everyone who uses the public key that corresponds to the compromised private key receives the key revocation in time.

9.4.3 Security for Session Layer

At the session layer of an OSI stack Secure Socket Layer (SSL) or Transport Layer Security (TLS) can be seen. The SSL was originally developed by Netscape Communications Corporation to provide privacy and reliability between two communicating applications at the Internet session layer. The SSL uses public key encryption to exchange a session key between the client and server. This session key is used to encrypt a HyperText Transfer Protocol (HTTP) transaction (prefixed by "http:\\"). Each transaction uses a different session key. Even if someone manages to decrypt a transaction the session itself is still secure (just one transaction is violated).

TLS is an IETF standard (RFC 2246) that has been developed from SSL. There is also a specific modified version called "Wireless Transport Layer Security" (WTLS), which is optimised for the wireless environment. It is used to secure Wireless Application Protocol (WAP) sessions.

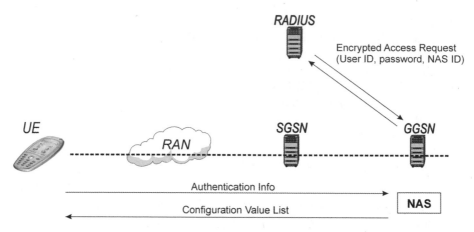

Figure 9.15 RADIUS protocol data flow

9.4.4 AAA Mechanisms

The Remote Authentication Dial In User Service (RADIUS) is described in RFC 2138 as a protocol for carrying authentication, authorisation and configuration information between Network Access Servers (NASs) and a shared authentication server (Figure 9.15). It provides a means of allowing multiple dial-in NAS devices to share a common authentication database. It was originally developed by Livingston Inc. for their Portmaster line of NAS products and has since been widely deployed.

A client–server model is used for an NAS to manage user connections. An NAS operates as a client of RADIUS, which is responsible for passing user information to designated RADIUS servers. RADIUS servers are responsible for receiving user connection requests, authenticating the user and then returning all configuration information (type of service such as SLIP, PPP and Login User-ID values to deliver the service) necessary for the client to deliver the service to the user.

The IETF's Authorisation, Authentication and Accounting (AAA) working group has developed a new protocol called "Diameter". This is used in 3GPP Release 5 for the Cx interface between the S-CSCF and HSS, and should gradually replace RADIUS as the dominant AAA mechanism.

9.5 Lawful Interception

One crucial goal in security is to encrypt information in such a way that it is only available to the correct receiver. So far in this chapter we have presented mechanisms that show how this is done in 3G networks.

On the other hand, there are many countries where local authorities and laws set the limits for encryption, thus restricting the security level the network is able to offer. In addition/parallel to this, local regulations may set a requirement for authorities to have a way of accessing sensitive information and monitoring subscriber numbers (i.e., the authorities must be able to listen to calls or monitor data traffic, both CS and PS). In

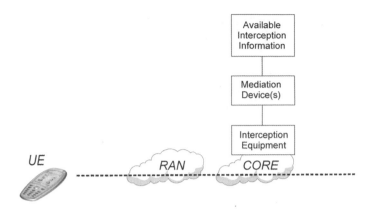

Figure 9.16   Lawful interception arrangement—rough principle

the GSM, this kind of arrangement was later added on top of the existing system, but with the advent of the 3G this localised requirement was taken into account at the very beginning.

As a whole the lawful interception arrangement consists of three parts: interception equipment/functionality, mediation devices and interception information (Figure 9.16). Interception equipment/functionality collects the items shown in Tables 9.1 and 9.2, with local requirements defining what is needed and what is not. This filtering is taken care of by local mediation devices, which then show only the information that local regulations specify. This filtered information is called "interception information". Full details of lawful interception can be found in 3GPP TSs 33.106, 33.107 and 33.108.

In addition to the items listed in Tables 9.1 and 9.2 there are some other regulations related to the position of a subscriber. For example, in some countries it will be mandatory for the system to show the position of a subscriber in case of an emergency call down to an accuracy of 50 metres. This degree of accuracy cannot be achieved using normal MM methods; instead, the 3G network contains a special device called a "positioning system" to bring about the desired accuracy. In this book, the most common positioning methods were presented in Chapter 8.

Table 9.1   Items collected for lawful interception of circuit-switched transactions

Collected item	Explanation
Observed MSISDN	Target identifier with the Mobile Subscriber ISDN (MSISDN) of the target subscriber (monitored subscriber)
Observed IMSI	Target identifier with the IMSI of the target subscriber (monitored subscriber)
Observed IMEI	Target identifier with the IMEI of the target subscriber (monitored subscriber). It should be checked for each call over the radio interface
Event type	Describes the type of event that is delivered: establishment, answer, supplementary service, handover, release, SMS, location update, Subscriber Controlled Input (SCI)
Event date	Date of event generation in the 3G Mobile Switching Centre (MSC)
Event time	Time of event generation in the 3G MSC
Dialled number	Dialled phone number before digit modification, Intelligent Number (IN) modification, etc.
Connected number	Number of the answering party
Other party address	Directory number of the other party for a Mobile Originated Call (MOC) or a calling party for a Mobile Terminated Call (MTC)
Call direction	Information as to whether the monitored subscriber is calling or called (e.g., MOC/MTC or originating/terminating party)
Correlation number	Unique number used to correlate between each call and intercept-related information
Network element identifier	Unique identifier for the relevant network element (e.g., MSC)
Location information	The service area identity and/or location area identity that is present at the 3G MSC at the time of event record production
Basic service	Information about the teleservice or bearer service
Supplementary service	Supplementary services used by the target (e.g., CF, CW, ECT)
Forwarded-to number	Forwarded-to number at CF
Call release reason	Call release reason of the target call
SMS initiator	SMS indicating whether the SMS is MO, MT or undefined
SMS message	The SMS content with header that is sent using the Short Message Service
Redirecting number	The number that invokes call forwarding to the target. This is provided if available
SCI	Non-call-related SCI which the 3G MSC receives from the ME

Table 9.2 Items collected for lawful interception of packet-switched transactions

Collected item	Explanation
Observed MSISDN	MSISDN of the target subscriber (monitored subscriber)
Observed IMSI	IMSI of the target subscriber (monitored subscriber)
Observed IMEI	IMEI of the target subscriber (monitored subscriber). It should be checked for each call over the radio interface
Event type	Describes the type of event that is delivered: PDP attach, PDP detach, PDP context activation, start of intercept with PDP context active, PDP context deactivation, SMS, cell and/or Receiver Address (RA) update
Event date	Date of event generation in the 3G SGSN and/or GGSN
Event time	Time of event generation in the 3G SGSN and/or GGSN
PDP address	The PDP address of the target subscriber. Note that this address might be dynamic
Access Point Name (APN)	The APN of the access point (typically the GGSN of the other party)
Location information	Location information is the Service Area Identity (SAI), Routing Area Identity (RAI) and/or Location Area Identity (LAI) that is present at the GGSN at the time of event record production.
Old location information	Location information of the subscriber before RA update
PDP type	The used PDP type (e.g., PPP, IP or X.25)
Correlation number	Unique number used to correlate communication content and intercept-related information
SMS	SMS content and header that is sent using the Short Message Service. The header also includes the SMS centre address
Network element identifier	Unique identifier for the relevant network element (e.g., SGSN)
Failed attach reason	Reason for failed attach of the target subscriber
Failed context activation reason	Reason for failed context activation of the target subscriber
IAs	Observed interception areas
Session initiator	The initiator of the PDP context activation, deactivation or modification request (i.e., either the network or UE)
Initiator	SMS indicating whether the SMS is MO or MT
Deactivation/termination cause	The termination cause of the PDP context
QoS	Indicates the Quality of Service associated with the PDP context procedure
Serving system address	Information about the serving system (e.g., serving SGSN number or serving SGSN address)

Part Three

10

UMTS Protocols

Ari Ahtiainen

In previous chapters the Universal Mobile Telecommunication System (UMTS) network has been discussed in a subnetwork- and network-element-centric manner. The functional split between network elements was presented by allocation of major system-level management functions to network elements. At the same time the interfaces between the different kinds of network elements were introduced.

In this chapter we now take an interface-centric view of the UMTS network by focusing on system protocols. UMTS protocols are used to control the execution of network functions in a coordinated manner across system interfaces.

The protocols for the UMTS network proper (i.e., UTRAN and CN) are first discussed in Sections 10.1–10.6. Due to the access-independent nature of the IP Multimedia System (IMS) the protocols for IMS networking and services are discussed separately in Section 10.7.

10.1 Protocol Reference Architectures at 3GPP

Since protocol specification work was divided between various groups in the Third Generation Partnership Project (3GPP) technical organisation it was quite natural that each of the groups developed its own protocol reference architecture for those protocols for which it was given the mandate to carry out the specification work. This resulted in three major areas, each with protocol reference models of its own.

Before presenting a combined model for all UMTS network protocols, which will be followed throughout this chapter, let us have a look at each of these three reference models. It appears that among them some major protocol architectural concepts have been introduced and we can make good use of them when deriving the combined protocol architecture.

10.1.1 The Radio Interface Protocol Reference Model

The UMTS Terrestrial Access Network (UTRAN) radio interface is based on Wideband Code Division Multiple Access (WCDMA) radio technology, which has some

UMTS Networks Second Edition   H. Kaaranen, A. Ahtiainen, L. Laitinen, S. Naghian and V. Niemi
© 2005 John Wiley & Sons, Ltd   ISBN: 0-470-01103-3

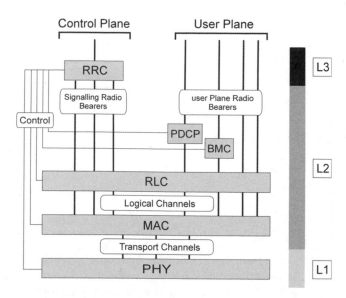

Figure 10.1 Radio interface protocol reference model

fundamental differences from all 2G radio access technologies. The key aspect to be controlled by radio interface protocols is the multiplexing of traffic flows of different kinds and different origins. To ensure effective control of multiplexing a strict layering of duties has been applied, thus resulting in the three-layer protocol reference model illustrated in Figure 10.1.

The layers are simply named according to their position in the architecture as radio interface layer 1, layer 2 and layer 3. Although the well-known Open System Interconnection (OSI) principle of layering has been applied in this design, these three layers cannot as such be considered as the three lowest layers of the OSI model. The layers have well-defined responsibilities and interfaces with each other, which makes it possible to name them as follows:

- L1 – radio physical layer.
- L2 – radio link layer.
- L3 – radio network layer.

As shown in Figure 10.1, the physical layer provides its services as a set of WCDMA transport channels. This makes the physical layer responsible for the first multiplexing function: to map the flows from transport channels to WCDMA physical channels and vice versa. The mapping of transport channels onto physical channels has been explained in Chapter 4.

The radio link layer is another multiplexing layer, one that makes a major contribution to dynamic sharing of capacity in the WCDMA radio interface. Instead of the wide variety of L1 transport channels this layer allows the upper layer to see only a set of radio bearers, along which different kinds of traffic can be transmitted over the radio link. This UTRAN radio bearer is part of the UMTS system-wide bearer architecture, which has already been explained in Chapter 1.

The Medium Access Control (MAC) sublayer controls the use of the transport block capacity by ensuring that capacity allocation decisions (done at the UTRAN end) are executed promptly at both ends of the radio interface. The Radio Link Control (RLC) sublayer then adds regular link layer functions onto the logical channels provided by the MAC sublayer. Due to the characteristics of radio transmission some special ingredients have been added to RLC sublayer functionality.

For L3 control protocol (signalling) purposes the RLC service is adequate as such, but for domain-specific user data additional convergence protocols may be needed to accomplish the full radio bearer service. For CS domain data (e.g., transcoded speech) the convergence function is null, but for the PS domain an additional convergence sublayer is needed. This Packet Data Convergence Protocol (PDCP) sublayer makes the UMTS radio interface applicable to carry Internet Protocol (IP) data packets. Another convergence protocol (BMC) has been specified for message broadcast and multicast domains. The scheduling and delivery of cell broadcast messages to User Equipment (UEs) is the main task of this protocol.

Separation of control signalling from user data is another key design criterion that has long been applied in protocol engineering. In this way the protocols used for control purposes become part of the system-wide control plane whereas protocols carrying end-user data belong to the user plane.

The L3 control plane protocol is the Radio Resource Control (RRC) protocol. As shown in Figure 10.1 an RRC protocol entity at both the UE and UTRAN ends has control interfaces with all other protocol entities. Whenever the protocol entity to be controlled by RRC is located in another UTRAN network element, there is a need to support this control mechanism using standardised protocols. In all other cases, control interfaces are internal to a single UE or UTRAN element and, hence, not precisely standardised, but their existence is crucial for the RRC sublayer to carry out its task as executor of radio resource management decisions.

10.1.2 UTRAN Protocol Reference Model

The access network, UTRAN, has overall responsibility of WCDMA radio resources. UTRAN carries out radio resource management and control among its own network elements and creates the radio access bearers that allow communication between UEs and the Core Network (CN) across the whole UTRAN infrastructure. This communication is structured according to the generic UTRAN protocol reference model illustrated in Figure 10.2. The model is generic in the sense that the architectural elements— though, not the specific protocols—are the same for all the UTRAN interfaces: Iu, Iub and Iur. The specific protocols are referred to in Figure 10.2 only using their 3GPP specification numbers, which also underlines the idea of sharing a common protocol architecture.

The main elements of the UTRAN protocol architecture are layers and planes. The general layering principle is applied in the UTRAN design primarily to distinguish between the two major parts of the protocol stack. Therefore, all the lower layer

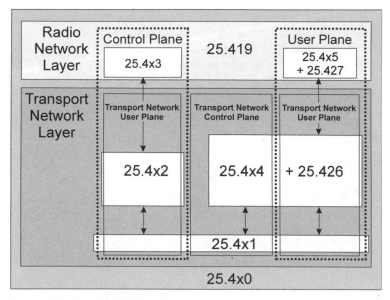

Legend to the 3GPP Specifications:
x = 1   Iu protocols
x = 2   Iur protocols
x = 3   Iub protocols

Figure 10.2   Generic UTRAN protocol reference model

protocols together comprise what is called the "transport network layer" of UTRAN. The name "transport" is used here pretty much in its OSI sense (i.e., to cover the set of protocols allowing non-adjacent network nodes to communicate across an internetwork composed of possibly many different kind of subnetworks). Another, more pragmatic design aspect has been to include in the transport network layer all those protocols that were selected from existing protocol suites instead of having to design them specially for UMTS purposes.

The other set of protocols on top of the transport network layer is—consistently with the radio interface protocol model—called the "radio network layer". These protocols have been carefully designed for the UMTS system and the usual task for them is to control the management and use of radio access bearers across the various UTRAN interfaces.

Figure 10.2 also distinguishes between control-plane and user-plane protocols. This design aspect, which has already been discussed on radio interface protocols, has been applied over all UTRAN interfaces. Within the transport network layer, protocols have been selected on the basis of keeping in mind the various properties required to support both the control- and user-plane protocols of the radio network layer in the most appropriate way. For control-plane protocols this means reliability as a major selection criterion and for user-plane protocols, on the other hand, Quality of Service (QoS) support from the transport network has been targeted for protocol selection. The control protocols for the radio network layer have been specially designed to meet radio and bearer control requirements. The control protocols for the UTRAN radio network layer are commonly named "UTRAN Application Part (AP) protocols". On the other hand, the user-plane protocols within the radio network layer that are responsible for efficient transfer of user data frames are commonly known as "UTRAN frame protocols".

The UTRAN AP and frame protocols together form the UMTS access stratum, which covers all those communication aspects that are dependent on the selected radio access technology. Within generic UMTS radio access bearers, non-access stratum protocols are then used for direct transfer of signalling and transparent flow of user data frames between UEs and the CN. This is achieved by encapsulation of the higher layer payload into UTRAN protocol messages.

UTRAN control-plane protocols have been designed to follow the client–server principle. Regarding the Iu interface UTRAN plays the role of a radio access server and the CN behaves as a client requesting access services from UTRAN. The same is true in the Iub interface, where the Base Station (BS) is the server and its Controlling Radio Network Controller (CRNC) is a client. To some extent this applies also to the Iur interface, where it is the Drifting RNC (DRNC), as a server, that actually provides the Serving RNC (SRNC) with the possibility of controlling remotely located BSs as well. In the client–server-based protocol design the behaviour of a server protocol entity is specified in terms of which actions it should take when receiving a service request from its client. On the other hand, the conditions under which the client generates such service requests are more variable.

The UTRAN protocol architecture is discussed in more detail by Holma and Toskala (2004).

10.1.3 The CN Protocol Reference Model

The CN consists of the network elements that provide support for network features and end-user services. The support provided includes such functionality as the management of user location information, control of network features and services and the transfer (switching and transmission) mechanisms for signalling and for user-generated information.

Within the CN the 3GPP protocol architecture is derived from that of the Global System for Mobile Communications/General Packet Radio Service (GSM/GPRS) system. Therefore, five main protocol suites can be distinguished:

- Non-access stratum protocols between UEs and CN.
- Network control signalling protocols between serving and home networks.
- Packet data backbone network protocols.
- Transit network control protocols.
- Service control protocols.

As a legacy of GSM/GPRS, these protocols have quite different specification backgrounds and origins.

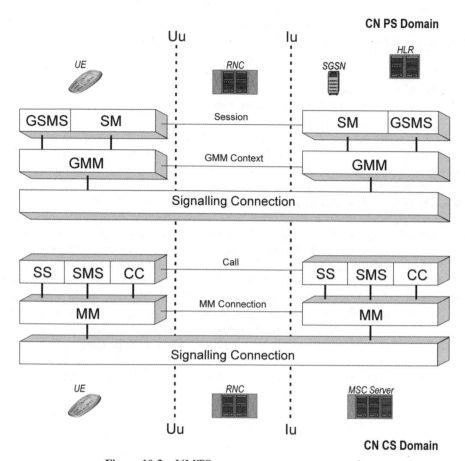

Figure 10.3   UMTS non-access stratum protocols

A typical example of this is that the UMTS CN terminates the *non-access stratum protocols* from UEs. Among these the GSM/GPRS background is very evident. This is reflected by the 3GPP specification TS 24.008, which is an evolved version of its famous "ancestor" GSM 04.08. The design goal for non-access stratum protocols has been to maintain compatibility between GSM/GPRS and UMTS systems in order to share the same CN and to allow dual-mode UEs.

UMTS non-access stratum protocol stacks are shown in Figure 10.3. All of these protocols are carried over the signalling connection, which is established between the UE and the CN at the initial access and signalling connection establishment phase. In both Packet Switched (PS) and Circuit Switched (CS) domains two sublayers can be distinguished in non-access stratum protocols. The lower sublayer has to do with Mobility Management (MM); in the CS domain this protocol is called the "MM" protocol and in the PS domain it is called the "GMM" protocol to reflect the fact that it deals with the GPRS MM. On top of this common sublayer the more service-specific Communication Management (CM) protocols operate. The CM protocols and their control functions are as follows:

- Session Management (SM) protocol, which controls the establishment and release of Packet Data Protocol (PDP) contexts (sessions) for packet data transfer in the CN PS domain.
- Call Control (CC) protocol, which controls the establishment and release of circuit-switched calls in the CN CS domain.
- Supplementary Service (SS) protocol, which controls the activation and deactivation of various call-related and non-call-related supplementary services.
- Short Message Service (GSMS/SMS) protocol, which controls the delivery of short text messages to and from UEs.

Network control signalling between serving and home networks utilises the Mobile Application Part (MAP) protocol suite, which was originally designed for controlling GSM CS services. With the introduction of PS services by the GPRS subsystem it became necessary to extend the MAP protocol to the interfaces between GPRS support nodes and home network nodes. Of the 3G CN interfaces, which were illustrated in Figure 6.4, the following ones are controlled by the 3G version of the MAP protocol:

- Interfaces C, D, E, F and G, which originate from the GSM system.
- Interfaces Gc, Gr, Gf and Gs, which originate from the GPRS system.

This means that both Serving GPRS Support Node (SGSN) and Gateway GPRS Support Node (GGSN) must normally be equipped with the MAP protocol implementation for control-plane purposes.

The MAP protocol follows a transaction-oriented communication scheme. Each transaction (e.g., registering the subscriber's location at the HLR) is executed as a dialogue between CN nodes. This communication structure is created for the MAP sublayer by the Transaction Capabilities Application Part (TCAP) sublayer directly below it. The TCAP then utilises a Signalling System #7 (SS7)-based signalling transport network as the backbone for the signalling network across network operator boundaries.

The protocol suite for the *packet data transfer* within the CN PS domain has also been adopted from the GPRS backbone network. It follows the internetwork protocol paradigm and IP protocol suite, which had fully established itself by the time GPRS specification was started. Actually, only one GPRS-specific protocol needed to be added. This GPRS Tunnelling Protocol (GTP) controls the communication across the PS domain backbone network in the Gn and Gp interfaces (Figure 6.4).

The GTP protocol can also be modelled as two subprotocols: GTP-C for control signalling, which operates between the GGSN and SGSNs, and the user-plane part GTP-U, which extends from the GGSN across the Iu interface to the UTRAN end. This termination of GTP-U communication at the RNC (instead of at an SGSN) differs from the original GPRS specifications.

Any UMTS network has to interwork with external telecommunications and data communications networks. This *interworking* takes place *across the transit network*, which in the case of CS services is the international Public Switched Telephone Network/Integrated Services Digital Network (PSTN/ISDN) backbone or in the case of PS services most often another IP backbone network. The protocols in the transit network boundary must therefore be aligned with those used in the transit network backbone. In the telephone network case the ISDN User Part (ISUP) protocol or Bearer Independent Call Control (BICC) protocol is the obvious choice, and, correspondingly, IPv4 or IPv6 is used in interworking with external IP data networks.

Within the UMTS CN, value-added services should be available to subscribers not only within their home networks but also in visited serving networks. Therefore, the service control protocol Customised Applications for Mobile network Enhanced Logic (CAMEL) Application Part (CAP) has been carried over from the GSM infrastructure, as described in network evolution in Chapter 2. The CAP protocol is another transaction-based protocol and can therefore use the TCAP service and international SS7 signalling backbone just like the MAP protocol discussed above.

10.2 UMTS Protocol Interworking Architecture

The number and wide variety of UMTS system protocols described in Section 10.1 may look confusing—and due to the legacy considerations which had to be taken into account in the protocol selection and design—it cannot be taken as a uniform and homogeneous protocol suite. In this section a combined protocol architecture model for the UMTS network is elaborated on in order to help us in our later examination of individual protocols in this chapter.

While traversing through radio interface, UTRAN and CN protocols, we have already familiarised ourselves with two major protocol design aspects:

- Separation of (generic) transport aspects from (UMTS-specific) mobile networking aspects by layering.
- Separation of network control aspects from user data aspects by planes.

Before we start elaborating on this basis, let us take a closer look at two additional fundamental design considerations in building network-wide protocol architectures:

- Protocol interworking.
- Protocol termination.

Protocol interworking deals with those protocols that belong to the same layer but extend across multiple network elements and, therefore, comprise a set of protocols, which contribute to (distributed) execution of a common system-wide function. As examples of such functions one might consider radio access bearer set-up or location updating. The first requires interworking from the UE up to the CN element and coordinated actions from the RRC protocol and the AP protocols within UTRAN. For location updating, the "chain" of interworking protocols consists of the MM protocol (between the UE and the CN) and the MAP protocol between the CN nodes. Such a chain of distributed actions is called a (system) procedure among protocol designers.

Protocol interworking is often designed as an extension to the above-mentioned client–server model. Once an event is triggered at one end of the network the first protocol entity to discover it becomes the client, and starts the procedure by requesting some action from its server. In a UMTS network the server is seldom able to satisfy such a request on its own: it has to assume the role of another client and request other services from its server(s). This is how the procedure works, until those servers—often at the end of the network opposite to which the triggering event took place—are reached that cannot delegate the task any further, but have to obey the commands themselves. In the radio access bearer set-up the procedure is initiated by a CN element and finally executed by the BS and the UE that set up the radio link between them. When location updating it is the UE that initiates the procedure, and, finally, the Home Location Register (HLR) has to record the new location in its database.

More complete examples of UMTS system-level procedures are shown in Chapter 11, but this interworking of protocols within a layer is the key design aspect in setting up a layered view on UMTS protocols in this chapter as well.

Protocol termination can now be taken as the end point of a protocol chain, which interworks in the manner described above. Within a network element, where a protocol terminates, a system function or algorithm can be identified as the source/sink of data or commands. For the MM protocols mentioned above in the location update example the termination points are therefore the UE and the HLR. Termination is sometimes discussed with respect to a single protocol, but then the termination points are simply the two end points of that single protocol. Therefore, from the system-wide protocol architecture point of view, the termination of a set of interworking protocols is much more significant.

Throughout the rest of this chapter UMTS protocols will be described by dividing the UMTS protocol suite into three different protocol interworking layers, each of which extends across multiple network elements and can be seen as a logical network of its own. This decomposition does not strictly follow the structure of 3GPP specifications, but rather takes a network-wide, protocol-oriented view, where layering and end-to-end interworking are the main structuring principles.

First, the UMTS protocol model is divided horizontally into layers. The protocols within each of the layers operate across multiple interfaces and belong together in the sense of interworking and termination described above.

Three layers can be distinguished as shown in Figure 10.4:

- Transport network layer.
- Radio network layer.
- System network layer.

In this division the transport network (layer) is responsible for providing a general-purpose transport service for all UMTS network elements, thus making them capable of communicating across all the interfaces described earlier. UMTS functionality is then distributed among the network elements by using the protocols within the radio and system network layers, which are by definition UMTS-specific. Radio network (layer) protocols ensure interworking between the UE and the CN on all radio access bearer-related aspects. System network (layer) protocols extend from the UE until the transit

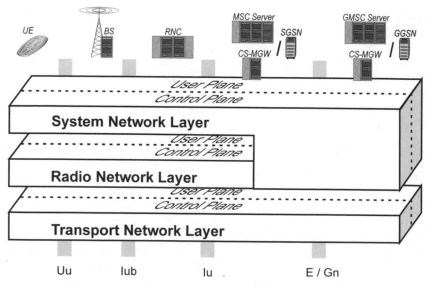

Figure 10.4   UMTS protocol internetworking architecture

network edge of the UMTS CN. They ensure interworking on UMTS communication service-related aspects.

Within all three layers it is then possible to distinguish between control aspects and user data transfer aspects, which create the other structuring dimension of the UMTS protocol model shown in Figure 10.4. Those protocols that only deal with control aspects belong to the control plane, while those dealing with user data transfer belong to the user plane. The distinction between control and user planes is most visible within the radio and system networks, but within the transport network many protocols are common to the control and user plane. For optimisation reasons some transport network protocols are however specially selected to carry control or user data only.

Control-plane protocols in different UMTS interfaces interwork with each other to ensure system-wide control of communication resources and services. In a similar way, user-plane protocols interwork with each other to ensure the end-to-end flow of user data.

Based on the characteristics of user traffic two different system domains are distinguished. In the CS domain, user-plane capacity is allocated in circuits for the whole duration of the service (e.g., a phone call). In the PS domain, user-plane capacity is allocated to data packets. The different requirements of these two domains may also be reflected in selection of transport network protocols and in the design of system network protocols; therefore, in this chapter they are illustrated separately for both domains, whenever necessary.

With this separation of functionalities the three different protocol subsystems can be more readily upgraded when new technologies become available. In fact, such evolution has already taken place in the introduction of alternative transport network protocols and the new IMS in Release 5 of 3GPP specifications.

10.3 Transport Network Protocols

Although the communications capacity of the user plane and control plane should be ideally physically separate, it would lead to wasteful use of radio and terrestrial capacity in the network. Therefore, the protocols providing general-purpose transport service for both the user and control plane were designed as much as possible as a subsystem of their own, referred to as the "transport network".

As its name suggests the UMTS transport network is actually a network within the UMTS network. A more detailed look at transport network technology further reveals that it is not just a single network, but actually consists of a number of networks that, when used in a UMTS system-specific way, can be seen as a single (logical) UMTS transport network.

Physical transmission of bits is done by a (cellular) transmission network, which is based on digital transmission network technology. The (optical) trunk-line capacity of such a transmission network is shared by statistical multiplexing, which is controlled by transmission network nodes. This transmission capacity is then utilised by the "second-degree" network on top of the transmission network. In 3GPP systems the standardised technology choices for this switching network are Asynchronous Transfer Mode (ATM) cell switching and IP packet switching. This means that the data cells or packets traverse through multiple switching nodes before reaching their final destination. It is only at the edges of this switching network that we can can find actual UMTS network elements.

The principle of composing an end-to-end transport network from a number of interconnected subnetworks is well known from IP architecture. Similarly, the UMTS transport network is seen as an "internetwork", although it does not follow such a uniform protocol structure as IP internetworks. Instead, the transport service within the different parts of the UMTS system—radio interface, RAN and CN—is designed and optimised for each part separately. As discussed below, there is a drive towards such a harmonised transport network architecture, which is based on IP networking with real-time transport capability.

Since the transport network itself provides switching and routing capabilities, it may also require an embedded control plane for the creation of transport network circuits or for distributing routing information. These kinds of embedded control protocols are selected according to the transport network technology itself and they are not discussed in any more detail here.

The key requirement for the UMTS transport network is in-sequence delivery of user data packets. In the case of ATM this is guaranteed by ATM cell-switching itself, but in the case of IP transport a well-engineered IP network is needed. From the 3GPP specifications' point of view this is not an issue since the UMTS transport network is considered to be an operator-specific, dedicated network that is not shared with any public IP traffic.

10.3.1 Transport Network Protocol Architecture

The transport network consists of the lowest layers of the UMTS protocol architecture, thus providing facilities to transport and route both control and user traffic across all

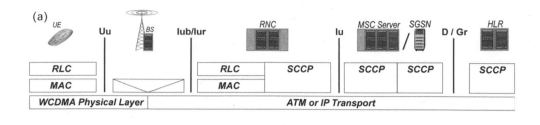

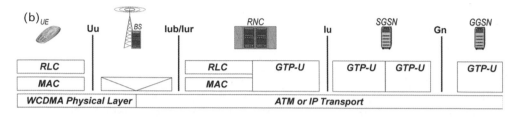

Figure 10.5 Transport network protocols: (a) control-plane protocols in the transport network; (b) user-plane protocols in the transport network (PS domain only)

UMTS network interfaces. Figure 10.5 presents the high-level protocol architecture for the signalling transport network and the packet transport network.

The physical layer controls the physical media through which both control signalling and user data traffic are transferred. The physical layer of the UMTS radio interface is based on WCDMA radio technology, as described in Chapter 4. The purpose of the physical layer protocol is to provide the upper layers with a set of WCDMA transport channels.

For terrestrial interfaces both ATM transport and IP transport networking have been specified by 3GPP. In Release 99, ATM with its Adaptation Layers (AAL2/AAL5) was the most popular choice of transport, while IP transport was only specified for Iu-PS as the only alternative for the user plane and as an option for the control plane. From Release 5, ATM transport and IP transport are equally valid choices for all user-plane and control-plane UTRAN interfaces.

The selection of physical layer transmission media and appropriate data link layers needed to support ATM and IP transport are left open in order not to limit the use of transmission network technologies available in operator networks. However, it is important to note that the underlying assumption in 3GPP specifications for IP transport is that UTRAN is considered as a private network dedicated only for an operator's own traffic. This helped to overcome security and QoS issues, which would have otherwise required much more detailed specifications on IP networking protocols and technical choices. A well-engineered IP network with adequate overprovisioning is needed as an operator's private network for UTRAN transport.

The transport network protocol stack for the UTRAN radio interface, as shown in Figure 10.5, is intentionally simplified. Due to UTRAN's internal protocol structure and functional division the protocols MAC and RLC are terminated at the RNC. Therefore, there is a need to carry MAC/RLC frames over the Iub and Iur interfaces.

The frame protocols used for carrying MAC/RLC frames are not shown in this figure, but will be discussed in Section 10.4.2.2. These frame protocols are also running on top of either ATM or IP transport. Within UTRAN, it is the SRNC that is responsible for radio interface-related activities for a UE at the WCDMA transport channel level, while BSs actually only maintain WCDMA physical channels.

10.3.2 WCDMA Physical Layer in the Uu Interface

The physical layer protocol controls the use of WCDMA physical channels at the Uu interface (Figure 10.6). It provides the upper layer with a set of WCDMA transport channels that characterise the different ways in which data can be transmitted over the WCDMA radio. The mapping between transport channels and physical channels is done by the physical layer protocol.

The physical layer provides a bandwidth-on-demand service, which means that the transport channels support variable bit rates. As the lowest layer, the physical layer is responsible for multiplexing at the WCDMA frame level. To manage this kind of multiplexing the physical layer controls the transport formats used on transport channels. The *transport format* is a format for delivery of a transport block set during each transmission time interval on a transport channel. *Transport block sets* are the physical layer SDUs composed by the MAC protocol entities that use the physical layer service. The *transmission time interval* determines how often the physical layer can accept data from the MAC layer. The transport format attributes are listed in Table 10.1.

The physical layer implements the multiplexing of transport blocks from different transport channels. In order to maintain the frame synchronisation of WCDMA radio transmission, the physical layer also segments transport blocks into radio frames. Other

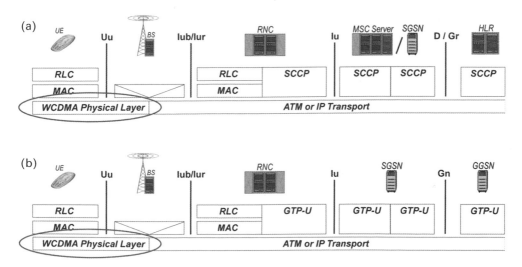

Figure 10.6 Radio transport in Uu interface: (a) control-plane protocols in the transport network; (b) user-plane protocols in the transport network (PS domain only)

Table 10.1   Transport format attributes for WCDMA-FDD*

Dynamic attributes	Transport block size
	Transport block set size
Semi-static attributes	Transmission time interval
	Error protection scheme: type, coding rate, rate matching
	Size of CRC

* Frequency Division Duplex.

functions of the physical layer are channel coding, interleaving and rate matching, all of which are done before radio frames are mapped for transmission over different physical channels.

As a comparison to the physical layer protocol in the GSM radio system it is worth noting that there is no ciphering at the WCDMA physical layer.

The physical layer protocol also attaches a Cyclic Redundancy Check (CRC) checksum (0, 8, 12, 16 or 24 bits) to transport blocks for transmission over the radio interface. At the receiving end the physical layer delivers the transport blocks to the higher layer together with the possible error indication resulting from CRC checksum calculation.

Services provided by the WCDMA physical layer protocol are described in 3GPP specification TS 25.302, and the detailed specification of the WCDMA physical layer protocol is covered by 3GPP specifications TS 25.211–25.215. A detailed presentation of the WCDMA physical layer can be found in Holma and Toskala (2004).

10.3.3 Backbone Networking in Other Interfaces (Figure 10.7)

Unlike the case of the radio interface, the transport network protocols for UMTS terrestrial interfaces have not been designed specifically for UMTS purposes. Instead, the 3GPP standardisation bodies have selected them from among all existing protocol suites. As in the case of all transport network protocols, the focus in this selection has been on the ability to multiplex traffic from different end-users and—to be more precise—from different UMTS bearers taking into account their QoS characteristics.

This—together with some standardisation politics, which inevitably is also present—has led to the selection of two major protocol suites: one from broadband tele-communications (the ATM protocol family) and the other from the Internet data communication networks (known as the IP protocol family). In 3GPP Release 99, ATM was the dominating transport network technology at the UTRAN end; the same was true for IP-protocol technology in the PS domain at the CN end. Within the CN CS domain the transport network continued to be based on the time-slotted PCM trunking network. In 3GPP Release 5, both ATM and IP transport were made equally valid options in all interfaces including speech transmission over the Iu-CS interface as well.

The use of ATM or the IP network as the UMTS transport network requires in both cases a set of adaptation and convergence protocols to be used for running UMTS-

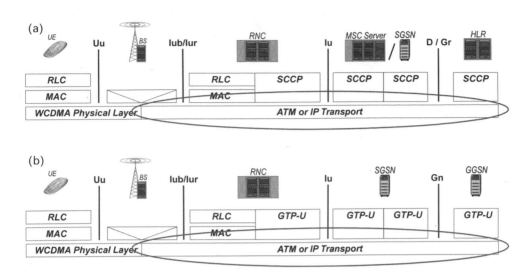

Figure 10.7 Transport networking in other interfaces: (a) control-plane protocols in the transport network; (b) user-plane protocols in the transport network (PS domain only)

specific transport protocols on top of these general-purpose protocols. The complete protocol stacks standardised by 3GPP Release 5 for different interfaces are shown in Figure 10.8.

10.3.3.1 ATM Transport

The basic idea behind ATM is to split the information flow to be transferred into small pieces (packets), attach address tags to those packets and then transfer the packets through the physical transmission path. The receiving end collects received packets and forms an original-like information flow from the contents of the packets. The packet containing transmitted information is called the *ATM cell.*

An ATM cell (see Figure 10.9) consists of two parts, a 5-byte-long header (address information) and payload (transmitted information). Compared with "conventional" protocols and messages, the header is very short. This sets some limitations on what can be done, but, on the other hand, information transfer effectiveness is high: the addressing overhead is $5/(5+48) \sim 9.5\%$. Another aim has been to establish a very lightweight transmission system without any extra "bureaucracy". Because of this, the payload of an ATM cell is *not* protected by checksum method(s). Nowadays, this is possible because the transmission networks carrying ATM traffic are of high quality and the terminals used are able to perform error protection themselves if required.

The header of an ATM cell contains some address information (see Figure 10.10). The most essential items are:

- VPI (Virtual Path Identifier): the identifier for a Virtual Path (VP) or, more generally, an identifier for a constantly allocated semi-permanent connection.

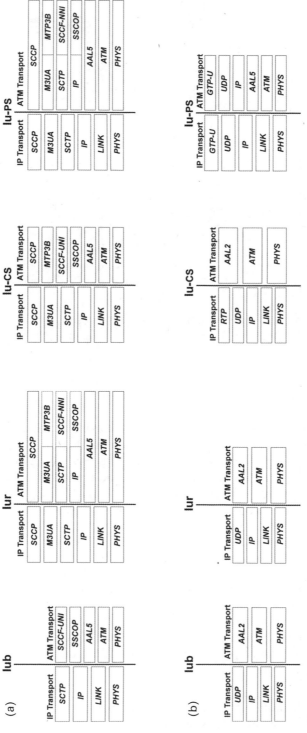

Figure 10.8 Protocol stacks for ATM and IP transport: (a) control-plane protocols in the transport network; (b) user-plane protocols in the transport network

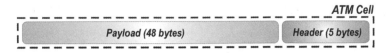

Figure 10.9 ATM cell structure

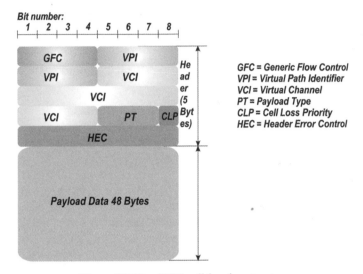

Figure 10.10 ATM cell header structure

- VCI (Virtual Channel Identifier): an identifier for a Virtual Channel (VC). This field is long because there may be thousands of channels to be identified within one VP (e.g., multimedia applications may require several VCIs simultaneously, one VC per multimedia component).
- PT (Payload Type): this indicates whether the 48-byte payload field carries user data or control data.
- CLP (Cell Loss Priority): this is a flag indicating whether this ATM cell is "important" or "less important". If CLP = 1 (low priority/less important) the system may lose this ATM cell if it has to.
- HEC (Header Error Control): in ATM, the ATM cell header is error-protected. This is because a failure in the ATM cell header is more serious than in the payload (e.g., since the ATM cell may be delivered to the wrong address as a result of header error). The error correction mechanism used is able to detect all errors in the header and one failure can be corrected.

Figure 10.11 illustrates the transmission path of ATM. One ATM transmission path may consist of several virtual paths, which further contain virtual channels.

A virtual path is a semi-permanent connection simultaneously handling many virtual connections/channels. Actual data are transferred in ATM cells over the virtual channels. From the point of view of the UMTS system, an ATM transmission path

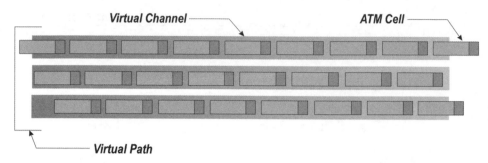

Figure 10.11 Virtual Path (VP) and Virtual Channel (VC)

would be, say, between the BSs and the RNC. If a loop transmission is the matter in question, the transmission path contains many virtual paths (one per BS) and the virtual channels in the virtual path are set up on a per-call basis. The bandwidth of the virtual channel varies depending on the bearer service used.

The ATM layer as such consists of fairly simple transport media and in theory is suitable for transmission purposes as such. In practice, the ATM layer must be adapted to higher protocol layers and the lower physical layer. The International Telecommunication Union (ITU-T) has defined what are called "ATM service classes" using AALs. The original idea was that each service class from A to D should correspond to one AAL from 1 to 4. In course of time, the original idea disappeared, as can be seen from Figure 10.12.

The service classes of the ATM are:

- Constant Bit Rate (CBR).
- Unspecified Bit Rate (UBR).
- Available Bit Rate (ABR).
- Variable Bit Rate (VBR).

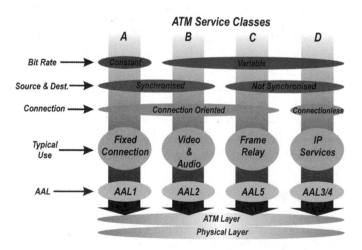

Figure 10.12 ATM Adaptation Layers (AALs)

Any transparent data transfer may use the CBR and the resources are allocated on a peak data rate basis. The UBR uses free bandwidth when available. If there are no resources available, queuing may occur. The ABR is used when the user service has a minimum bit rate defined. Otherwise, the bandwidth is used as in UBR. The VBR provides a variable bit rate based on statistical traffic management.

The adaptation layers for ATM are as follows:

- AAL1 offers synchronous mode, connection-oriented connection and constant bit rate for services requiring this kind of adaptation.
- AAL2 offers synchronous mode, connection-oriented connection with variable bit rate for services using this adaptation.
- AAL3/4 offers connectionless and asynchronous connection with a variable bit rate.
- AAL5 offers asynchronous mode, connection-oriented connection with a variable bit rate.

As shown in Figure 10.13, AAL is divided into two sublayers: the Convergence Sublayer (CS) and the Segmentation And Re-assembly (SAR) sublayer. The CS adapts AAL to upper protocol layers and the SAR sublayer splits the data to be transmitted into suitable payload pieces and, in the receiving direction, collects payload pieces and assembles them back to the original dataflow. Depending on the case, the CS may be divided further into smaller entities.

From the point of view of the UMTS transport network, AAL2 and AAL5 are the most interesting alternatives. AAL2 is seen as a suitable option for CS user-plane connections and AAL5 is seen as suitable for control protocol exchange. A more detailed list of these AAL variants in different UTRAN protocol interfaces is given in Table 10.2.

Within UTRAN the ATM adaptation layer AAL5 is used to carry all control protocols as well as the PS domain user data at the Iu interface. In contrast, AAL2

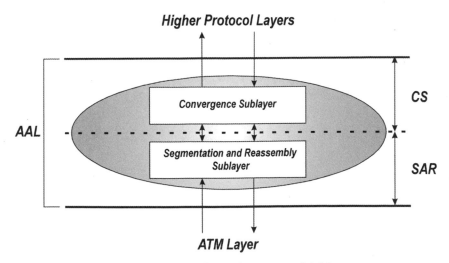

Figure 10.13   General structure of AAL

Table 10.2 ATM adaptation layers used in UTRAN interfaces

AAL5	Iu: CS C-plane, PS C/U-plane, internal C-plane of the transport network
	Iur: C-plane, internal C-plane of the transport network
	Iub: C-plane, internal C-plane of the transport network
AAL2	Iu: CS U-plane
	Iur: U-plane
	Iub: U-plane

C-plane = Control plane; U-plane = User plane.

is used as the common carrier for user data in all interfaces, except the PS domain user data at the Iu interface.

The sublayers of the AAL protocol stack in different UTRAN interfaces, as shown in Figure 10.8, are not described in detail here. The key observation in the selection of these convergence protocols is that, in all Release 99 control-plane protocols for the PS domain, convergence could already be accomplished in two alternative ways: either the convergence protocols specified by ITU-T for signalling transport over AAL5 were selected or IP-over-AAL5 was used. In addition, user-plane convergence for the PS domain at the Iu interface was already part of Release 99 and based on the User Datagram Protocol (UDP) over IP. In 3GPP Release 5 this led to even wider utilisation of IP-based transport in all UTRAN and CN interfaces, as described next.

10.3.3.2 IP Transport

IP-based transport started to gain more momentum towards the end of the 3GPP R99 specification campaign. In order to create a basis for full-scale utilisation of IP inter-networking technology as the common transport for the future evolution of UMTS networks, it became necessary to study the IP option for both signalling and user data transport in all UTRAN interfaces.

Ever since 3GPP R5, user data can be transported across all UTRAN interfaces by using the IP protocol stack as well (see Figure 10.8). The common part of the IP stack used for user data transport is UDP/IP: on the Iub and Iur interfaces, radio frames are transported as plain UDP datagrams; on the Iu-PS interface, the tunnelling protocol (GTP-U) is used to carry user data packets; and on the Iu-CS interface, the Real Time Protocol (RTP) is used on top of UDP/IP to carry speech frames.

3GPP specifications for the UTRAN transport network require all UTRAN nodes to support IPv6. Support for IPv4 is optional and the specifications do not preclude implementation of just IPv4. However, in order to support the transition from IPv4 to IPv6, it is recommended to use dual-stack (IPv4 and IPv6) implementations in UTRAN nodes.

Although 3GPP specifications do not mandate any single link layer protocol to be used in the IP transport network, there is a minimum requirement to support the Point-to-Point Protocol with High-level Data Link Control (HDLC) framing according to the Internet Engineering Task Force's (IETF) Requests for Comments (RFCs) 1661 and 1662. Other data link protocols that fulfil the same requirements can also be used. One key requirement of the UTRAN transport network is in-sequence delivery of data packets.

The IP transport bearers in UTRAN interfaces are identified by pairs of UDP port numbers and IP addresses (for source and destination nodes). For QoS purposes every IP node in the UTRAN transport network have to support Differentiated Services Code Point (DSCP) marking according to IETF RFC 2474.

In order to maintain harmonisation between IP transport and ATM transport of UMTS signalling both stacks should provide the same transport service. Therefore, it was decided to "surface" both stacks using the Signalling Connection Control Part (SCCP) protocol entity, which always delivers a harmonised signalling transport service (the structures of the resulting protocol stacks were shown in more detail in Figure 10.8).

Since the UMTS CN is based on the GSM/GPRS network subsystem, so signalling transport protocols are likewise inherited from GSM/GPRS. The 3GPP R99 implementation still includes a protocol stack based on common channel SS7. Ever since 3GPP R4/5, the CN signalling transport solution can be harmonised with the UTRAN transport network by also running CN signalling protocols on top of ATM or IP, instead of plain SS7.

In the original case of SS7 signalling network the SCCP protocol is running on top of the Message Transfer Part Layer 3 (MTP3) protocol, which is responsible for routing signalling messages between signalling points. MTP routing is based on Signalling Point Codes (SPCs) and can only route signalling messages within a single SPC addressing space, which is typically administered by a single network operator. In order to make it possible to route signalling across network boundaries the SCCP protocol is needed. SCCP uses global title addressing, which makes it possible, for example, for a Visitor Location Register (VLR) within a visited network to reach the HLR in the subscriber's home network by using the Global Title (GT) of the HLR for addressing.

The adaptation of the SCCP protocol on top of the IP protocol stack was achieved by using two convergence protocols, which are developed and standardised by the SIGTRAN Working Group at the IETF. These protocols and the corresponding IETF documents are:

- Stream Control Transport Protocol (SCTP), RFC 2960.
- MTP3 User Adaptation (M3UA) layer (Internet draft document).

The purpose of these two sublayers is to make the SCCP layer look as though it is sitting on top of the regular MTP3 service, as has been the case in telecommunication signalling networks for more than two decades.

SCTP assumes that it is running over an IPv4 or IPv6 network. Even more importantly, it assumes it is running over a well-engineered IP network. This, in practice, means that there is a diverse routing network underneath so as to avoid a single point of failure. SCTP takes into consideration multi-homed end points (i.e., end points with more than one IP address/port number tuple) for additional reliability. Furthermore, it provides a Maximum Transmission Unit (MTU) discovery function to determine the MTU size of the data path; this is to avoid IP-level fragmentation.

The purpose of the SCTP is to provide a robust and reliable signalling bearer. To achieve this, SCTP provides appropriate congestion control procedures, fast retransmit in the case of message loss and enhanced reliability. It also provides additional security against blind attacks and will be used to increase security in connecting the UMTS networks of different operators.

The purpose of the M3UA protocol is to support SCCP signalling, so that the SCCP lower interface does not need to be modified. M3UA needs to manage the use of SCTP streams and to match the performance of its SS7 counterpart, the Message Transfer Part-3 Broadband (MTP-3B) protocol, while running on top of an IP stack. M3UA provides mapping of SS7 addressing to IP addressing. It provides failover support as well as the ability to load-share among end points.

M3UA has two architectural modes: signalling gateway to IP Signalling Point (IPSP) and IPSP-to-IPSP. It is assumed that IPSP-to-IPSP will be not only the most likely but the simplest as well.

10.3.4 UMTS Transport Network Protocols

The protocols running on top of the transport backbone network (ATM or IP) described above provide addressing and routing of signalling messages from one UMTS network node to another. They also allow multiplexing of traffic from different users to the common transport network capacity. Besides the WCDMA physical layer, which is itself capable of some multiplexing of traffic onto the WCDMA physical channels, multiplexing is a task given to the layers above the physical layer in the transport network.

10.3.4.1 Transport Network Protocols in the Uu Interface

The transport network protocols, which carry all control signalling and user data—for both PS and CS traffic—over the Uu interface are:

- MAC protocol.
- RLC protocol.

These two protocols work together as the radio link of the transport network.

10.3.4.1.1 Medium Access Control (MAC) Protocol

The MAC protocol is active at UE, BS and RNC entities (Figure 10.14). MAC is a layer 2 protocol and, as part of the common transport network, provides services to both the control and user plane. In the design of the MAC protocol both CS and PS traffic, as well as signalling traffic considerations, were taken into account from the very beginning. This is a major difference from 2G packet radio protocols, which can be seen as packet data-specific add-ons on top of the normal CS (time-slotted) 2G radio carriers.

MAC provides its service as a set of logical channels that are characterised by what

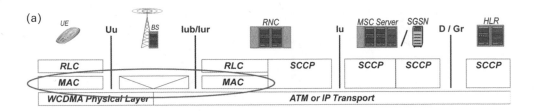

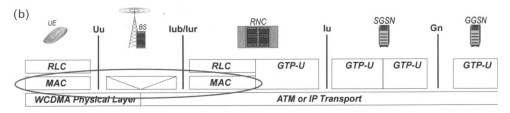

Figure 10.14   Transport network—Medium Access Control (MAC): (a) control-plane protocols in the transport network; (b) user-plane protocols in the transport network (PS domain only)

type of data is transported on them. There are four kinds of control channels and two kinds of traffic channels among these logical channels (as described in Chapter 4).

MAC has overall responsibility for controlling communications over the WCDMA transport channels provided by the physical layer. In order to be able to share the capacity of the transport channels among the set of logical channels, MAC uses transport blocks as units of transmission. The possible transport block sizes for each transport channel are offered to MAC by the WCDMA physical layer as part of the transport format set definition.

According to the type of transport channels provided by the physical layer MAC is split into three entities: *MAC-Common*, *MAC-Dedicated* and *MAC-Broadcast*.

The *MAC-Common* entity resides in the CRNC and the UE and is in charge of control over common and shared transport channels. It is capable of multiplexing Protocol Data Units (PDUs) from higher layers into transport block sets carried over these transport channels. Identification of UEs on common and shared transport channels is performed by a temporary identity (allocated by the RNC), which can be the Cell-Radio Network Temporary Identity (C-RNTI, 16 bits), or the UTRAN Radio Network Temporary Identity (U-RNTI, 32 bits) for common channels or the Downlink Shared Channel (DSCH) Radio Network Temporary Identity (DSCH-RNTI, 16 bits) for shared transport channels.

The *MAC-Dedicated* entity resides in the SRNC and the UE and provides control over dedicated transport channels. On these channels MAC is capable of multiplexing by composing transport block sets from higher layer PDUs. In this case all blocks in the set must belong to the UE for which the transport channel has been dedicated, and multiplexing is only possible if the QoS parameters are identical to those of the services supported by the Dedicated Channel (DCH).

The *MAC-Broadcast* entity can be found in the BS and the UE. This entity handles a single broadcast transport channel in each cell.

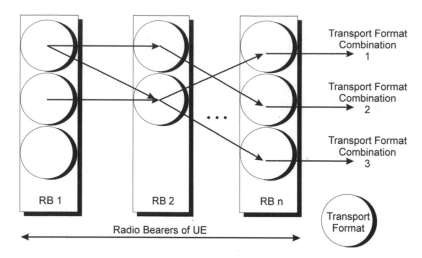

Figure 10.15   Transport format combinations in MAC

Before initiating data transmission on any transport channel, the MAC entity selects the appropriate transport format for the given transport channel and takes into account the data transmission requirements of the other transport channels. This combination of different transport formats from each transport channel is based on the instantaneous bit rate of the traffic offered at any moment together with power control considerations and results in the transport format combinations shown in Figure 10.15.

Due to the fact that MAC is also part of the user plane, it is a real-time protocol. MAC has to meet the tight timing requirements of the physical layer and must be ready to send a new set of transport blocks according to the transmission time intervals indicated by the physical layer. During each time interval MAC must select the optimal transport format combination based on the characteristics of the data to be transmitted and on the current traffic situation.

The basic service of the MAC layer is to provide data transfer of MAC SDUs between peer MAC entities. From the upper layers' point of view it is worth noting that this service is given without any acknowledgement or segmentation.

Switching between common, dedicated and shared transport channels is also executed by the MAC layer, despite the commands to switch the configuration coming from the RRC layer.

MAC also collects statistical information about the traffic to be used by the RRC layer. These MAC measurements are provided per logical channel and include local measurements, such as buffer occupancy, variance and average. A MAC status indication about underflow or overflow detected for each transport channel and measurement mode (event-triggered, periodic) is reported to the RRC.

The MAC layer performs a ciphering algorithm of transparent-mode RLC data. MAC Signalling Data Units (SDUs), which are sent on dedicated logical channels (DCCH and DTCH) and mapped onto dedicated transport channels, are ciphered with a UE-specific key by using the block cipher algorithm KASUMI as described in Chapter 9.

MAC has only one PDU called a "data-PDU", which consists of a MAC SDU and MAC protocol header. The header is constructed according to the transport channel to be used. In some cases (e.g., when the UE has only one DCH without any logical channel multiplexing) the MAC header may be left out.

In 3GPP R5 specifications, MAC also supports High Speed Downlink Packet Access (HSDPA). As part of this extension a new *MAC-High-Speed* entity was introduced. The MAC-High-Speed entity is responsible for control of the new transport channel: High Speed Downlink Shared Channel (HS-DSCH). This is a shared channel, one of which exists in every HSDPA-capable cell. This channel between the UE and the BS is controlled by MAC-High-Speed peer entities on both sides. Between the BS and the SRNC, high-speed traffic is carried over so-called "MAC-Dedicated Flows"—yet another MAC extension in Release 5. From the MAC-Dedicated point of view, MAC-Dedicated Flow is considered to be another type of dedicated transport channel (similar to DCH).

Due to the unidirectional nature of the HS-DSCH transport channel (it only exists in the downlink direction), MAC-High-Speed peer entities perform asymmetrical functionalities. MAC-High-Speed at the BS executes functions related to data transmission, such as scheduling and priority handling between users and their priority queues, and transport formation selection as well. MAC-High-Speed at the UE takes care of data reception, which includes reordering of data in the queues according to priorities and disassembly of MAC-High-Speed PDUs.

In order to increase the reliability of data transmission, MAC-High-Speed provides fast retransmission of erroneously delivered data PDUs. Every MAC-High-Speed PDU is acknowledged by the receiver through associated uplink signalling. Based on the received acknowledgement, the sender may initiate retransmission. However, whenever the MAC-High-Speed receiver detects an error it does not discard the received packet, rather it stores it in the soft buffer and then coherently combines the stored packet with its retransmissions until it becomes decodable. MAC-High-Speed in the UE is capable of soft-combining. Soft buffer capacity is configured by the RRC layer.

Fast retransmission and soft-combining is facilitated by an *n*-channel Hybrid Automatic Repeat Request (HARQ) functional entity. There is one HARQ entity in MAC-High-Speed at the UE and one per cell in MAC-High-Speed at the BS. There are up to eight stop-and-wait channels in every entity.

The MAC-High-Speed entity in the BS receives data from MAC-Dedicated flows and orders them in priority queues. At every Transmission Time Interval (TTI, which can be as short as 2 ms) offered by the physical layer for transmission, data from only one queue can be handled. The scheduling rate for each queue (i.e., how often it gets the right to transmit/retransmit on HS-DSCH) depends on the scheduling algorithm. Therefore. there is a need for flow control functionality between MAC-High-Speed in the BS and MAC-Dedicated in the SRNC to avoid queue overflow. Flow control is provided with the help of the HS-DSCH frame protocol capacity allocation procedure.

The complete specification of the MAC protocol is given in 3GPP specification TS 25.321 and a more detailed description of the MAC protocol can be found in Holma and Toskala (2004).

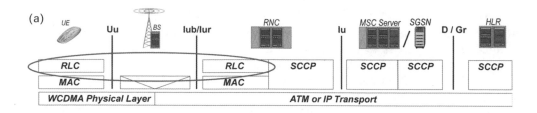

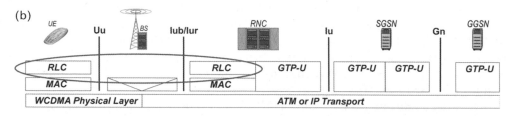

Figure 10.16 Transport network—Radio Link Control (RLC): (a) control-plane protocols in the transport network; (b) user-plane protocols in the transport network (PS domain only)

10.3.4.1.2 Radio Link Control (RLC) Protocol

The RLC protocol runs both in the RNC and the UE and implements the regular (data) link layer functionality over the WCDMA radio interface (Figure 10.16). The RLC is active in both the control and user plane simultaneously and provides data link services for both CS and PS connections.

The operational environment of the RLC depends on the plane discussed; in the case of the user plane the RLC is used by PDCP and in the case of the control plane the RLC is used by the RRC protocol. The lower layer protocol for the RLC is MAC, which provides data transfer services as MAC SDUs over logical radio channels. Because MAC data transfer mode is unacknowledged, the RLC becomes responsible for the delivery of higher layer PDUs for which reliability is needed. Also, because the MAC layer is not able to segment larger SDUs according to the transport formats available, the RLC takes care of such functionality. Therefore, transport formats are visible at the RLC layer as shown in Figure 10.17.

The RLC provides the data transfer service of higher layer PDUs and does so as RLC SDUs. This service is called the "radio bearer service". Three modes of operation have been defined for the RLC: transparent (Tr), unacknowledged (UM) and acknowledged (AM).

In *transparent mode* the RLC transmits SDUs without adding any protocol information. SAR is still possible, but has to be negotiated at the RRC layer during radio bearer set-up. This mode can be used for streaming class services.

In *unacknowledged mode* the RLC transmits SDUs without guaranteeing delivery to the peer entity. This mode is used by some RRC control procedures, where acknowledgement is taken care of by the RRC protocol itself.

In *acknowledged mode* the RLC transmits SDUs and guarantees delivery to the peer entity. Guaranteed delivery is ensured by means of retransmission. In case RLC is

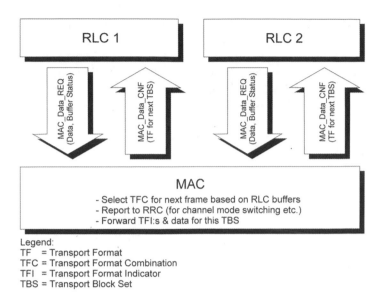

Legend:
TF = Transport Format
TFC = Transport Format Combination
TFI = Transport Format Indicator
TBS = Transport Block Set

Figure 10.17   RLC interoperation with MAC on dedicated channels

unable to deliver the data correctly, the user of the RLC at the transmitting end is notified. This mode is used for PS data transfer over dedicated logical channels.

The RLC has many functions to perform on its layer:

- SAR.
- Concatenation.
- Padding.
- Error correction.
- In-sequence delivery of SDUs.
- Duplicate detection.
- Flow control.
- Sequence number check.
- Protocol error detection and recovery.
- Suspend/resume functionality.
- SDU discard.
- Ciphering.

SAR involves adjusting variable length higher layer PDUs into/from RLC PDUs. The RLC PDU size is adjustable to the actual set of transport formats.

Concatenation is used when the contents of an RLC SDU do not fill an integer number of RLC PDUs. Thus, the first segment of the next RLC SDU may be put into the RLC PDU chained to the last segment of the previous RLC SDU. When concatenation is not applied and the remaining data for transmission do not fill an entire RLC PDU of a given size, the remainder of the data field will be filled with padding bits. Padding can be replaced by piggybacked status information for the reverse link.

Many RLC functions have to do with error detection and correction. Bit errors may actually be detected by a physical layer CRC check, but the RLC layer is responsible for error recovery. The most effective error recovery is provided by retransmission in acknowledged data transfer mode. The RLC protocol can be configured by the RRC to obey different retransmission schemes (e.g., Selective Repeat, Go Back N or Stop-and-Wait).

In-sequence delivery of SDUs preserves the order of higher layer PDUs that were submitted using the acknowledged data transfer service. If this function is not used, out-of-sequence delivery may happen.

The RLC is able to detect duplicates of received RLC PDUs and can ensure that the resultant higher layer PDU is delivered only once to the upper layer. Flow control allows an RLC receiver to control the rate at which the peer RLC entity may send information.

The sequence number check function may be used in unacknowledged mode to guarantee the integrity of reassembled PDUs. This provides a mechanism for the detection of corrupted RLC SDUs by checking the sequence number of RLC PDUs when they are reassembled into an RLC SDU. A corrupted RLC SDU will be discarded.

The protocol error detection and recovery functionality detects and recovers from errors caused by operation of the RLC protocol. The RLC reset procedure is used to recover from an error situation. In case of an unrecoverable error the RLC entity notifies the RRC layer.

The RLC may suspend and resume data transfer on request from the RRC.

The SDU discard function facilitates the removal of any remaining RLC PDUs of the SDU in question from the buffer at the transmitter end, when transmission of an RLC PDU does not succeed for a long time. The SDU discard function allows buffer overflow to be avoided. There are several alternative operation modes for the RLC SDU discard function.

Ciphering is performed in the RLC layer for those radio bearers that use unacknowledged or acknowledged mode. Ciphering is done by means of a UE-specific key by using the block cipher algorithm KASUMI as described in Chapter 9.

The complete specification of the RLC protocol is given in 3GPP specification TS 25.322 and a more detailed description of the RLC protocol can be found in Holma and Toskala (2004).

10.3.4.2 Transport Network Protocols in Other Interfaces

10.3.4.2.1 Signalling Transport (SCCP)

The common signalling transport protocol in both UTRAN and CN interfaces is SCCP (see Figure 10.18).

SCCP is derived from SS7 and provides both a connectionless and connection-oriented service. The connection-oriented service is used to support signalling bearers.

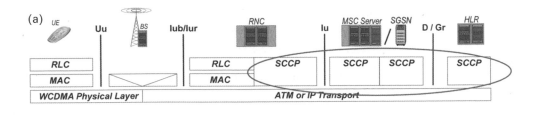

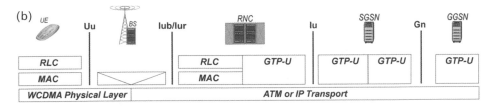

Figure 10.18 Transport network—signalling transport: (a) control-plane protocols in the transport network; (b) user-plane protocols in the transport network (PS domain only)

10.3.4.2.2 IP Transport for User Data (GTP-U)

The major user-plane protocol providing the transport service in the CN PS domain and across the Iu interface is GTP-U for the packet data user plane (see Figure 10.19). As its name suggests, this protocol has been adopted from the 2G GSM/GPRS system, but with a significantly different architectural design choice. In 2G, GTP-U is only used between 2G GSN network elements, but in 3G it was extended to reach RNC across the Iu-PS interface as well.

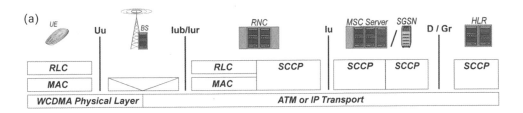

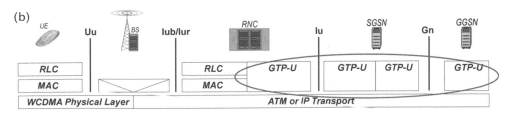

Figure 10.19 Transport network—GPRS Tunnelling Protocol for User Plane (GTP-U): (a) control-plane protocols in the transport network; (b) user-plane protocols in the transport network (PS domain only)

GTP-U is a transport network protocol supporting user-plane data transfer in the UMTS network. GTP-U belongs to the PS domain in the UMTS network and is located in the RNC at the UTRAN end, as well as in the SGSN and GGSN at the CN end.

GTP-U operates on three different interfaces: Gn, Gp and Iu-PS. The Gn interface is between GSNs (i.e., SGSNs and GGSNs) belonging to the same UMTS packet domain network. The Gp interface is used for network interworking between two SGSNs or between an SGSN and GGSN belonging to different UMTS packet domain networks. The interface between the RNC and SGSN is the Iu-PS.

Over all interfaces GTP-U operates on top of the UDP/IP protocol family. UDP provides connectionless message transfer between two IP network nodes. UDP addressing is used to identify GTP-U end points within source and destination network elements. UDP also provides a checksum mechanism to detect transmission errors inside data packets. As is generally known, the main service of the IP is the routing of messages between source and destination network elements. This routing is based on IP network addresses (both IPv4 and IPv6 can be used within the UMTS transport network).

GTP-U provides connectionless data transfer services for upper layers and allows multi-protocol user data packets to be tunnelled across the Iu-PS, Gn and Gp interfaces. Because of encapsulation and tunnelling mechanisms it is possible to transfer user data packets, although different routing protocols are used in the packet domain IP backbone network. In addition, UMTS network elements and protocols transferring user data packets between the UTRAN and GGSN have no need of being aware of different addressing mechanisms at the IP layer to be able to associate user data packets with specific PDP contexts.

Since the main purpose of GTP-U is to transfer user data, it has been optimised to that specific task. GTP-U multiplexes packets received from one interface (e.g., Gi at the GGSN) and addressed to several destinations (different UEs). It receives the user data packet from the external packet network, interprets the destination of the packet and passes the packet onto the next node along the path. Tasks for setting up the tunnels between GTP-U end points have been excluded from GTP-U, and control-plane protocols are used for these purposes. On the Gn interface, GTP-C (GTP for the control plane) and, on the Iu-PS interface, Radio Access Network Application Part (RANAP) protocols are used to control the set-up of GTP-U tunnels.

GTP-U entities have to interwork with other protocol entities as the data packets are transported. In the GGSN a GTP-U entity communicates with the packet relay functionality located at the edge of the UMTS network (i.e., the Gi interface). At the Iu-PS end of an SGSN a GTP-U entity interworks with another GTP-U entity at the Gn end. Within the RNC a GTP-U entity terminates the tunnel and forwards the packets to the frame protocol entity at the radio network layer.

Interworking between GTP-U and the neighbouring protocol entities is quite simple. After all, each parameter related to user data packet transfer has been negotiated in advance by the control plane. User data packets are simply passed to the next entity, which is then able to handle the packet according to its predefined PDP context parameters.

The main functions of GTP-U are:

- Data packet transfer.
- Encapsulation and tunnelling.
- Data packet sequencing.
- Path alive check.

The maximum size of a user data packet is 1,500 octets. User data packets containing 1,500 octets or less should be transmitted as one packet. In case the user data packet received from the external network by GGSN exceeds the 1,500-octet limit, the GGSN will fragment or discard the user data packet depending on PDP type. Since IP fragmentation is inefficient and does not tolerate transmission errors, fragmentation should be avoided. Thus, all links between network elements containing GTP-U should have MTU values exceeding the 1,500 octets in addition to the size of GTP-U, UDP and IP headers.

Each user data packet is encapsulated before being transmitted to the UMTS packet domain network. Encapsulation adds a GTP header containing tunnelling information to every user data packet. Tunnelling information includes a 32-bit Tunnel Endpoint Identifier (TEID), which serves two important purposes. First, TEID is used to address a PDP context inside a tunnel end point. Hence, it also indirectly refers to different UEs that have one or more PDP contexts active in the UMTS network. Second, tunnelling enables multiplexing of user data packets destined for different addresses into a single path which is identified by two IP addresses.

As an optional feature GTP-U may preserve the order of user data packets between the RNC and GGSN. During PDP context establishment at the control plane, reordering can be negotiated. In case user data packet reordering is applied, the GTP header will contain a 16-bit sequence number, which GTP-U uses to determine whether user data packets are received in the correct order or whether it has to wait for missing packets. However, in case the missing user data packet is lost, the data packet sequencing mechanism does not provide a mechanism to recover a lost packet. It is up to the other layers to recover the lost user data packet.

GTP may check at regular time intervals whether a peer GTP is alive by sending an echo request message to the peer GTP. The peer GTP will respond to an echo request message with an echo response message. On the Gn interface, one may think that the path alive check procedure belongs to GTP-C, but, on the Iu-PS interface, GTP-U is the only GTP element and, thus, handles the path alive check procedure. On the Gp interface, the path alive check is similar to the Gn interface.

The main procedures of the UMTS network where GTP-U is used are primarily user data packet transfer between the UTRAN and SGSN as well as between the SGSN and GGSN. When a routing area update procedure needs to switch a GTP-U tunnel from one SGSN to another, GTP-U is used to relay user data packets not yet sent to the UE from the old SGSN to the new SGSN. GTP-U is also used as part of the SRNS relocation procedure to tunnel user data packets not yet sent to the UE from the source RNC to the target RNC via the pair of SGSNs attached to the RNCs.

GTP-U is defined in 3GPP specification TS 29.060.

10.4 Radio Network Protocols

In the UMTS internetwork protocol architecture (see Figure 10.4), the radio network protocols comprise the next layer on top of the transport network protocols discussed above. The radio network extends from the UE across the whole UTRAN and terminates at the edge nodes of the CN. The protocols in the radio network are needed to control the establishment, maintenance and release of radio access bearers and to transfer user data between the UE and CN along these radio access bearers.

10.4.1 Radio Network Control Plane

The control-plane protocols in the radio network layer execute all the control needed for management of Radio Access bearers (RABs). According to the client–server approach applied in the overall design of the UTRAN control plane, these protocols implement radio resource management decisions and maintain the RABs for all UEs, even when they are moving within the UTRAN area.

10.4.1.1 RANAP—Radio Access Network Application Protocol

The Iu interface connects the UMTS Radio Access Network (RAN) and the CN. The same Iu interface is used to connect both CS service and PS service domains to the UTRAN (see Figure 10.20). The Iu interface has been designed to support the independent evolution of UTRAN and CN technologies. In addition, it was made possible for CS and PS service domains inside the CN to develop independently and still use the same Iu interface to connect the domains to the UTRAN.

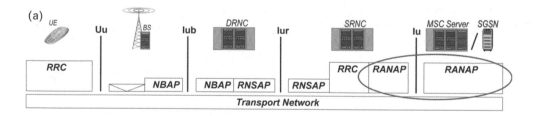

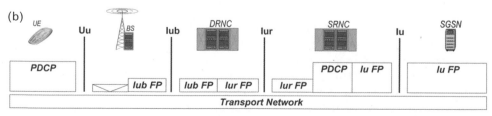

Figure 10.20 Radio network—RANAP: (a) control-plane protocols in the radio network; (b) user-plane protocols in the radio network (PS domain only)

From UTRAN's perspective, it is desirable that connections to the CS and PS service domains are as similar as possible. This is the reason a single signalling protocol between the UTRAN and CN was defined. This protocol is called Radio Access Network Application Part (RANAP) and is defined in 3GPP specification TS 25.413. Both CS and PS domains use RANAP to access services provided by the UTRAN.

RANAP is the protocol that controls resources in the Iu interface. One RANAP entity resides in the RNC and another peer entity resides in the MSC server or the SGSN.

RANAP is located on top of Iu signalling transport layers and uses the signalling transport service to transfer RANAP messages over the Iu interface. The transport protocol stack consists of several protocol layers, but the topmost protocol is always SCCP. In spite of the two possible transport options (ATM or IP) it is always an SCCP entity that offers RANAP an interface to the services of the transport protocol stack.

RANAP has certain requirements of the transport layers that lie below it. Basically it is assumed that each message (PDU) that RANAP sends to the peer entity will reach the destination without any errors. In other words it is the responsibility of the transport protocol stack to provide reliable data transfer across the Iu interface for RANAP PDUs.

Each PDU of dedicated control services should be sent during a unique signalling connection. In the Iu interface, signalling connection is brought about by an Iu signalling bearer. Transport layers will provide RANAP with the means to dynamically establish and release signalling bearers for the Iu interface when RANAP requests it. Each active UE will have its own Iu signalling bearer. It is the responsibility of transport layers to maintain the bearers. If the Iu signalling bearer connection is interrupted for some reason, SCCP will inform RANAP about it.

RANAP offers services to upper layer protocols: on the MSC server and SGSN side, non-access stratum protocols make use of RANAP services. By interworking with the RRC protocol in the RNC, RANAP contributes to the transfer of the signalling messages of these upper layer protocols between the UE and CN.

RANAP services are divided into two groups: general control services and dedicated control services. General control services are related to the whole Iu interface between the RNC and CN. Dedicated control services support the separation of each UE in the Iu interface and always relate to a single UE. The majority of RANAP services are dedicated services. All the RANAP messages of the dedicated control services are sent during a dedicated connection. This connection is called a "signalling connection".

Overall RAB management is one of the main services RANAP offers. RANAP provides the means for the CN to control the establishment, modification and release of RABs between the UE and CN. Moreover, RANAP supports UE mobility by transferring an RAB to a new RNS when the UE moves from the area of the SRNS to another area. This service is called "SRNS relocation". Another RANAP service controls the security mode in the UTRAN. All these services are dedicated control services.

General control services are only needed in exceptional situations. RANAP provides, for example, a means of controlling the load in the Iu interface if the amount of user traffic gets too large. In the case of a fatal failure at either end of the Iu interface

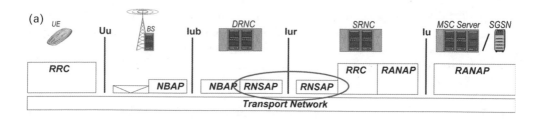

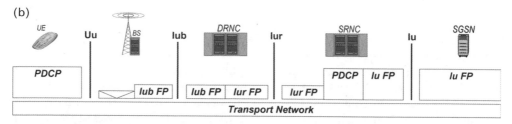

Figure 10.21 Radio network—RNSAP: (a) control-plane protocols in the radio network; (b) user-plane protocols in the ratio network (PS domain only)

RANAP offers a reset service. This initialises the whole Iu interface and clears all active connections.

The detailed specification for RANAP is given by 3GPP specification TS 25.413.

10.4.1.2 RNSAP—Radio Network Subsystem Application Protocol

RNSAP is a control-plane protocol at the radio network layer (Figure 10.21). It provides control signalling exchange across the Iur interface. RNSAP is run by two RNCs, one of which plays the role of SRNC and the other that of DRNC. It is the SRNC that facilitates RANAP signalling connection with the CN.

RNSAP always operates on top of SCCP. Just like RANAP protocol there are two signalling transport options below SCCP: either ATM with an AAL5, or IP-based transporting which the SCTP and M3UA protocols are used as the adaptation layer. As a signalling transport SCCP provides two different modes for RNSAP: connection-oriented and connectionless data transfer service.

RNSAP is responsible for bearer management signalling across the Iur interface. RNSAP is used for setting up radio links and allowing the SRNC to control these radio links using dedicated resources in a DRNC.

RNSAP uses one signalling connection per DRNC and UE when the UE has one or more active radio links for the transfer of layer 3 messages. The information transferred over the Iur interface can be categorised as follows:

- Radio and mobility control signalling: the Iur interface provides the capability to support radio interface mobility between RNSs for those UEs having a connection with UTRAN. This capability includes the support of handover, radio resource control and synchronisation between RNSs.

- Iub/Iur DCH data streams: the Iur interface provides the means for transport of uplink and downlink Iub/Iur DCH frames carrying user data and control information between the SRNC and a remote BS via the DRNC.
- Iur DSCH data streams: an Iur DSCH data stream corresponds to the data carried on one DSCH transport channel for one UE. A UE can have multiple Iur DSCH data streams. In addition, the interface provides a means for the reporting of queues to the SRNC and for the DRNC to allocate capacity to the SRNC.
- Iur Random Access Channel/Common Packet Channel (RACH/CPCH) data streams: in WCDMA FDD variant only.
- Iur Forward Access Channel (FACH) data streams.
- Iur USCH data streams: in WCDMA TDD (Time Division Duplex) variant only.
- Iur HS-DSCH data streams: an Iur HS-DSCH data stream corresponds to the data carried by one MAC-Dedicated flow for one UE. A UE can have multiple Iur HS-DSCH data streams.

The main UTRAN control functions supported by RNSAP exchange are:

- Transport network management.
- Traffic management of common transport channels (e.g., paging).
- Traffic management of dedicated transport channels (e.g., radio link set-up, addition and deletion as well as measurement reporting).
- Traffic management of downlink shared transport channels (e.g., radio link set-up, addition and deletion as well as capacity allocation).
- Measurement reporting of common and dedicated measurement objects.
- SRNS relocation—this function coordinates activities when the SRNS role is to be taken over by another RNS.

RNSAP participates in the radio link set-up procedure when it is used for establishing the necessary resources in the DRNC for one or more radio links. The connection-oriented service of the signalling bearer will be established in conjunction with this procedure.

Other UTRAN-wide procedures that may require protocol activities from RNSAP entities are soft and hard handovers, cell update, UTRAN Registration Area (URA) update and paging as well as downlink power control actions and RRC connection release and re-establishment whenever it needs to be performed over the Iur interface.

The complete specification of RNSAP is given in 3GPP specification TS 25.423.

10.4.1.3 NBAP—Node B Application Protocol

NBAP is a radio network layer protocol that maintains control-plane signalling across the Iub interface and, thus, controls resources in the Iub interface and provides a means for the BS and RNC to communicate (Figure 10.22). A peer entity of NBAP resides in the BS and for each BS another entity resides in the RNC that controls the BS (i.e., in the CRNC).

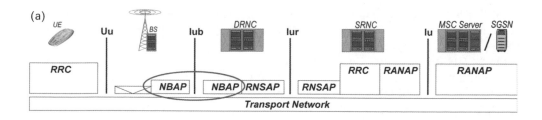

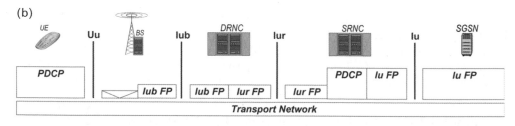

Figure 10.22 Radio network—NBAP: (a) control-plane protocols in the radio network; (b) user-plane protocols in the ratio network (PS domain only)

NBAP resides on top of Iub transport layers and uses their services to transfer NBAP messages over the Iub interface. ATM is 3GPP R99's choice of underlying transport technology to carry NBAP signalling over the Iub interface. ATM connection, together with the necessary ATM adaptation layers, provides a signalling bearer for NBAP as defined in 3GPP specification TS 25.432. 3GPP R5 specifications define two choices of underlying transport technology: the first is ATM connection as in R99; and the second is SCTP over the IP stack. The 3GPP R5 standard allows operators to choose one of two protocol stacks for transporting NBAP messages.

The signalling bearer for NBAP provides reliable point-to-point connection. This means that NBAP assumes that each message it sends to its peer entity will reach its destination without any errors. There may be multiple point-to-point links for NBAP signalling at the Iub interface.

The BS does not make any decision of its own about radio resource or RAB management but, instead, obeys commands from the RNC. NBAP offers its services to the control units of RNC and BS. NBAP also has an interface with the RRC protocol such that it can provide the RRC entity in the BS with information, which should be mapped to the logical Broadcast Control Channel (BCCH). NBAP has no other interfaces to the radio network layer protocols of UTRAN and all interactions with them are carried out via the RNC control unit.

Figure 10.23 illustrates the operational environment of NBAP, in which all the interfaces of NBAP are depicted as bold black arrows.

All resources physically implemented in a BS logically belong to and are controlled by the RNC. Therefore, the physical resources of a BS are seen as logical resources by the CRNC. This RNC controls the level of logical resources, while the physical implementation of resources and control procedures in the BS is left unspecified. Physical procedures performed in the BS change the conditions of the logical resources

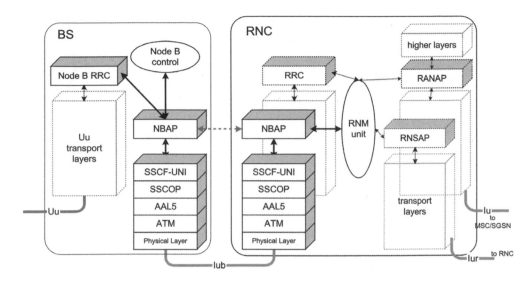

Figure 10.23 Operational environment of NBAP

owned by the RNC, and this requires some information exchange between the RNC and the BS to update the RNC of any changes. Such information exchange is referred to as the Logical Operation and Maintenance (Logical O&M) of the BS. Logical O&M constitutes an integral part of NBAP signalling.

Other important responsibilities of NBAP are the establishment and maintenance of a control-plane connection over the Iub interface, initiation of the set-up and release of dedicated user plane connections across the Iub interface and commanding the BS to activate resources for new radio links over the Uu interface. All NBAP signalling functions are divided into common procedures and dedicated procedures.

Common NBAP procedures are not related to any particular UE, but to common resources across the Iub interface. Signalling related to the Logical O&M of the BS constitutes the major part of common NBAP procedures. They include procedures for configuration management of logical resources and procedures that enable the BS to inform the RNC about the status of logical resources in the BS. Common NBAP procedures also allow the RNC to initiate specific measurements in the BS and the latter to report the results of measurements. Besides Logical O&M, another common NBAP procedure is used to deliver the information to be transported on the broadcast channel from the RNC to the BS. There is also a common NBAP procedure that initiates the creation of a new UE context for each specific UE in BS by setting up the first radio link for that UE and yet another procedure to maintain control-plane signalling connection across the Iub interface. It is for this reason that there is always one signalling link for common NBAP procedures at the Iub interface, which terminates at the common control port of the BS.

Dedicated NBAP procedures are always related to a specific UE that already has a UE context in the BS. They include procedures for management and supervision of existing radio links, whose UE is connected to the BS. Dedicated NBAP procedures allow the RNC to command the BS to establish or release some radio links for the UE context. A dedicated NBAP procedure also provides the BS with the ability to report failure or restoration of transmission on radio links. There are also dedicated procedures for radio link reconfiguration management and to support measurements on dedicated resources and corresponding power control activities that allow the RNC to adjust the downlink power level on radio links. Dedicated NBAP signalling is carried out via one of the communication control ports of the BS logical model. There can be several dedicated signalling links for dedicated NBAP procedures across the Iub interface.

The complete specification of NBAP is given in 3GPP specification TS 25.433.

10.4.1.4 Radio Resource Control (RRC) Protocol

The RRC protocol is the key radio resource control protocol within the UTRAN (Figure 10.24). It supports the Radio Resource Management (RRM) functionality discussed in Chapter 5 by coordinating the execution of resource control requests that result from the decisions made by RRM algorithms. With the help of the RRC protocol, the effect of RRM decisions can be extended to all UTRAN network elements affected by such decisions. RRC protocol messages also carry in their payload all the signalling belonging to non-access stratum protocols.

The RRC protocol operates between the UE and the RNC. RRC protocol entities use the signalling bearers provided by the RLC sublayer to transport signalling messages. All three modes of the RLC are available as different kinds of Service Access Points (SAPs) as shown in Figure 10.25. Transparent mode TM-SAP is used, for example, whenever a UE has to communicate with an RNC prior to establishment of a full RRC connection (e.g., initial access or cell/URA update with a new cell). It is

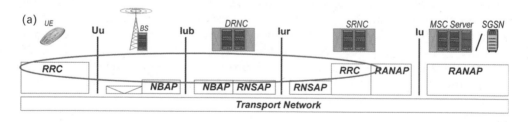

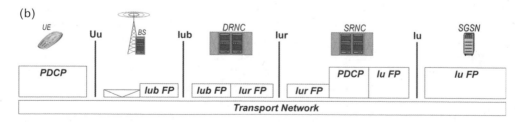

Figure 10.24 Radio network—RRC: (a) control-plane protocols in the radio network; (b) user-plane protocols in the radio network (PS domain only)

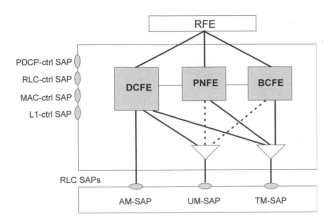

Figure 10.25 Logical structure of an RRC protocol entity in WCDMA-FDD

also used for frequently repeated messages, like paging, to avoid unnecessary over-heads. Acknowledged mode AM-SAP is used for control signalling that is specific to one UE whenever reliability of message exchange is required. Unacknowledged mode UM-SAP, on the other hand, is used to avoid the potential delay present in acknowledged-mode signalling (e.g., when releasing an RRC connection the release message is repeated many times—quick repeat function—via UM-SAP to increase the probability of the UE receiving it).

Figure 10.25 also shows the logical structure of an RRC protocol entity. The Dedicated Control Function Entity (DCFE) is used to handle all signalling specific to one UE. In the SRNC there is one DCFE entity for each UE having an RRC connection with this RNC. The Paging and Notification Function Entity (PNFE) handles paging messages sent to idle-mode UEs. In a CRNC there is at least one PNFE entity for each cell to be controlled. The Broadcast Control Function Entity (BCFE) handles system information broadcasting on BCCH and FACH logical channels. There is at least one BCFE entity for each cell in a CRNC. Besides these functional entities a special Routing Function Entity (RFE) is also modelled on top of Figure 10.25. Its task is to route non-access stratum messages to different MM/CM entities at the UE end and to different CN domains at the RNC end.

As key executors of radio resource allocation, the RRC entities in the UE and RNC have control over the L1, MAC and RLC entities at their end. In order to execute control commands on these lower layer protocol entities, special control SAPs are available to the RRC protocol entity as shown in Figure 10.25. These control SAPs are also used for reporting measurements and exceptional conditions detected by the lower layers.

The major function of the RRC protocol is to control the radio bearers, transport channels and physical channels. This is done by set-up, reconfiguration and release of different kinds of radio bearers. Before such actions can take place, RRC protocol communication must itself be initiated by using a number of (minimum four) signalling radio bearers. The resulting signalling connection, together with any other subsequently established bearers, is called *RRC connection*. During RRC connection, set-up, reconfiguration and release of other radio bearers for user-plane traffic may be executed by exchanging commands and status information between peer RRC entities over signalling radio bearers. RRC connection will continue to exist until all user-plane bearers have been released and the RRC connection between the UE and RNC is explicitly released.

The security mechanisms applied to radio bearers are activated and deactivated under the control of the RRC protocol. Besides activation of confidentiality protection by ciphering, the RRC protocol can also guarantee the integrity of all higher layer signalling messages as well as most of its own signalling messages. This property, which is discussed more thoroughly in Chapter 9, is achieved by attaching to every RRC PDU a 32-bit message authentication code, which is calculated and verified by the RRC entities themselves. This ensures that the receiving RRC entity can verify that signalling data have not been modified and that the data genuinely originated from the claimed peer entity. In addition, the operation to change the keys used by these functions is performed by the RRC protocol under instructions that originally come from the CN.

UE MM at the UTRAN level is controlled by RRC signalling. Such MM functions executed by the RRC protocol are cell update, URA update and active set update, all of which were discussed in Chapter 5 and some examples of which will be shown in Chapter 11.

Control and reporting of radio measurements is taken care of by the RRC protocol. A total of seven different categories of measurements can be activated at the UE, among them measurement of UE location. Each measurement can be controlled and reported independently by means of further measurements.

Besides the above-mentioned tasks the RRC protocol also controls the broadcast of system information and paging of UEs and exchange parameters for power control purposes.

As a protocol, RRC is fairly complex: altogether it has some 40 different procedures and more than 60 different kinds of PDU.

The RRC protocol is specified in 3GPP specification TS 25.331.

10.4.2 Radio Network User Plane

10.4.2.1 PDCP—Packet Data Convergence Protocol

As its name suggests PDCP is designed to make WCDMA radio protocols suitable for carrying the most common user-to-user packet data protocol, TCP/IP. As shown in Figure 10.26, PDCP entities can be found at both ends of the WCDMA radio interface: the RNC and the UE. The key functionality of PDCP is its ability to compress the headers of payload protocols, which—if sent without compression—would waste the invaluable radio link capacity (e.g., without header compression the header size for an RTP/UPD/IP header is at least 40 bytes for IPv4 and at least 60 bytes for IPv6. Since payload size in such services as IP voice is about 20 bytes or less, the overhead without compression is obvious.

In the 3GPP radio interface protocol model, PDCP belongs to the radio link layer

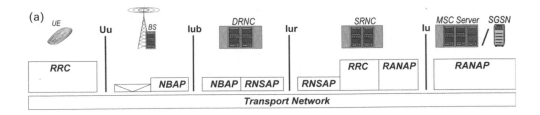

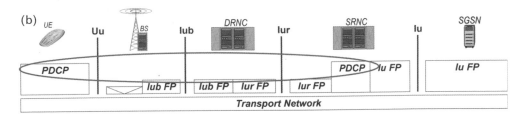

Figure 10.26 Radio network user plane—PDCP: (a) control-plane protocols in the radio network; (b) user-plane protocols in the radio network (PS domain only)

(L2), the topmost sublayer of which is specially designed for user-plane radio bearers carrying packet data.

PCDP makes use of RLC-layer services in three different modes. Conceptually, these services are available at three different SAPs:

- Data transfer is acknowledged (AM-SAP).
- Data transfer is unacknowledged (UM-SAP).
- Data transfer in transparent mode (TM-SAP).

PDCP assumes that SAR is taken care of by the RLC layer below it. When lossless SRNS relocation is supported, PDCP uses an AM-SAP and assumes in-sequence delivery of RLC SDUs.

PDCP uses an AM-SAP whenever reliable data transfer is requested. In this case the RLC layer is only allowed to discard some of the RLC SDUs as long as it is configured in such a way, but is expected to be able to indicate the PDCP of each of the discarded SDUs at both the transmitter and receiver. These indications are made necessary by the fact that—as a compression protocol—PDCP avoids sending explicit sequence numbers over the radio and, instead, makes use of a virtual sequence numbering scheme. This makes it further necessary to inform PDCP about possible resetting of RLC connection and radio bearer reconfigurations, especially when lossless SRNS relocation is carried out.

As far as TM-SAP is concerned, the size of PDCP PDUs is expected to be fixed, which means very strict requirements either for the PDU size of the upper layer (e.g., IP protocol) or for applied IP header compression. In a generic case neither the upper layers in the PS domain nor the currently specified IP header compression method can satisfy these requirements and, therefore, TM-SAP should be used very carefully and in well-studied cases only.

A PDCP entity running at the RNC has to interwork with UTRAN frame protocols in order to relay data packets in the downlink direction from Iu to Iub/Iur and in the uplink direction from Iub/Iur towards Iu. In normal cases this is achieved simply by copying packets from one interface to another and performing the specified header compression/decompression along the way. Things get more complicated whenever an intersystem handover or lossless SRNS relocation needs to be performed. PDCP is therefore equipped with special interworking rules with both Iu frame protocols and the RRC protocol, which ensure that packets are not lost or duplicated and that packet sequence numbering can be restored after the handover or relocation at the target system.

The only IP header compression algorithm specified in 3GPP R99 is RFC 2507 (defined by the IETF), but the generic protocol structure of PDCP will allow the use of other compression algorithms in the future. In 3GPP R4/R5 another header compression algorithm called Robust Header Compression (ROHC), as defined in IETF RFC 3095, was added to the PDCP specification.

The complete specification of PDCP is given in 3GPP specification TS 25.323 and more information can also be found in Holma and Toskala (2004).

10.4.2.2 UTRAN Frame Protocols

UTRAN frame protocols are radio network layer user-plane protocols that carry UMTS user data over the common UTRAN transport network (Figure 10.27). They are active across the UTRAN interfaces Iu, Iur and Iub. The transport network service

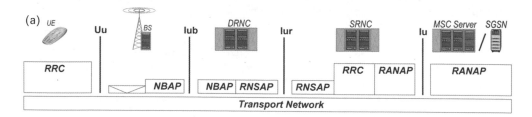

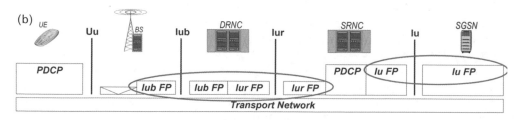

Figure 10.27   Radio network user plane—UTRAN frame protocols: (a) control-plane protocols in the radio network; (b) user-plane protocols in the radio network (PS domain only)

used by frame protocols is based on either ATM or the IP protocol stack, as discussed in Section 10.3.3.

10.4.2.2.1 Frame Protocol at Iu

The frame-handling protocol in the Iu interface has been designed to meet the needs of both CS domain and PS domain user data traffic. The transport services for these two domains are different and, therefore, two modes of Iu frame protocol are used:

- *Transparent mode*, which is used on RABs that do not require any special protocol activities, like framing. Therefore, user data in this mode are simply passed to the transport network stack without adding any protocol header information. Transparent mode hence becomes a null protocol. It can be used for carrying non-real-time data packets in their plain GTP-U format from the RNC to the SGSN and vice versa.
- *Support mode*, which takes the form of a real protocol with control PDUs and framing of user data. These protocol formats are applied to RABs, which can carry AMR-coded speech data in the CS domain. Therefore, support-mode control PDUs can take care of rate control and time alignment, which are needed to support real-time speech transport. In this case the support-mode protocol directly uses the AAL2 service or the RTP session provided by the underlying transport network.

10.4.2.2.2 Frame Protocol for Dedicated Channels at Iur and Iub

There is a common frame protocol for DCHs over the Iub and Iur as illustrated in Figure 10.28. With this protocol the SRNC can exchange user data frames with UEs being serviced by its own as well as by remote BSs. Besides normal uplink and downlink data transfer, this frame protocol is capable of performing many control functions, which deal with timing adjustment and synchronisation. Also outer loop power control

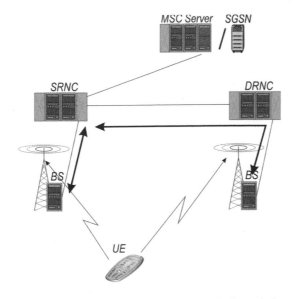

Figure 10.28   Frame protocol for Iub/Iur dedicated channels

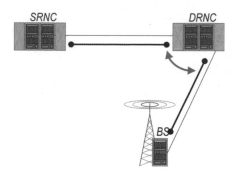

Figure 10.29   Frame protocol for Iub/Iur common transport channels

commands can be sent to BSs in DCH frame protocol messages as well as other updates
to radio interface parameters.

10.4.2.2.3 Frame Protocols for Common Transport Channels at Iur and Iub

For sending and receiving data frames over common transport channels there are
two different frame-handling protocols: one for the Iub and another for the Iur.
Therefore, interworking between these protocols is done at the DRNC as illustrated
in Figure 10.29.

Besides uplink and downlink data transfer, which is to some extent specific to
different transport channel types, these frame protocols also carry out control tasks.
On the Iur, frame-handling flow control based on credits is applied. The frame-handling
functions on the Iub interface include synchronisation and timing adjustment as well as
flow control for the HS-DSCH.

10.5 System Network Protocols

Once radio network protocols make it possible to communicate across the UTRAN
subnetwork by maintaining the communication path to mobile terminals, it is the
system network protocols that then create the communication services to the users of
these terminals. System (network) protocols operate on top of the radio network and
within the UMTS CN itself.

10.5.1 Non-Access Stratum Protocols

The non-access stratum refers to the group of control-plane protocols that control
communication between UEs and the CN. The common name "non-access stratum"
refers to the fact that these protocols are carried transparently through the RAN.

The protocols in this group belong to two sublayers of the system network as shown
in Figure 10.3. The MM sublayer operates on top of the signalling connection provided
by the radio network. Besides its own functionality, the MM sublayer also acts as a
carrier to CM sublayer protocols.

10.5.1.1 MM Protocol for the CS Domain

The MM protocol operates between the UE and the MSC/VLR (Figure 10.30). This control-plane protocol provides basic signalling mechanisms for controlling MM and authentication functions between UEs and the CN CS domain. The MM protocol is supported by UEs operating either in PS/CS operation mode or in CS-only operation mode.

The MM protocol makes use of the signalling connection provided by radio network layer protocols. This signalling connection is sometimes referred to as the Radio Resource (RR) connection, which consists of RRC protocol connection over the signalling radio bearer and RANAP connection over the Iu signalling bearer. Before any MM protocol activity can take place the signalling connection must first be established. Inside the CN the MM protocol entity at the Mobile service Switching Centre (MSC)/VLR end has to interwork with the MAP protocol to maintain location registration in the HLR.

The MM protocol controls the MM state of the UE and the corresponding peer entity at the network end. The major states of these protocol entities indicate the location update status of the UE, which has a major impact on the UE's ability to use the services of the CM sublayer. This is also the reason the MM protocol is used to carry messages of CM sublayer protocols between the UE and the MSC/VLR.

The MM protocol has three basic categories of procedures:

- MM connection procedures.
- MM common procedures.
- MM specific procedures.

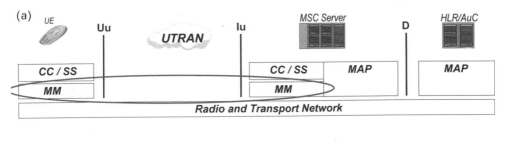

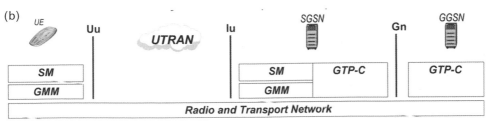

Figure 10.30 System network non-access stratum—MM protocol for the CS domain: (a) control-plane protocols in the system network (CS domain); (b) control-plane protocols in the system network (PS domain)

MM connection procedures are used to establish and release MM connections as well as for transferring CM sublayer messages. The concept of MM connection is established without exchange of any special MM protocol initiation messages. Instead, an MM connection is established and released implicitly on request from a CM entity.

When MM connection establishment is initiated by the UE, the MM protocol entity issues a CM SERVICE REQUEST, which contains the following parameters:

- Identity of the UE (e.g., TMSI).
- Mobile station classmark 2.
- Ciphering key sequence number.
- CM service type identifying the requested type of transaction, which can be mobile originating call establishment, emergency call establishment, SMS or supplementary service activation.

The MM protocol entity receiving this request at the CN end will analyse these parameters and decide on approval and continuation of the service request.

On the other hand, the request to establish MM connection may come from the CM entity at the network end. This is then accomplished by making the UE establish MM connection via the paging procedure.

After MM connection has been established, it can be used by the CM sublayer entity for information transfer. Each CM entity will have its own MM connection. These different MM connections are identified by the Protocol Discriminator (PD) and, additionally, by the Transaction Identifier (TI).

An established MM connection can be released by the local CM entity. The release of the connection will then be done locally in the MM sublayer (i.e., no MM messages are sent over the radio interface for this purpose).

MM common procedures can be initiated at any time while MM connections are active. These procedures are used mainly for security functions when the identity and location of the UE is known by the CN and MM connection exists from the CN to the UE. Mandatory MM common procedures are:

- *TMSI reallocation*, which is used to provide confidentiality of user identity (i.e., to protect a user against being identified and located by an intruder). In this procedure a Temporary Mobile Subscriber Identity (TMSI) is allocated by the MSC/VLR for use within radio interface procedures instead of the permanent and, therefore, traceable IMSI identification. The procedure is initiated by the CN and is usually performed in ciphered mode.
- *Authentication*, which permits the CN to check whether the identity provided by the UE is acceptable or not, and provide parameters enabling the Universal Subscriber Identity Module (USIM) to calculate new ciphering and integrity keys. In a UMTS authentication challenge, the authentication procedure is extended to allow the UE to check the authenticity of the CN (see Chapter 9) (e.g., this allows detection of a false BS). The authentication procedure is always initiated by the CN.
- *Identification*, which is used by the CN to request the UE to provide specific identification parameters to the network, like the International Mobile Subscriber Identity (IMSI) and International Mobile Equipment Identity (IMEI).

- *IMSI detach*, which is initiated by the UE when the UE is deactivated or if the USIM is detached from the Mobile Equipment (ME). Each UMTS network can broadcast a flag within the system information block to indicate whether the detach procedure is required by UEs. When receiving an MM IMSI DETACH INDICATION message, the network may set an inactive indication for this IMSI. No response is returned to the UE. After this the network will locally release any ongoing MM connections, and start the normal signalling connection release procedure.
- *Abort*, which can be used by the CN to abort the ongoing establishment of an MM connection or one that is already established due to network failure or having found an illegal ME.

MM-specific procedures handle messages used in location update procedures. An example message flow of a location update procedure is described in Chapter 11. There are three variants of the basic location update procedure:

- *Normal location updating*, which is used in situations where the UE has discovered that its location area has changed and, therefore, needs to inform the CN about the new location. Detection is done by checking the Location Area Identity (LAI) value broadcast by the network against the current LAI value stored in the USIM. The normal location updating procedure will also be started if the network indicates that the UE is unknown in the VLR in response to a service request. The UE has to maintain a list of "forbidden location areas for roaming" on the USIM based on failed location update requests.
- *Periodic location updating*, which is used to notify the CN periodically of the availability and location of the UE according to a timer in the UE. The timeout value for this timer is taken from the broadcast system information block.
- *IMSI attach*, which is performed when the UE enters (or is switched on in) a network in the same location area where it was previously detached (i.e., when the LAI value broadcast in the current cell is found to be equal to the LAI value on the USIM).

A common, characteristic feature of MM-specific procedures is that they can only be initiated when no other MM-specific procedure is running or when no MM connection exists. Note also that any of the MM common procedures described above (except IMSI detach) can be initiated during an MM-specific procedure.

Since the MM protocol is a direct extension of its GSM predecessor, a reader wanting more information about it should consult any of the GSM textbooks listed in the Bibliography. The complete specification of the MM protocol together with UMTS-specific details is given in 3GPP specification TS 24.008.

10.5.1.2 GPRS Mobility Management (GMM) for the PS Domain

GMM operates between the UE and SGSN (Figure 10.31). The name "GPRS mobility management" is used because the protocol has basically evolved from the corresponding protocol for PS services in the GSM/GPRS system. This control-plane protocol provides basic signalling mechanisms for MM and authentication functions between

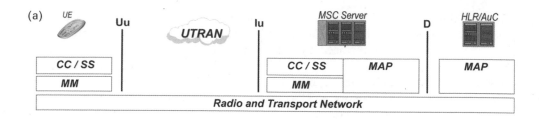

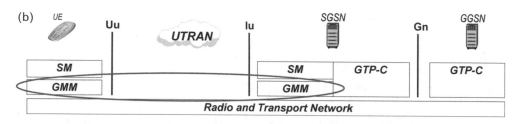

Figure 10.31 System network non-access stratum—mobility management protocol for the PS domain: (a) control-plane protocols in the system network (CS domain); (b) control-plane protocols in the system network (PS domain)

UEs and the CN PS domain. The GMM protocol is supported by UEs working either in PS/CS operation mode or in PS operation mode.

Like the MM protocol, the GMM protocol makes use of a signalling connection between the UE and the SGSN. This signalling connection—which is sometimes called a "PS signalling connection"—is provided by radio network layer protocols using RRC over the signalling radio bearer and RANAP over the Iu signalling bearer. The GMM protocol entity at the SGSN end has to interwork with a GTP-C entity for location information exchange.

It is possible for a UE working in PS/CS operation mode to have signalling connections both with the MSC/VLR and the SGSN, simultaneously. For optimisation reasons it has been preordained that in such circumstances the GMM entities in both the UE and the SGSN can perform so-called "combined" procedures. The idea behind combined GMM procedures is to avoid sending both MM and GMM messages when the CN PS domain can simply inform the CN CS domain about such combined activities inside the CN. The optional Gs interface is used to carry out this action at the CN end.

From the packet data services' point of view, it is important to distinguish between the state when the UE together with its identity and location is known by the CN PS domain and the state when it is not known. The importance of this stems from the fact that packet user data are protected by encryption, effected by user-specific keys. This separation is highlighted by the *GMM context* concept. When the user is known to the CN PS domain network the GMM context exists, otherwise it does not.

GMM procedures for explicit establishment and release of the GMM context are:

- *GPRS attach and combined GPRS attach,* which are used to indicate user identity (typically P-TMSI) and current routing area to the network and to establish a GMM context for this user. This procedure is typically carried out when a packet-mode UE is switched on or the USIM is inserted into it. In the case of the combined GPRS attach procedure, activities for normal IMSI attach are carried out.

- *GPRS detach* or *combined GPRS detach*, which are invoked when the UE is switched off or the USIM is removed from the UE or a network failure has occurred. The procedures cause the UE to be marked as inactive in just the CN PS domain or in the combined GPRS detach procedure as well for the CN CS domain.

The following GMM procedures are used for location management when a GMM context already exists:

- Normal *routing area updating* and *combined routing area updating*, which are performed when the UE recognises that the routing area has changed and the UE needs to inform the CN PS domain about the event. In combined routing area updating the location area is also included for the CN CS domain.
- *Periodic routing area updating*, which can be used to notify the CN periodically of the availability of the UE, according to a timer in the UE.

The following GMM common procedures for security purposes can be initiated when a GMM context for the user has been established:

- *P-TMSI re-allocation*, which provides a packet-TMSI (P-TMSI) for use within radio interface procedures instead of the permanent and, therefore, traceable IMSI identification. Normally, the P-TMSI re-allocation procedure is performed in conjunction with another GMM procedure like routing area updating.
- *GPRS authentication and ciphering*, always initiated by the CN, permits the CN to check whether the identity provided by the UE is acceptable. It also provides parameters enabling the USIM to calculate UMTS ciphering and integrity keys and permits the UE to authenticate the CN. See Chapter 9 for more details about these security functions.
- *GPRS identification*, which is used by the CN PS domain to request the UE to provide the network with specific identification parameters, like IMSI and IMEI.

One more GMM procedure is needed to provide services to the CM sublayer on top of the GMM protocol:

- *Service request*, which is initiated by the UE, is used to establish secure connection with the network and/or to request the bearer establishment to send data. The procedure is typically used when the UE has to send a signalling message that requires security protection (e.g., an SM or paging response message).

Further details about the GMM protocol can be found in 3GPP specification TS 24.008.

10.5.1.3 Call Control (CC) Protocol

The CC protocol provides basic signalling mechanisms for establishing and releasing CS services (e.g., voice call or multimedia call) (Figure 10.32). The CC protocol operates between an MSC/VLR and UEs operating in CS/PS or CS-only mode. The CC protocol uses the connection service provided by the MM sublayer to carry CC protocol messages.

A normal CS voice call is controlled in pretty much the same way as in ISDN telephony. Therefore, the CC protocol entity in the GMSC has to interwork with the ISUP protocol for establishment of a call path to an external ISDN network.

A UMTS CS multimedia call is based on 3G-324M. The 3G-324M is a variant of ITU-T recommendation H.324. For details see 3GPP specification TS 26.111.

Call establishment procedures establish CC connection between the UE and the CN and activate the voice/multimedia codec and interworking functions, if needed. The CC protocol entity in the MSC/VLR also interworks with the RANAP protocol entity in order to establish an RAB for the CS call. This means that the QoS requirements received by the CC protocol entity in the CC SETUP message have to be mapped onto a RAB request for the RANAP protocol entity in the same MSC/VLR. Only after the UTRAN has confirmed the assignment of an RAB can call establishment proceed.

Established calls can be Mobile Originating (MO) or Mobile Terminating (MT). For multimedia CS calls the normal call establishment procedure applies with some extra information elements for CC protocol messages. The call release procedure, initiated by the UE or the CN, is used to release resources for the call.

Call collisions as such cannot occur at the network end, because any simultaneous MO or MT calls are dealt with separately by assigned, different transaction identifiers in the CC protocol.

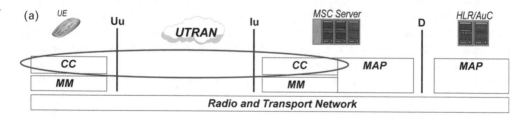

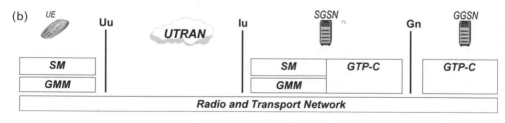

Figure 10.32 System network non-access stratum—CC protocol: (a) control-plane protocols in the system network (CS domain); (b) control-plane protocols in the system domain (PS domain)

CC as a CN function was discussed in Chapter 6. An example of message flows for CS call establishment and release are given in Chapter 11.

Since the CC protocol is a direct extension of its GSM· predecessor, the reader wanting further information about it should consult any of the GSM textbooks listed in the Bibliography. The full specification of the CC protocol together with UMTS-specific details is given in 3GPP specification TS 24.008.

10.5.1.4 Session Management (SM) Protocol

The protocol entities running the SM protocol are located in the UE and SGSN (Figure 10.33). This protocol is a counterpart to the CS CC protocol in the sense that it is used to establish and release packet data sessions. To underline the possibility of using several packet data protocols within the UMTS network these sessions are called "PDP contexts", as discussed in Chapter 6. In practice, the single most important packet data protocol to be supported is IP (especially IPv6).

The main purpose of the SM protocol is to support PDP context handling of the UE. In the SGSN the SM protocol entity interworks with the GTP-C protocol entity at the Gn interface end in order to extend PDP context management to the GGSN. The PDP context contains the necessary information for routing user-plane data packets from the GGSN to the UE, and vice versa. A UE may have more than one PDP context activated simultaneously, in which case each is controlled by an SM protocol entity of its own.

SM procedures require the existence of a GMM context for the identified user. If no GMM context for the user has been established, the MM sublayer has to initiate a GMM procedure to establish the GMM context (see the GMM protocol in Section

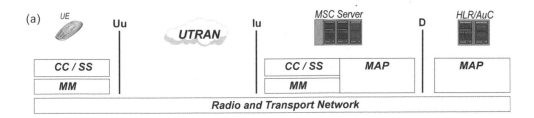

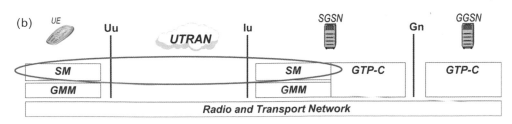

Figure 10.33 System network non-access stratum—SM protocol: (a) control-plane protocols in the system network (CS domain); (b) control-plane protocols in the system network (PS domain)

10.5.1.2). After GMM context establishment, the SM protocol uses services offered by the MM sublayer. Ongoing SM procedures are suspended during GMM procedure execution.

SM procedures can be initiated either by the GGSN or by the UE. The basic SM procedures are:

- *PDP context activation*, which establishes a new PDP context at the UE, SGSN and GGSN with the necessary routing and QoS parameters for the session.
- *PDP context modification*, which is invoked when a change in the session parameters, like QoS, or a traffic flow template is needed during the session. Modification can be initiated either by the UE or the GGSN.
- *PDP context deactivation*, which destroys PDP contexts in UE and CN elements when the corresponding session is no longer needed.

A more detailed description of the SM protocol is given in 3GPP specification TS 24.008.

10.5.1.5 Supplementary Services (SS) Protocol

Supplementary services are additional services that have been defined to be used over CS connections (Figure 10.34). This concept was inherited from the 2G GSM system and, therefore, a UMTS terminal operating in CS mode is able to provide the same supplementary services as a GSM mobile phone.

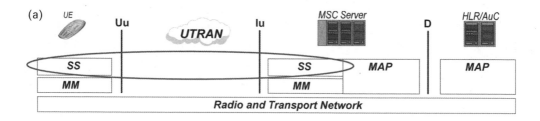

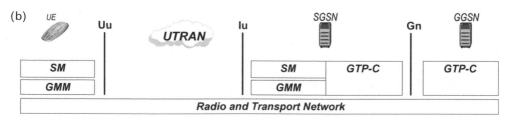

Figure 10.34 System network non-access stratum—SS protocol: (a) control-plane protocols in the system network (CS domain); (b) control-plane protocols in the system network (PS domain)

The 3GPP implementation supports all the SSs as far as CS traffic is concerned. Some typical supplementary services are:

- Call-forwarding services with different triggering conditions: no reply, busy, unconditional, not reachable.
- Call-barring services for all incoming, all outgoing, all outgoing international or roaming calls.
- In-call services like call wait and call hold.

A complete list and description of supplementary services can be found in 3GPP specification TS 24.080.

SSs are controlled by the SS protocol, which is part of the CM sublayer. SS protocol entities at the UE and MSC/VLR end communicate by using an MM connection. This communication has to do with activation and deactivation of SSs. At the CN end the SS protocol interworks with the MAP protocol, which is used whenever the SS command requires, say, access to subscriber information in the HLR.

10.5.2 Control Plane between CN Nodes

Section 10.5.1 briefly introduced the control-plane protocols in the non-access stratum between the UE and the CN. CN entities also have to perform some signalling functions with each other (especially with the registers located in the home network part of the CN).

The MAP and CAP protocols have already been mentioned in Section 10.1.3: a reader wanting more detailed descriptions of these protocols should consult the GSM textbooks listed in the Bibliography. The 3G version of the MAP protocol is defined in 3GPP specification TS 29.002.

10.5.2.1. GTP-C—GPRS Tunnelling Protocol for the Control Plane

GTP-C belongs to the PS domain in the CN and is located in the SGSN and the GGSN (Figure 10.35). GTP-C is a control-plane protocol specifying tunnel management and control procedures so that the SGSN and GGSN can undertake user data packet transfer. GTP-C is also used to transfer GMM signalling messages between SGSNs, which corresponds to the MAP/G interface at the CS end.

GTP-C operates on two different interfaces, Gn and Gp. The Gn interface is shared between GSNs (i.e., SGSNs and GGSNs) belonging to the UMTS PS domain network. The Gp interface is used for UMTS network interworking between two UMTS PS domain networks (see Figure 6.4).

On both the Gn and Gp interfaces GTP-C operates on top of the UDP/IP protocol family. UDP provides connectionless message transfer (i.e., there is no need to establish a connection). The main service provided by IP is message routing between source and destination network elements.

In the SGSN the GTP-C protocol entity not only interworks with the GTP-U protocol entity at the Gn end but also with GMM and SM protocol entities. In the

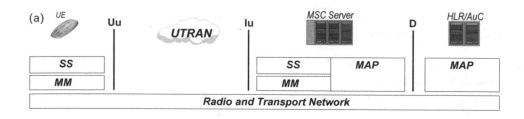

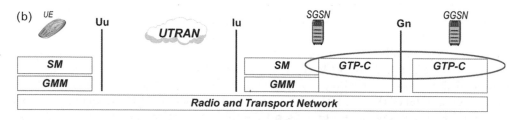

Figure 10.35 System network control plane—GPRS Tunneling Protocol (GTP-C): (a) control-plane protocols in the system network (CS domain); (b) control-plane protocols in the system network (PS domain)

GGSN the GTP-C entity interworks with the GTP-U protocol entity located at the Gn interface and with the external network interworking function at the Gi interface (e.g., to allocate dynamic addresses for PDP contexts).

The GTP-C protocol has to be able to transfer its signalling messages reliably over the Gn and Gp interfaces. Therefore, its design had to solve the problem of unreliable message transfer provided by UDP/IP transport layers. Thus, GTP-C includes the timers and message sequence numbers needed to control and detect message loss and message retransmission.

10.5.2.1.1 GTP-C Procedures between the SGSN and GGSN

GTP-C is used to create, delete and modify those GTP tunnels that are used to transfer user data packets. GTP-C also checks whether the path toward the peer GTP-C is alive.

When a PDP context is activated, GTP-C in the SGSN initiates GTP tunnel establishment towards the GGSN. The purpose of the GTP tunnel establishment procedure is to negotiate parameters related to user data transfer between the SGSN and GGSN: such parameters are QoS and those associated with the activated address, like the traffic flow template.

When a PDP context is deactivated, GTP-C removes GTP tunnelling information from the SGSN and GGSN.

The PDP context modification procedure provides a flexible and dynamic way to react to changed conditions. When a PDP context is modified (e.g., due to changed QoS attributes or load conditions), GTP-C initiates a GTP tunnel modification procedure, in which both the SGSN and the GGSN are able to modify almost all the parameters negotiated during the GTP tunnel establishment procedure. The PDP context modification procedure is also performed as part of the inter-SGSN routing area update procedure and SRNS relocation procedure, which involves two SGSNs notifying the GGSN of the new SGSN to serve the UE.

As far as the network is concerned, requested PDP context activation is supported, but the GGSN does not include the necessary protocols to communicate with the HLR across the Gc interface; instead, GTP-C is used to transfer messages to another GSN which converts messages between GTP and MAP. The MAP protocol is then used to communicate across the Gc interface towards the HLR. The reasons a GGSN needs to communicate with the HLR are to discover whether the UE is attached to the network and to find the address of the SGSN serving the UE. This information is needed by the GGSN before it can request the SGSN to initiate a PDP context activation procedure.

10.5.2.1.2 GTP-C Procedures between SGSNs

GTP-C is used in the communication between two SGSNs, to exchange information that relates to a UE. Communication is initiated when the UE makes either a GPRS attach procedure or an inter-SGSN routing area update procedure or when UTRAN decides to perform the SRNS relocation procedure involving two SGSNs.

The use of GTP-C in exchanging UE-related information between SGSNs minimises the need to use a radio interface. For example, when an inter-SGSN routing area update procedure or an SRNS relocation procedure involving two SGSNs is performed, the new SGSN serving the UE receives almost all the necessary information about the UE from the old SGSN. Such information consists of GMM-related parameters, like IMSI and GMM context, and SM-related parameters, like active PDP contexts.

The use of GTP-C in the exchange of UE-related information between SGSNs also strengthens system security. For example, when a UE undertakes a GPRS attach procedure, the IMSI may be retrieved from the old SGSN and not from the UE over the radio interface.

The GTP-C protocol is defined in 3GPP specification TS 29.060.

10.5.3 The User Plane in the System Network

As far as the CS domain is concerned, the role of user-plane protocols is taken by the speech codec functions that are active in the UE and serving MSC/VLR (Figure 10.36). The predefined speech codec is the Adaptive Multi Rate (AMR) codec. Besides transcoding, interworking functions (already discussed in Chapter 6) have to be taken care of between the UMTS and the external PSTN/ISDN network.

As for the PS domain, the UE and GGSN exchange data packets according to some point-to-point or multicast packed data protocol. It seems likely that the most common user-plane protocol will continue to be the IP protocol.

10.6 Summary of UMTS Network Protocols

Having now finished our exploration of all the major UMTS network protocols, we can summarise the overall composition of PS domain protocol stacks (see Figures 10.37 and 10.38).

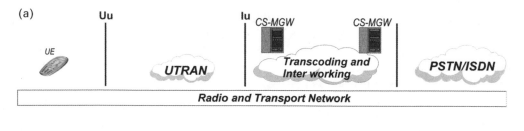

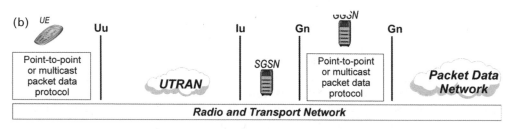

Figure 10.36 User plane in the system network: (a) user plane in the system network (CS domain); (b) user plane in the system network (PS domain)

Figure 10.37 illustrates the control-plane protocol stacks in all PS domain network elements. It shows those control-plane protocols that are active during a typical system network-level signalling procedure (e.g., PDP context activation). The signalling bearer is established and maintained by the UTRAN control protocols—RANAP, RNSAP and NBAP—and the RRC protocol, which interwork with each other in order to carry out distributed radio resource management. System network-level control signalling (e.g., SM protocol messages) is relayed through UTRAN elements within RRC signalling messages, which further on are embedded into RLC/MAC and Dedicated Channel Frame Protocol (DCH-FP) messages. Relaying of RLC/MAC messages is illustrated in Figure 10.37 by showing two DCH-FP protocol stacks at the CRNC element. Transport may be realized across all UTRAN interfaces either by IP or the ATM transport network, as discussed earlier.

At the CN end the SM protocol interworks with the GTP-C protocol to control the PDP context. In a similar way the GMM protocol interworks with the MAP protocol to carry out mobility management for the UE. The GTP-C protocol runs on top of IP transport, but the MAP protocol can run on top of either ATM or IP transport.

Figure 10.38 illustrates the PS user-plane protocol stacks in all PS domain network elements during a typical packet data session (e.g., when running any of the IMS procedures). The user IPv6 packet flow between the UE and the GGSN is carried by UTRAN user-plane protocols and by the GTP-U tunnelling protocol. At the SRNC the GTP tunnelling protocol interworks with PDCP and user-plane RLC protocols to adapt IP packets for transmission over the WCDMA radio interface.

In Figure 10.38 we assume that packet data flow makes use of the HSDPA enhancement of WCDMA radio transmission. Therefore, MAC-d messages are transported by the High Speed Downlink Shared Channel Frame Protocol (HS-DSCH-FP or HS-FP

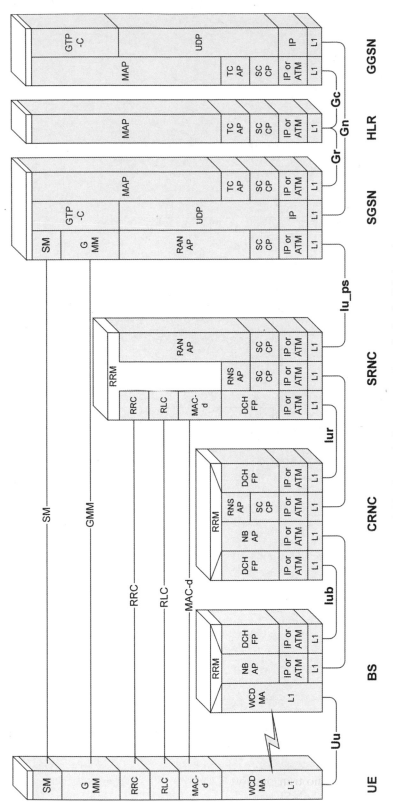

Figure 10.37 Control-plane protocol stacks in the PS domain

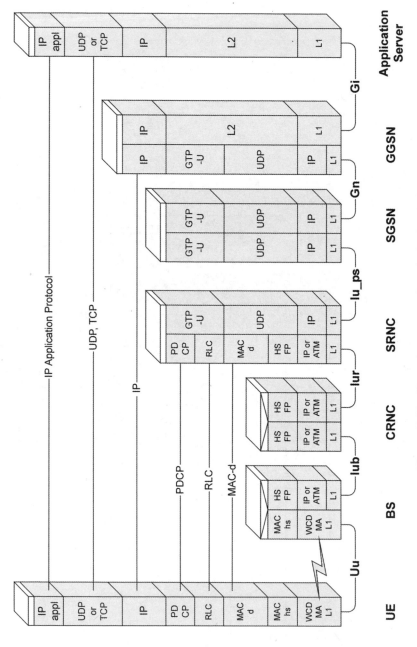

Figure 10.38 User-plane protocol stacks in the PS domain

for short) from the SRNC to the BS, where the actual high-speed channel is maintained by the MAC-hs protocol across the Uu interface.

On the PS CN the GTP tunnel makes use of IP transport across Iu as well as CN interfaces. IP packets are transported over the Gi interface according to the IP transport stack selected for IMS transport.

10.7 Overview of IMS Protocols

This section briefly describes the main protocols used by the IMS and lists their usage over various reference points. At the time of writing, Release 6 work is still ongoing and, therefore, Table 10.3 can only summarise stable IMS reference points and list their associated protocols.

3GPP chose to use a single session control protocol, Session Initiation Protocol (SIP), as defined by RFC 3261. SIP is an application-layer protocol that is used for establishing, modifying and terminating multimedia sessions in an IP network. It is part of the multimedia architecture whose protocols are continuously being standardised by the IETF. Its applications include, but are not limited to, voice, video, gaming, messaging, call control and presence. The base SIP protocol itself is not sufficient for IMS system operations. The IMS requires the support of many other SIP-related RFCs (2327, 3262, 3264, 3265, 3311, 3312, 3313, 3325, 3327, etc.). For more information, please refer to 3GPP TS 24.229 or Poikselkä et al. (2004).

For database queries and charging purposes a protocol other than SIP is needed. 3GPP chose to use Diameter when IMS entities need to communicate with the HSS or charging entities.

Diameter is the Authentication, Authorisation and Accounting (AAA) protocol developed by the IETF. It is used to provide AAA services for a range of access technologies. Instead of being built from scratch, Diameter has its roots loosely embedded in the Remote Authentication Dial In User Service (RADIUS), which is defined in RFC 2865 and has been used to provide AAA services in the dial-up and terminal server access environments. Diameter actually includes a mode in which backward-compatible operation with RADIUS is possible.

Diameter is actually split into two parts: the Diameter base protocol and Diameter applications. The base protocol is needed for delivering Diameter data units, negotiating capabilities, handling errors and providing for extensibility.

A Diameter application defines application-specific functions and data units. Each Diameter application is specified separately. Currently, in addition to the base protocol defined in RFC 3588, there are a few Diameter applications that either have been defined or are in the throes of being defined. In addition to this, the AAA transport profile (defined in RFC 3539) includes discussions and recommendations on the use of transports by AAA protocols, and 3GPP Release 5 has also been allocated a specific set of Diameter command codes (defined in RFC 3589).

The Diameter base protocol uses both TCP and SCTP as transport. However, SCTP is preferred, because of the connection-oriented relationship that exists between Diameter peers. Both IP Security (IPSec) (RFC 2401) and Transport Layer Security (TLS) (RFC 2246) are used to secure these connections.

Table 6.3 Summary of IMS reference points

Name of reference point	Involved entities	Purpose	Protocol
Gm	UE, P-CSCF	This reference point is used to exchange messages between the UE and CSCFs	SIP
Mw	P-CSCF, I-CSCF, S-CSCF	This reference point is used to exchange messages between CSCFs	SIP
ISC	S-CSCF, I-CSCF, AS	This reference point is used to exchange messages between the CSCF and AS	SIP
Cx	I-CSCF, S-CSCF, HSS	This reference point is used to communicate between the I-CSCF/S-CSCF and HSS	Diameter
Dx	I-CSCF, S-CSCF, SLF	This reference point is used by I-CSCF/S-CSCF to find a correct HSS in a multi-HSS environment	Diameter
Sh	SIP AS, OSA SCS, HSS	This reference point is used to exchange information between SIP AS/OSA SCS and HSS	Diameter
Si	IMS-SF, HSS	This reference point is used to exchange information between the IMS-SF and HSS	MAP
Dh	SIP AS, OSA, SCF, IMS-SF, HSS	This reference point is used by AS to find a correct HSS in a multi-HSS environment	Diameter
Mm	I-CSCF, S-CSCF, external IP network	This reference point will be used for exchanging messages between the IMS and external IP networks	Not specified
Mg	MGCF → I-CSCF	MGCF converts ISUP signalling to SIP signalling and forwards SIP signalling to the I-CSCF	SIP

	Entities	Description	Protocol
Mi	S-CSCF → BGCF	This reference point SIP is used to exchange messages between the S-CSCF and the BGCF	SIP
Mj	BGCF → MGCF	This reference point SIP is used to exchange messages between the BGCF and MGCF in the same IMS network	SIP
Mk	BGCF → BGCF	This reference point SIP is used to exchange messages between BGCFs and different IMS networks	SIP
Mr	S-CSCF, MRFC	This reference point SIP is used to exchange messages between the S-CSCF and MRFC	SIP
Mp	MRFC, MRFP	This reference point SIP is used to exchange messages between the MRFC and MRFP	H.248
Mn	MGCF, IMS-MGW	This reference point allows control of user plane resources	H.248
Ut	UE, AS (SIP AS, OSA SCS, IMS-SF)	This reference point enables the UE to manage information related to his or her services	HTTP
Go	PDF, GGSN	This reference point allows operators to control QoS in a user plane and exchange charging correlation information between the IMS and the GPRS network	COPS
Gq	P-CSCF, PDF	This reference point is used to exchange policy-decision-related information between the P-CSCF and PDF	Diameter
Ro	AS, MRCF, S-CSCF, OCS	This reference point is used by AS/MRFC/S-CSCF for online charging towards the OCS. *Note*: there might exist an inter-working function between the S-CSCF and OCS	Diameter
Rf	P-CSCF, S-CSCF, I-CSCF, BGCF, MGCF, AS, MRFC, CCF	This reference point is used by IMS functions for offline charging towards the CCF	Diameter

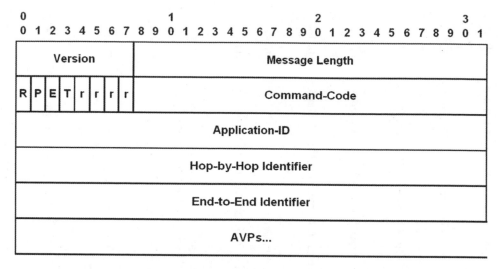

Figure 10.39   Diameter header structure

Diameter is a peer-to-peer protocol, as a result of any Diameter node being able to initiate a request. Diameter has three different types of network nodes: clients, servers and agents. Clients are generally a network's edge devices that carry out access control. A Diameter agent provides either relay, proxy, redirect or translation services. A Diameter server handles AAA requests for a particular domain or realm. Diameter messages are routed according to the Network Access Identifier (NAI) (RFC 2486) of a particular user. Diameter messages consist of a Diameter header, followed by a number of Diameter Attribute Value Pairs (AVPs). The header comprises binary data and, as such, resembles an AP header or a TCP header (see Figure 10.39).

AVPs contain encapsulated AAA information elements, as well as the routing, security and configuration information elements that are relevant to the particular Diameter request or answer message.

In 3GPP's IMS, Diameter is used for location and service management, user data handling, user authentication and authorization in reference points Cx, Dx, Sh and Dh, as well as for offline and online accounting in reference points Rf and Ro.

For IP policy control purposes the Common Open Policy Service (COPS) (RFC 2748) is used. COPS is an IETF protocol used for the general administration, configuration and enforcement of policies. It defines a simple query and response protocol for exchanging policy information between a policy server (such as a Policy Decision Function, or PDF) and its clients (GGSN). The reference point between PDF and GGSN uses COPS to perform media authorisation and charging correlation. It also employs COPS usage for policy provisioning extensions (RFC 3048). Moreover, 3GPP has defined something called a policy information base, which identifies policy provisioning data. To transmit IMS policy-control-specific data, 3GPP has defined its own policy information base, the details of which can be found in 3GPP TS 29.207.

Table 10.3 depicts the IMS architecture in a simplified form. For the sake of clarity, it is impossible to include everything in one table, so please note that the table does not show:

- Charging-related functions or reference points.
- The different types of Application Servers (ASs).
- The user-plane connections between different IMS networks and the AS.

Figure 10.40 illustrates the IMS system architecture with the reference points. However, for the sake of clarity, some details have been omitted:

- The figure does not show the SEG at the Mm, Mk, Mw reference points.
- The dotted line between the entities indicates a direct link.
- ISC, Cx, Dx, Mm, Mw terminate at both the S-CSCF and the I-CSCF.

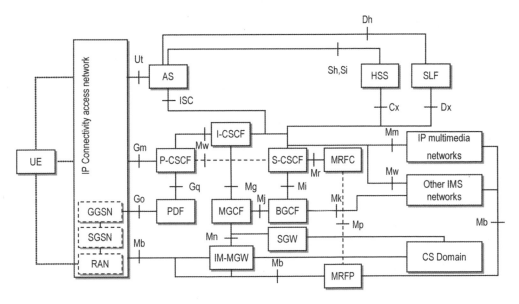

Figure 10.40   IMS architecture

11

Procedure Examples

Heikki Kaaranen, Ari Ahtiainen and Siamäk Naghian

For the reader to get a network-wide view of some of the most important Universal Mobile Telecommunication System (UMTS) functionalities discussed in the preceding chapters we present a few examples of UMTS system procedures. These procedures illustrate the way in which the actions carried out by all network entities are coordinated as well as the role of UMTS protocols in controlling this coordination.

11.1 Elementary Procedures

In this chapter a basic model for system-wide procedures is presented by modelling each system procedure as a kind of communication-intensive transaction. The concept of transaction is used to underline the fact that these system-wide procedures are executed into well-defined completion and that subsequent procedures represent fairly independent instances of communication.

The layering and interworking of protocols as described in Chapter 10 allows us to identify a set of elementary procedures that can be used as building blocks in the design of network transactions. Depending on the transaction and its type—whether it is Mobile Terminating (MT) or Mobile Originating (MO)—some parameters and messages are known to vary within elementary procedures and certain elementary procedures may or may not be used.

The UMTS protocol interworking model (see Figure 10.4) is useful here in the sense that different procedures make use of different layers in the UMTS network. The transport network is used in every procedure: signalling transport is always required and user data transport is required by many transactions. Radio network functions are needed whenever access network services are required within an elementary procedure. The overall control of system transactions is carried out by system network protocols, which invoke the elementary procedures in a stepwise manner and determine how the transaction proceeds through the different steps.

Basically, any network transaction can be divided into the eight steps presented in Figure 11.1. For each step an elementary procedure can be distinguished.

UMTS Networks Second Edition   H. Kaaranen, A. Ahtiainen, L. Laitinen, S. Naghian and V. Niemi
© 2005 John Wiley & Sons, Ltd   ISBN: 0-470-01103-3

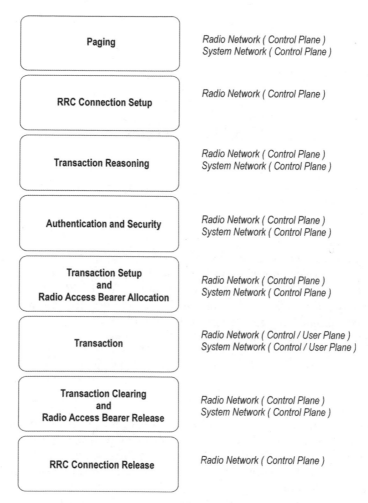

Figure 11.1   Basic model of UMTS network transactions

Paging is a Mobility Management (MM) procedure that is used when searching for a given subscriber within the network coverage area. This procedure is only executed if the transaction originates from the network side. The remaining seven steps are the same irrespective of whether the transaction is originated or terminated by the User Equipment (UE).

Radio Resource Control (RRC) connection setup is an elementary procedure containing both activities and a message flow to establish radio control connection between the terminal and Radio Access Network (RAN). The details of this procedure vary depending on the case.

Transaction reasoning is an elementary procedure in which the terminal indicates to the Core Network (CN) the kind of transaction requested. Based on reasoning information the CN may decide whether to proceed with the transaction or decide to terminate execution of the transaction.

After these steps, *authentication and security procedures* normally take place. This elementary procedure mutually authenticates the UMTS subscriber and the network and later activates the necessary security mechanisms for access network connection.

Transaction setup and Radio Access Bearer (RAB) allocation is an elementary procedure that allocates the actual communication resources for the transaction. Therefore, the details of this procedure depend on the type of transaction (circuit-switched or packet-switched).

The elementary procedure *transaction* is the phase when there exists a user-plane connection (i.e., the UE has a UMTS bearer connection active across the whole UMTS network).

Transaction clearing and RAB release is an elementary procedure used for releasing network resources related to the transaction.

RRC connection release is an elementary procedure containing mechanisms by means of which the radio control connection between the UE and the access network can be released.

11.1.1 Paging

The paging procedure is needed whenever a UE-terminating transaction is to be carried out. Unlike the case for the Global System for Mobile Communications (GSM), the UMTS contains two types of paging methods, Paging Type 1 and Paging Type 2. Paging Type 1 is used by CN domains and is the "conventional" way to use paging (see Figure 11.2). Paging is part of the radio network control plane and is delivered over the Iu Interface by using a RANAP PAGING message. The CN domain that initiates paging addresses it to those Location Areas/Routing Areas (LAs/RAs) to which the desired UE has last reported its location (i.e., by carrying out either location update or routing area update).

A RANAP PAGING message contains two mandatory parameters, the requesting CN domain and an International Mobile Subscriber Identity (IMSI). From a system security point of view, the unnecessary transfer of IMSI over the Uu reference point without encryption is not desired and, because of this, the Radio Network Controller (RNC) may or may not perform a number of conversions for the IMSI. For example, the RNC may issue an RRC PAGING TYPE I message containing a Radio Network Temporary

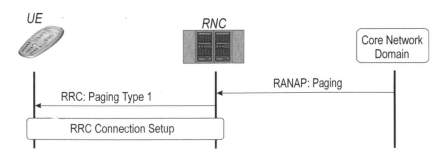

Figure 11.2   Paging Type 1

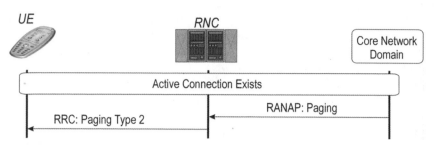

Figure 11.3   Paging Type 2

Identity (RNTI) instead of an IMSI. This value is assigned when the UE is attached to the network or has carried out the previous transaction.

Typically, receipt and recognition of an RRC PAGING TYPE I message by the UE leads to establishment of RRC connection, necessary to carry on the transaction, which was initiated by paging. In addition to the previously described use, RRC PAGING TYPE I has some RAN-specific uses. When the UE is not in idle mode and changes in system information parameters need to be made known to the UE, the RNC may use RRC PAGING TYPE I to "awaken" the UE and, thus, make it able to read revised system information or certain parts of it. If the UE has a Packet Switched (PS) connection in which no activity is taking place the RNC uses RRC PAGING TYPE I to inform the UE that the PS connection should be reactivated. In this case there are some packet data the CN needs to deliver to the UE. This forces the UE to perform cell update, after which the RNC is able to route data packets to the correct UE.

UMTS terminals are meant to be able to handle a number of connections simultaneously (i.e., they can have more than one connection at a time with CN domains). When the UE already has a connection with one CN domain and another connection is established with the same domain the CN sends RANAP PAGING as in previous cases. The paging sent to the UE by the Serving RNC (SRNC) is now RRC PAGING TYPE 2 (see Figure 11.3).

Separation between Paging Type 1 and Paging Type 2 is done in the RNC, where RANAP paging always looks the same. The difference between Paging Type 1 and Paging Type 2 is that the former is used for paging those UEs that are in idle mode, Cell Paging Channel (Cell_PCH) state or UMTS Terrestrial Access Network (UTRAN) Registration Area Paging Channel (URA_PCH) state and, hence, could either be used for certain UEs or for all UEs within cell(s). Paging Type 2, on the other hand, is used for paging those UEs that are in Cell Dedicated Channel (Cell_DCH) state or Cell Forward Access Channel (Cell_FACH) state and, therefore, is always dedicated and addressed to just the one UE.

11.1.2 RRC Connection Setup

Figure 11.4 illustrates the principle involved in establishing radio connection between the UE and the RNC over the Uu interface and the access domain internal interface Iub. RRC connection setup always starts at the UE with the message RRC CONNECTION

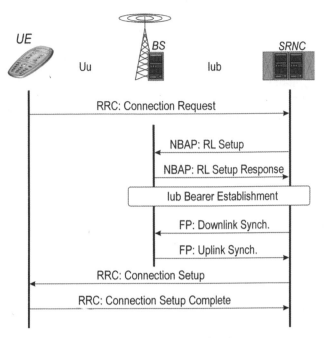

Figure 11.4   RRC Connection Setup

REQUEST sent over the Common Control Channel (CCCH). Recall from Chapter 4 that the CCCH in the uplink direction means the Random Access Channel (RACH) and that RRC CONNECTION REQUEST content actually comes from the UE over physical RACH as a result of the initial access procedure. This message reaches the RNC over the Iub RACH data port. Upon receiving this message the RRC entity at the RNC end changes its state from IDLE to CONNECTED Cell_FACH or CON-NECTED Cell_DCH. Then the RNC communicates with the UE over common control channels (of these, FACH and RACH are respectively used).

The RRC CONNECTION REQUEST message contains a lot of information related to requested radio connection, terminal and subscriber identity; in this message the UE sends to the network such information as the IMSI or Temporary Mobile Subscription Identity (TMSI), International Mobile Equipment Identity (IMEI), Location Area Identity (LAI) and Routing Area Identity (RAI). The RRC CONNECTION REQUEST must indicate how many of these values are inserted in the message and, accordingly, contain these values as well. In addition to these identities the RRC CONNECTION REQUEST contains the reason radio connection is requested. There are various reasons and the following list shows some of them:

- Originating conversational call.
- Originating streaming call.
- Originating interactive call.
- Originating background call.
- Terminating conversational call.

- Terminating streaming call.
- Terminating interactive call.
- Terminating background call.
- Emergency call.
- High-priority signalling.
- Low-priority signalling.
- Call re-establishment.

As can be seen from the list, the RRC CONNECTION REQUEST already indicates what kind of Quality of Service (QoS) will be requested when the transaction proceeds. An emergency call is itemised because it will be treated differently over the network than under normal transaction conditions. If the subsequent transaction will just be signalling, this is indicated, too.

Depending on the reason for the request, the RNC makes a decision about whether to allocate dedicated or common resources to this transaction. Based on this decision, the SRNC allocates RNTI and other resources to this transaction. The Iub interface is "opened" when the RNC sends an NBAP RADIO LINK SETUP message to the BS. This message contains the transport format description, power control information and code information, the latter being the uplink scrambling code for WCDMA-FDD. The Base Station (BS) acknowledges this message by sending an NBAP RADIO LINK SETUP RESPONSE message. This message informs the RNC about transport-layer addressing information (AAL2 address) and provides some reference information for establishment of an Iub bearer in the transport network.

The SRNC starts Iub bearer establishment according to information received from the BS. This procedure is carried out by the internal control plane of the transport network at the Iub interface. The established Iub bearer is bound together with the DCH assigned to the transaction. Afterwards, the frame protocol connection over the Iub is synchronised with message exchange. If we suppose that a dedicated radio connection using DCH is set up here, then it is the DCH Frame Protocol (DCH-FP) for the Iub. When Iub communication is ready, the RNC sends the RRC CONNECTION SETUP message to the UE over common control channels (FACH in this case). In this message the SRNC informs the UE of the transport format, power control and codes, which in the case of Wideband Code Division Multiple Access Frequency Division Duplex (WCDMA-FDD) is a downlink scrambling code. The UE confirms RRC connection establishment by sending the message RRC CONNECTION SETUP COMPLETE.

11.1.3 Transaction Reasoning

When RRC connection has been established, the UE sends the message RRC INITIAL DIRECT TRANSFER, which carries in its payload the first system network message of the transaction from the UE to the network (Figure 11.5).

Upon receiving this information the RNC adds some more parameters to the message and forwards this combination to the appropriate CN domain as a RANAP UE INITIAL message. This message contains in its payload the contents of the original RRC INITIAL DIRECT TRANSFER message together with the first system network message

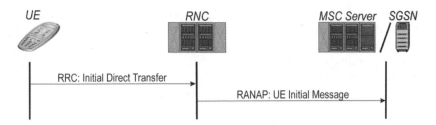

Figure 11.5 Transaction reasoning

generated by the UE. The contents of the RANAP UE INITIAL message provide the CN
with a lot of information about the transaction initiated by the UE. Among such
information is the (claimed) identity of the UMTS subscriber (TMSI or IMSI), his
or her current location area and the kind of transaction requested. All this information
is used by the CN node to decide how best to proceed with the transaction request.

11.1.4 Authentication and Security Control

In Chapter 9 we discussed security functions and algorithms. Figure 11.6 shows the
message flow related to access network security in the context of a transaction.

During RRC connection establishment the UE has already informed the RNC, using
the UE classmark parameter, about many of its capabilities, one of which being the
security algorithms it supports.

The UE and the network authenticate each other by sending the MM AUTHENTICA-
TION REQUEST message in the payload of RANAP AND RRC DIRECT TRANSFER messages
to the UE. After executing the authentication algorithms on the Universal Subscriber
Identity Module (USIM), the UE responds with an MM AUTHENTICATION RESPONSE

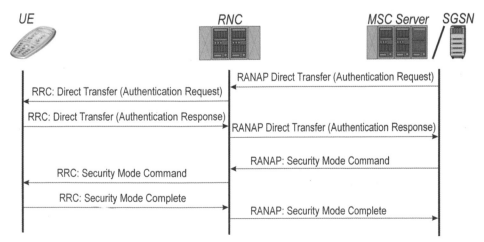

Figure 11.6 Authentication and security control

message, again in the payload of RRC AND RANAP DIRECT TRANSFER messages. In this dialogue the RNC acts as a relay that forwards the contents of RANAP DIRECT TRANSFER to RRC DIRECT TRANSFER and vice versa.

With a RANAP SECURITY MODE COMMAND message the related CN domain indicates to the UTRAN whether the transaction should be encrypted. This message indicates the selected security algorithms and delivers the integrity and encryption keys to the UTRAN.

Based on this information the RNC commands the UE to start encryption using the corresponding keys and algorithms by sending the RRC SECURITY MODE COMMAND. By issuing an RRC SECURITY MODE COMPLETE message the UE indicates that it has successfully turned on the selected integrity protection algorithm and the encryption algorithm, thus protecting further communication in this transaction.

The RNC still has to inform the related CN domain about procedure completion by sending the message RANAP SECURITY MODE COMPLETE.

11.1.5 Transaction Setup with Radio Access Bearer (RAB) Allocation

Up to now in our discussion, the contents of elementary procedures have been similar and, thus, independent of CN domain. Transaction setup with RAB allocation is the first step where the different nature of CN domains has to be taken into account.

In CS transactions the actual setup information is delivered through RRC/RANAP DIRECT TRANSFER messages. These messages can carry in their payload the CC SETUP message, as shown in Figure 11.7. This message identifies the transaction and indicates

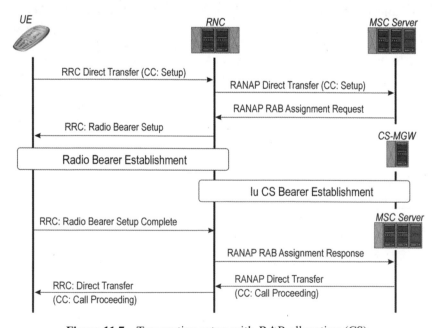

Figure 11.7   Transaction setup with RAB allocation (CS)

the QoS requirements; in other words, the type of bearer required for the service and the parameters it should contain:

- Transaction Identifier (TI).
- Stream identifier.
- Traffic class.
- Asymmetry indicator.
- Maximum bit rate.
- Guaranteed bit rate.

By means of the TI value the UE and the CN node are able to separate each other's calls. Each call has its own TI value. The stream identifier recognises the bearer used for this call. If the stream identifier value does not yet exist, this means that the CC protocol in the UE wants to establish a new bearer. The stream identifier value could be in use already, in which case the CC protocol will want to establish a multicall using an existing bearer (given by the stream identifier).

Upon receiving the CC SETUP message the MSC Server bursts into action. First, it checks whether the UE and current subscription are allowed to perform the requested operation. If yes, the MSC server starts RAB allocation by assigning a unique RAB ID and requesting the setup of an RAB with the requisite QoS parameters by sending the message RANAP RAB ASSIGNMENT REQUEST over the Iu interface.

When the RNC receives the RANAP RAB ASSIGNMENT REQUEST it commences radio bearer allocation by first checking whether there are enough resources available to satisfy the requested QoS. If so, the radio bearer is allocated according to the request. If not, the RNC may select either to continue allocation with lowered QoS values (e.g., the maximum bit rate for the bearer could be lowered) or it may choose to queue the request until radio resources become available. These special cases require some additional signalling to that shown in Figure 11.7.

The RNC informs the UE about bearer allocation by sending the message RRC RADIO BEARER SETUP to the UE. When the UE receives this message, it is then in a position to combine the information it originally sent to the network in its CC SETUP message with the received radio bearer identifier. Afterwards, the UE can route user data traffic (user plane) to the correct radio bearer. As soon as the UE is able to receive data from the new radio bearer it acknowledges this by sending the RRC RADIO BEARER SETUP COMPLETE message to the RNC. The RNC must also establish an Iu bearer for the new transaction. After this the RNC indicates that an RAB has now been allocated by sending the message RANAP RAB ASSIGNMENT RESPONSE to the MSC server. If the RNC made any changes to the QoS values requested by the MSC server, they are indicated in this message. Now, the RAB is established and the procedure continues at the CC protocol level.

When an RAB is to be established for a PS transaction, the procedure looks pretty much the same (see Figure 11.8) and any differences are within the messages. Instead of the CC protocol the UE now uses the Session Management (SM) protocol. The UE sends the SM ACTIVATE PDP CONTEXT REQUEST message to the network as an RRC DIRECT TRANSFER MESSAGE and the RNC relays this as a RANAP DIRECT TRANSFER message to the SGSN in the CN. Radio bearer allocation is similar to that of CS

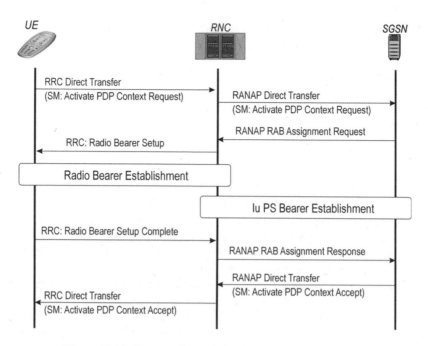

Figure 11.8   Transaction set-up with RAB allocation (PS)

traffic except that the parameters describing the QoS properties of the bearer are different. For example, guaranteed bit rate is an essential parameter for a PS RAB but is not significant for CS RABs. After an RAB has been established the CN domain confirms packet session establishment by sending an SM ACTIVATE PDP CONTEXT ACCEPT message. In terms of SM this means that the SM state is now active and that the UE and the CN domain are able to exchange PS data.

11.1.6 Transaction

In the most typical cases this phase of the transaction contains an active user-plane connection. Examples of this kind of transaction are given later in this chapter (e.g., see Section 11.4). However, if RRC connection was requested simply for a signalling purpose (e.g., an MM activity), no user-plane bearers are needed. Instead, a signalling connection is used to perform the MM activity (e.g., as shown in Section 11.3.3).

11.1.7 Transaction Clearing and RAB Release

If a transaction has an active user plane, then the user plane is disconnected first. The disconnect procedure depends on the transaction type. Figure 11.9 shows how a CS transaction is cleared. In this case, clearing is started by the UE, but the CN can also

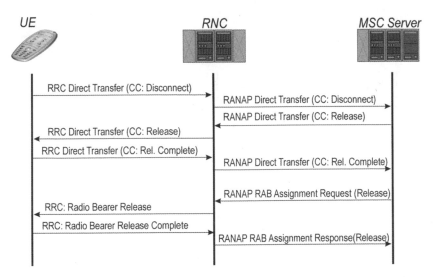

Figure 11.9   CS transaction clearing signalling with RAB release

start user-plane disconnection. More generally, either party can clear a normal CS call, and this does not depend on which party initiated the call.

When the user plane has been disconnected, the system releases the RAB by means of a separate subprocedure. When this is done using the RANAP RAB ASSIGNMENT REQUEST and RANAP RAB ASSIGNMENT RESPONSE, which indicate release as shown in Figure 11.9, signalling connection still remains. The UE still has an RRC connection to the UTRAN and any other RABs for the same UE may still exist. In the general case, when there is a single RANAP RAB Assignment Request message, the CN may cause a number of RABs to be created, deleted or modified.

If all the RABs between the UE and CN domain are released at the same time and the related radio resources are deallocated as well, this can be done by using a RANAP IU RELEASE command, as shown in Figure 11.10. Upon receiving this message the RNC initiates RRC RADIO BEARER RELEASE over the Iub interface. In addition to this, all possible radio bearers established through Drifting RNCs (DRNCs) are released. When the radio bearers are released, RRC connection is also released. When all the radio bearers and the RRC connection have been released, the RNC informs the CN domain with a RANAP IU RELEASE COMPLETE message.

In PS connections, the user plane and RAB(s) are cleared in the same way as done for CS connections. Again, any differences are at the message parameter level. Figure 11.11 illustrates the case where PS connection is closed by deactivation of the Packet Data Protocol (PDP) context. The closing of a PS connection is an SM protocol task where the SM changes its state from active to inactive. The signalling procedure used for this state transition is called "PDP context deactivation". When the UE wants to do this, it sends an RRC DIRECT TRANSFER message containing an SM DEACTIVATE PDP CONTEXT REQUEST to the SRNC. The SRNC further relays it as a RANAP DIRECT TRANSFER message to the CN PS domain. A PDP context deactivation request

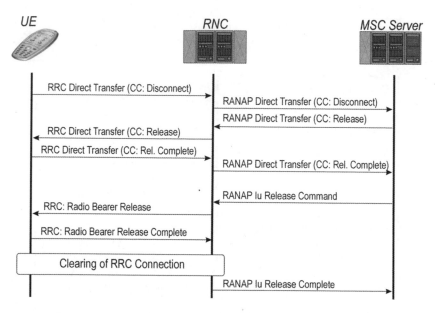

Figure 11.10   CS transaction clearing signalling with Iu Release

indicates to the SGSN that the RAB allocated for this PS connection is no longer needed and, thus, should be released. RAB release is done in the same manner as for CS connection.

When the RAB has been released, the SGSN indicates that PS connection is now

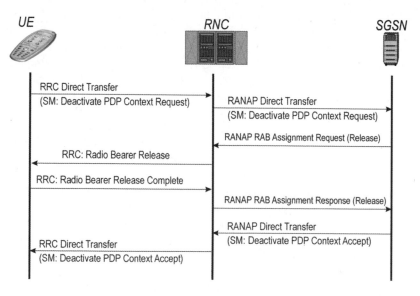

Figure 11.11   PS transaction clearing with RAB release

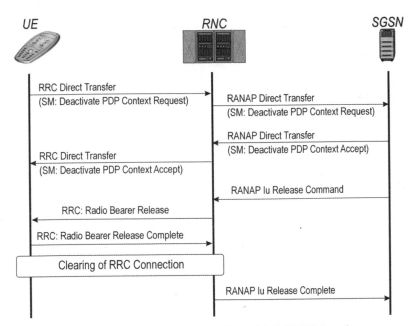

Figure 11.12   PS transaction clearing with RANAP Iu release

deactivated by sending the SM DEACTIVATE PDP CONTEXT ACCEPT message to the UE. After this procedure the UE still has RRC connection with the network and may have other CS and PS connections open. If all PS connections are to be terminated and related radio resources deallocated, the system uses the RANAP IU RELEASE command in the same way as it did for CS connections (see Figure 11.12).

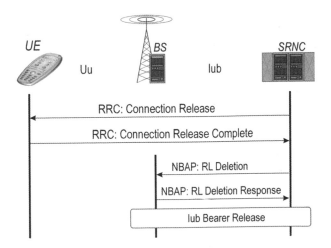

Figure 11.13   RRC connection release

11.1.8 RRC Connection Release

The RRC connection release procedure, which is shown in Figure 11.13, is always started by the RNC. The RNC identifies which RRC connection is to be released and then sends this information to the UE in the RRC CONNECTION RELEASE message. The UE acknowledges that RRC connection has been released by sending an RRC CONNECTION RELEASE COMPLETE message.

Afterwards, the RNC starts to clear Iub interface resources. This is done by exchange of NBAP RADIO LINK DELETION and NBAP RADIO LINK DELETION RESPONSE messages. When radio link deletion is agreed at the NBAP level, the transport data bearer at the Iub interface is released. Any radio links established over a Drifting RNC (DRNC) also have to be released by using Iur subprocedures (not shown in Figure 11.13).

11.2 RRM Procedure Examples

As described in Chapter 5, the RRM is a collection of algorithms needed to establish and maintain a good-quality radio path between the RNC and the UE. Thus, basically every algorithm the RRM contains is present in any transaction the UTRAN is involved in. The RRM algorithms are Hand Over (HO), Power Control (PC), Admission Control (AC), Packet Scheduling (PS) and Code Management (CM). Of these, PC, AC and CM are used "continuously" (i.e., from the very beginning of the transaction, they are involved in every phase). Packet scheduling is used for PS transactions and HOs are applied for both CS and PS transactions whenever needed. Since both CS and PS transactions are equal within the UTRAN, the need for HO(s) is defined by the connection QoS profile (see Chapter 8) and used radio channel type rather than the transaction type (CS/PS). According to the QoS profile, when the transaction is a real-time type, HO(s) are required. If the transaction uses one or more DCHs, HOs should be applied irrespective of whether the transaction type is CS or PS.

Those RRM entities that are located in the UE and the RNC communicate between themselves about RRM by making use of the RRC protocol. Therefore, establishment of an RRC connection, as described in Section 11.1.2, is a must for every transaction. However, it is important to bear in mind that, for any UE, one RRC connection with the UTRAN is enough no matter how many radio bearers are simultaneously open for the various system-wide transactions.

In this subsection we present some examples of HOs. The HO type that we explain first is soft HO; this follows the principles laid down in Chapter 5. After soft HO examples, we give an overview of the SRNS relocation procedure, where UEs are not involved in the procedure. The last RRM procedure example is the inter-system HO case where the UE carries out an HO from WCDMA to GSM radio access.

11.2.1 Soft Handover—Link Addition and Link Deletion

When the UE is using a service, RRC connection with the UTRAN exists and is active. In this case, the UE continuously measures the radio connection and sends measure-

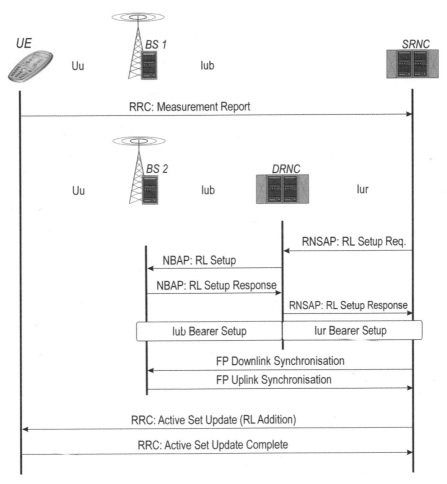

Figure 11.14   Soft HO—link addition

ment reports to the SRNC. The content of such a measurement report was presented in Chapter 5 in the context of RRM algorithms.

The HO algorithm located in the SRNC averages out and investigates the contents of the received measurement reports. Based on the results the SRNC realises that the UE has measured a cell located in BS 2 that has radio conditions fulfilling HO criteria defined in the SRNC. Based on radio network information stored in the SRNC database the SRNC discovers that the target cell in BS 2 does not belong to the same RNS.

The SRNC makes arrangements at the UTRAN end by requesting, by means of the Iur interface, the DRNC to set up a new radio link. This is done by sending a RNSAP RL SETUP REQUEST message. This triggers the DRNC to establish a radio link over the Iub interface between itself and BS 2 by means of NBAP protocol exchange. After these steps, Iub and Iur bearers are established and frame protocols are synchronised in the downlink and uplink directions between the SRNC and BS 2. The frame

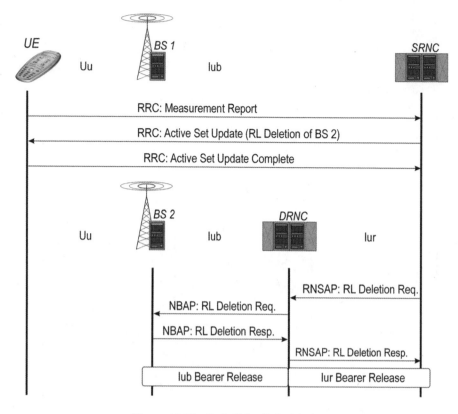

Figure 11.15 Soft HO—link deletion

protocols in the Iub and Iur interfaces implement the radio network user plane and carry the user data flow itself. In this example we assume the used service to be a voice call. Thus, the frame protocol used in this example is the Iub/Iur DCH-FP.

When the SRNC has achieved FP uplink synchronisation, it sends an RRC ACTIVE SET UPDATE message to the UE. In this message the SRNC indicates to the UE that a new radio link has been added to the active set for connection through the cell located in BS 2 and that the connection is ready for use. The UE acknowledges this by responding with RRC ACTIVE SET UPDATE COMPLETE.

From the bearer architecture point of view, this example illustrates a situation in which the UE has an active, CS RAB. This RAB is realised within the UTRAN by the Iu bearer located between the SRNC and CN and another going from the SRNC to the UE through a cell located in BS 1. The procedure described in Figure 11.14 is, in the sense of QoS and bearers, a case where the SRNC adds an additional radio bearer to the existing connection. The RAB and Iu bearer remain unchanged. When the frame protocols are synchronised the SRNC performs RAB mapping to the radio bearers. This situation and related protocols are illustrated at a general level in Figure 5.35.

When the UE moves in the network during the transaction, there comes a point when the SRNC discovers from received measurement reports that radio connection carrying the radio bearer through a cell located in BS 2 no longer fulfils the criteria set for radio connection. When this happens, the SRNC indicates to the UE that this particular radio connection can be removed from the active set. This is done by sending once again an RRC ACTIVE SET UPDATE message to the UE indicating that radio connection be removed. The UE acknowledges radio connection removal by sending an RRC ACTIVE SET UPDATE COMPLETE message to the SRNC.

Upon receiving the UE's acknowledgement, the SRNC can now initiate radio link deletion between itself and BS 2. This is done by using a RNSAP RL DELETION REQUEST over the Iur interface, which in turn triggers the DRNC to delete the radio link in the Iub interface by means of NBAP protocol exchange. When the radio link has been deleted both in the Iub and the Iur, the related Iub and Iur bearers are also released. The message flow related to soft HO by means of radio link deletion is illustrated in Figure 11.15.

From the bearer architecture point of view, this radio link deletion example illustrates the case when the SRNC removes one radio bearer, but an RAB between the CN and UE remains.

11.2.2 SRNS Relocation—Circuit Switched

The role played by the SRNC is one in which the RNC maintains the Iu bearer and performs RAB mapping to radio bearers with respect to a single active UE. Since the radio link components of radio bearers may continuously change due to soft HOs (see Section 11.2.1), the SRNC may end up in a situation where it no longer manages any radio links to its own BSs directly over the Iub. When this happens, the RAB mapping functionality should be changed to an RNC in a better position to perform it. Since RAB mapping requires the existence of the Iu bearer in the same RNC, the CN connection will change, too.

The RRM procedure in which Iu bearer termination is changed from one RNC to another (as are the radio bearers at the same time) is called SRNS relocation. The message flow of this procedure is illustrated in Figure 11.16. In this figure, RNC 1 is the original SRNC and RNC 2 is the RNC that will adopt SRNC functionality. The SRNS relocation procedure can be done in two different ways: either the UE is actively involved in the procedure or it is not. The example presented here illustrates the situation when the UE is not involved.

When the original SRNC (RNC 1) realises that SRNC functionality should be transferred to a better RNC candidate (RNC 2), it initiates this procedure by contacting the relevant CN domain with a RANAP RELOCATION REQUIRED message. This message contains the reason for relocation, target RNS identification and UE classmark information. Based on target RNS identification the CN domain is able to route this query further onto the target RNS. This re-routed query is the RANAP RELOCATION REQUEST message. In addition to other relocation-related information, the RANAP RELOCATION REQUEST contains information about bearer contexts, the CN's use of which at the same time effectively moves the RAB and RRC connection end point from the original SRNC (RNC 1) towards the target SRNC (RNC 2).

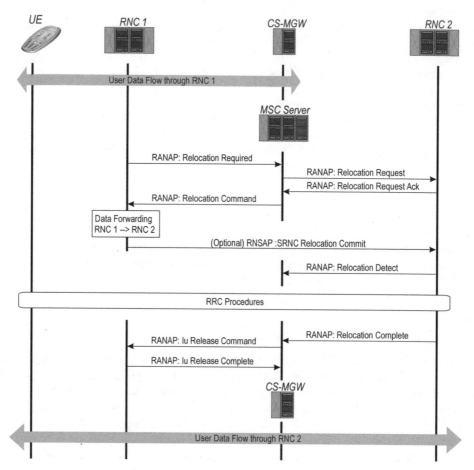

Figure 11.16   CS SRNS relocation—UE not involved

If the target RNC is able to provide resources to handle incoming SRNC function-ality it send an acknowledgement back to the CN, which relays this acknowledgement to the source RNS. From the RNC 1 point of view, this acknowledgement is a command to start relocation, since everything should now be ready in the target RNC. Upon receiving the RANAP RELOCATION COMMAND, the original SRNC (RNC 1) starts to forward data to the target SRNC (RNC 2). The expected SRNC (RNC 2) realises there are incoming data and lets the CN domain know that SRNS relocation has been detected with the message RANAP RELOCATION DETECT. Optionally, the source SRNC (RNC 1) may send a RNSAP SRNC RELOCATION COMMIT message to the target SRNC (RNC 2) over the Iur interface. In the case of SRNC relocation in which the UE is not involved, the UE is unaware of the message flow described above. If there is a need to carry out any RRC procedures, they are performed after the RANAP Reloca-tion Detect message. The UTRAN may, for instance, send an RRC UTRAN MOBILITY INFORMATION message to the UE. It is by means of this message that the UTRAN updates the C-RNTI and U-RNTI and possibly other relevant MM information at the UE.

When all relevant RRC procedures have been carried out, the new SRNC (RNC 2) informs the CN that the SRNC relocation procedure is now complete by sending the message RANAP RELOCATION COMPLETE. This triggers the CN to release resources related to the old SRNC (RNC 1) by issuing a RANAP IU RELEASE message, which is acknowledged by the old SRNC (RNC 1). Now, user data flows will go through the new SRNC and, depending on its role, it acts as the macrodiversity combination point for the UE and controls all RRC activities for this UE as well.

11.2.3 Inter-System Handover from UMTS to GSM—Circuit Switched

3GPP specifications define HO between two radio accesses to be a mandatory requirement of the system. From the UMTS point of view, this means that the system must be able to perform HO between the UTRAN and the GERAN. Inter System Hand Over (ISHO) between the UTRAN and the GERAN is a special case of the SRNS relocation procedure, as can be seen from Figure 10.17. Actually, RANAP carries much more than the same kind of information as the BSSMAP protocol in the GSM network. If HO is performed from the UTRAN to the GERAN, UTRAN-end messages are exactly the same as happens in SRNS relocation, but the contents of these messages vary. For example, the message RANAP Relocation Required contains the target RNC ID in the SRNC relocation case. When the target is at the GSM end, the target RNC ID is replaced with a cell global identity, which is a parameter more familiar to the GSM Base Station Controller (BSC).

Unlike SRNS relocation, ISHO is a procedure in which the UE is always involved, since the UE must perform Radio Resource (RR) activities in order to access the GERAN; note that the GERAN is not necessarily able to handle every kind of context information which is valid in the UTRAN. The UE must be able to measure the GSM cells surrounding the current UTRAN cell(s). This is made possible in the UTRAN by using the slotted mode procedure (see Chapter 5). Slotted mode gives the UE some time to perform measurements on the GSM band and, thus, detect any possible HO candidate(s). Information about these GSM cell candidates is delivered to the RNC with measurement reports just like that of the UTRAN cells.

As soon as the RNC realises that a GSM cell is the best candidate and no suitable UTRAN cells are available, the SRNC starts to require relocation information from the CN. The CN checks the contents of the RANAP RELOCATION REQUIRED message and discovers that the target cell for this HO belongs to a GERAN Base Station Subsystem (BSS). Thus, the CN sends the GSM BSSMAP HANDOVER REQUIRED message to the target BSC at the GSM network end. This triggers the target BSC to set up a Traffic Channel (TCH) so that the connection can be handed over to it. When TCH allocation has been successfully carried out within the GERAN BSS, an acknowledgement is sent back to the CN. The CN relays this message back to the SRNC as a RANAP RELOCATION COMMAND message.

The RANAP RELOCATION COMMAND starts the relocation procedure within the SRNC. Since this is the case in which the UE is involved the SRNC commands the UE to

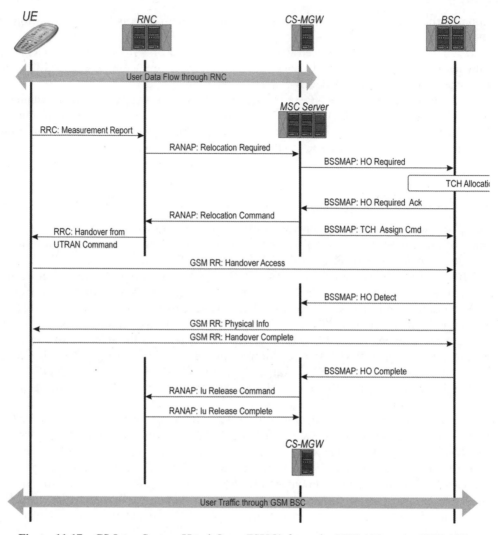

Figure 11.17   CS Inter System Hand Over (ISHO) from the UTRAN to the GERAN

perform ISHO by sending the RRC HANDOVER FROM UTRAN command. This message contains information about the target system and may also carry in its payload any additional information related to ISHO.

Upon receiving the RRC HANDOVER FROM UTRAN command the UE checks whether the message indicated any particular time for HO to be carried out (default is "immediately") and starts HO actions accordingly. Since the target radio access is the GERAN, the UE sends a GSM RR HANDOVER ACCESS message to the target cell in the GERAN BSS. When the GERAN BSS target cell detects this, it indicates to the BSC that the UE is accessing the GERAN. The BSC in turn informs the CN about the incoming UE by sending the message GSM BSSMAP HANDOVER DETECT.

As an acknowledgement to the GSM RR HANDOVER ACCESS message, the UE receives the GSM RR PHYSICAL INFO message from the GERAN BSS cell. This message contains information with which the UE is able to start GERAN radio access (e.g., channel descriptions are sent to the UE). Finally, when the UE has successfully accessed the target cell, the UE sends a GSM RR HANDOVER COMPLETE to the BSC. The BSC in turn relays the same information to the CN, thus indicating that the UE has now entered the GERAN BSS and ISHO is successfully completed in this respect. Since the UE no longer uses UTRAN resources, all resources related to the UE can be released and the CN issues a RANAP IU RELEASE command. This command in turn triggers the RNC to clear the RRC connection, and all the resources related to the UE are thus released. After this is done the RNC confirms the release to the CN with the message RANAP IU RELEASE COMPLETE.

11.3 MM Procedure Examples

Compared with the GSM, the MM behaves in much the same way in the UMTS, but there are some differences. MM as an entity covers all procedures, methods and identities required to maintain knowledge about the UE's location when it is moving in the network. As a result of the existence of PS transactions, a state model for MM has been introduced (see Chapter 6). In GSM networks, MM is completely handled between the MS and the Network Sub System (NSS). In UMTS networks, most MM functions are handled equally between the UE and the CN, but not all. The RNC partially handles the UE's movement within the RAN, using RRC procedures for this purpose. The MM activities handled by the RNC are cell and URA updates.

11.3.1 Cell Update

During a CS transaction, the CN knows the location of a UE down to the accuracy of an LA and the RNC knows the location of the UE to the accuracy of a cell or—to be more precise—cells if the UE is in soft HO state. The cell information is continuously changing during the transaction. The pace of change depends on radio network conditions and whether the UE is moving and at what speed.

In PS transactions the location information is distributed in such a way that the CN PS domain knows the location of the UE down to the accuracy of LA and RA and the RNC knows the location of the UE down to the accuracy of a cell. Because PS connection is discontinuous in nature, the cell information in the RNC will actually be out of date whenever data connection between the UE and the CN is temporarily closed. Hence, PS connection is enabled from the UE and the CN point of view, but the RNC is not able to route the data packets in this state. If the UE desires to send data packets and the current cell is different from its earlier location, the UE performs an RRC cell update procedure (Figure 11.18) to inform the RNC about the current cell. Afterwards, basically any RRC activity is possible (e.g., a radio bearer can be established).

If, on the other hand, it is the CN PS domain that has some data packets to be sent to the UE, the CN PS domain first issues a RANAP PAGING message to make the RNC send

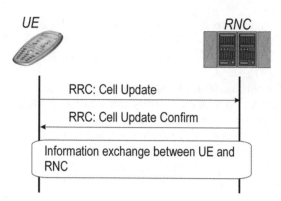

Figure 11.18   Cell update procedure

an RRC PAGING TYPE I message to the UE. Upon receiving this, the UE checks whether the cell update procedure is needed. If it is, the UE performs cell update, after which it is ready to start packet reception. To summarise, the various reasons for a UE to perform a cell update are as follows:

- Cell reselection.
- Periodic cell update.
- Uplink data transmission.
- Paging response.
- Re-entering the service area.
- Radio link failure.
- Unrecoverable RLC error.

In association with a cell update procedure, the RNC may allocate a new RNTI for RRC connection.

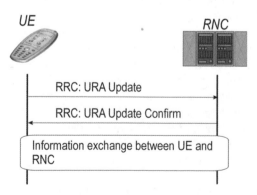

Figure 11.19   URA update procedure

11.3.2 URA Update

The RNC maintains the registration of the current URA for each UE. As described in Chapter 6, a URA consists of a number of cells belonging to either one RNC or several RNCs. Like LA and RA, a URA is one of those identities the UE stores in the USIM. When switched on, the UE continuously monitors the received URA identity. If the received URA differs from the one stored in the USIM, the UE performs an RRC URA UPDATE procedure (Figure 11.19).

Other conditions that trigger a URA update procedure are when the periodic URA update timer has expired or the UE has re-entered a network service area. As in the case of cell update, the RNC changes the RNTI allocated for the UE.

11.3.3 Location Update to the CN CS Domain

Figure 11.20 illustrates the signalling message flow for Location Update (LU) performed between the UE and the CN CS domain. If the UE does not have RRC connection with the UTRAN, RRC connection is established first. Actual LU is triggered by LA identity. The UE stores the LA in the USIM and if the LA identity received from the system information of the camped cell is different from the one stored in the USIM, the UE starts the LU procedure so that the CN knows its current location.

The MM LU REQUEST message is carried in the RRC INITIAL DIRECT TRANSFER envelope to the RNC where it is relayed to the CN within a RANAP UE INITIAL message. The MM LU REQUEST message contains an old LA identity and a new LA identity, and the subscriber identification information is also attached. In the normal case, this identity is a TMSI number that was previously allocated by the MSC server.

Visitor Location Register (VLR) functionality in the MSC server checks whether it has authentication vectors for this subscriber. If they are not present, the security parameters are fetched from the Authentication Centre (AuC), which sends a predefined number of them in the MAP SEND PARAMETERS message to the VLR. Upon receiving these parameters the VLR is able to initiate access-level security procedures for authentication and ciphering. If the old LA identity and the new LA identity differ, the VLR informs the subscriber's Home Location Register (HLR) about the new location. The HLR cancels the old location of the UE with a MAP CANCEL LOCATION message and starts a MAP INSERT SUBSCRIBER DATA operation. This Mobile Application Part (MAP) operation transfers the subscriber profile to the new VLR. The subscriber profile contains all the necessary identity and service provision details and possible restrictions concerning their use. When the VLR has updated the subscriber location information, the CN domain informs the UE of the successful completion of the procedure by sending an MM LU ACCEPTED message. This message also contains a new TMSI for the subscriber. The RNC relays this information by sending an RRC DIRECT TRANSFER message to the UE. The UE acknowledges LU acceptance and the new TMSI value by sending the MM TMSI REALLOCATION COMPLETE message to the MSC server.

When the VLR has received an acknowledgement of TMSI reallocation, it initiates a release-of-signalling connection. The Iu interface is released by means of the RANAP IU

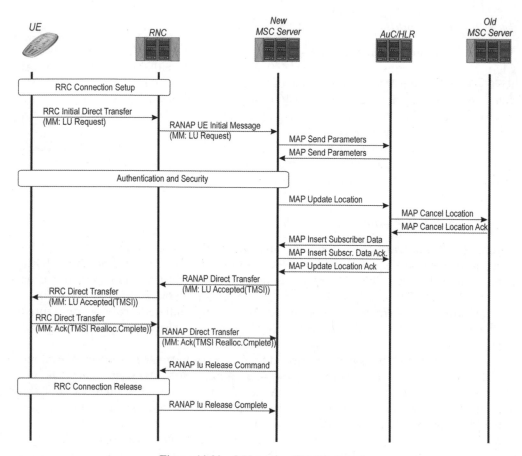

Figure 11.20 LU to the CN CS domain

RELEASE command, and when the RNC receives this it releases a RRC connection. When the connection is established, the RNC acknowledges the release by sending the RANAP IU RELEASE COMPLETE message to the MSC server.

11.3.4 Routing Area Update to the CN PS Domain

As already discussed in Chapter 6, the CN PS domain maintains its own mobility registrations for UEs. Although the PS domain recognises LAs, the more meaningful registration in this context is the Routing Area (RA). The UE stores the latest RA when carrying out RA update. From the system information broadcast by a camped cell, the UE receives its current RA identity. If this differs from the one stored in the USIM, the UE initiates a Routing Area Update (RAU) transaction (see Figure 11.21).

First, the UE sends, within a direct transfer message, the GMM RAU REQUEST message. This message contains the old RA and new RA identifier values. It arrives through the Iu interface to the new SGSN. The new SGSN knows the "MM environment" (i.e., the neighbouring RAs and their relationship with other SGSNs in the network). Based on

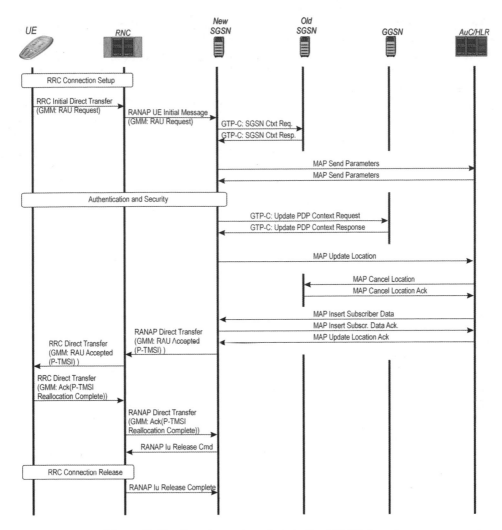

Figure 11.21   Routing area update to the CN PS domain

this information the new SGSN is able to determine the old SGSN, from which it requests information about the subscriber by sending the GTP-C SGSN CONTEXT REQUEST message. In other words, the new SGSN gets to know whether or not the UE performing RAU already has a PDP context.

Afterwards, the new SGSN is in a position to ask the subscriber's AuC to provide authentication vectors. The AuC calculates them and returns a set of vectors to the new SGSN within a MAP SEND PARAMETERS message. The new SGSN is now able to initiate authentication and security control for the UE over the UTRAN. On successful completion of authentication and security activities, the new SGSN informs the GGSN by sending a GTP-C UPDATE PDP CONTEXT message to the GGSN. The new SGSN informs the GGSN that the SGSN has now been changed and that PDP context information has been changed accordingly.

The new SGSN can now update location information with the HLR by sending a MAP UPDATE LOCATION message. Upon receiving this, the HLR cancels the UE's old location from the old SGSN. When the old location has been cancelled the HLR starts to transfer the subscriber profile to the new SGSN by means of a MAP INSERT SUB-SCRIBER DATA operation.

When the subscriber profile has been updated in the new SGSN it sends a GMM RAU ACCEPTED message to the UE. This parameter carries as a parameter the new P-TMSI number to the UE. The UE stores this P-TMSI on its USIM and acknowledges its receipt by sending a GMM ACKNOWLEDGEMENT message to the SGSN. Now that RA update is complete, the new SGSN releases the signalling connection used in this transaction. The message flow in Figure 11.21 presents RAU in the event the SGSN changes. The SGSN is not changed in every RA update. If the SGSN remains the same the procedure is simpler since the MAP dialogues between the SGSN and HLR are not needed.

11.4 CC Procedure Example

Figure 11.22 illustrates the signalling message flow of a UE-terminated CS call coming from a Public Switched Telephone Network (PSTN). PSTN signalling and the signalling used between MSC nodes is assumed to be ISDN User Part (ISUP).

Call signalling first enters the gateway MSC server from the PSTN by means of an ISUP INITIAL ADDRESS MESSAGE (IAM). This message contains the Mobile Subscriber ISDN (MSISDN) number of the called UMTS subscriber. At the same time this MSISDN number identifies the desired service. To set up a CS path between the UMTS Network (CS-MGW) and the PSTN the GMSC server responds with an ISUP ADDRESS COMPLETE MESSAGE (ACM).

The GMSC server asks the HLR to provide routing information for the subscriber by sending a MAP SEND ROUTING INFORMATION message. In this message the GSMC server sends to the HLR the received MSISDN number, from which the HLR is able to discover the subscriber's IMSI. It also discovers the latest reported location of this subscriber down to the accuracy of the current MSC server address. The HLR then issues a MAP PROVIDE ROAMING NUMBER REQUEST message to the MSC server in question, which allocates and returns a Mobile Station Roaming Number (MSRN) for CS path connection. When receiving this, the HLR relays the MSRN to the GMSC server.

The MSRN contains the necessary information for call-routing purposes, since every MSC server in the network allocates MSRNs from a certain numbering space and this numbering space is recognised in the CC entity of the Gateway MSC (GMSC) server. The GMSC server uses the received MSRN for call-routing to the MSC server using ISUP IAM and ISUP ACM messages, as described earlier.

The MSC server recognises the MSRN it allocated and determines that this call is to be terminated to one of the UEs under its own control. Thus, it sends RANAP PAGING messages to those RNCs maintaining the cells that belong to the LA where the addressed UE carried out its latest LU. The RNCs send RRC PAGING TYPE I messages to the Uu interface and the paged UE responds by initiating an RRC connection

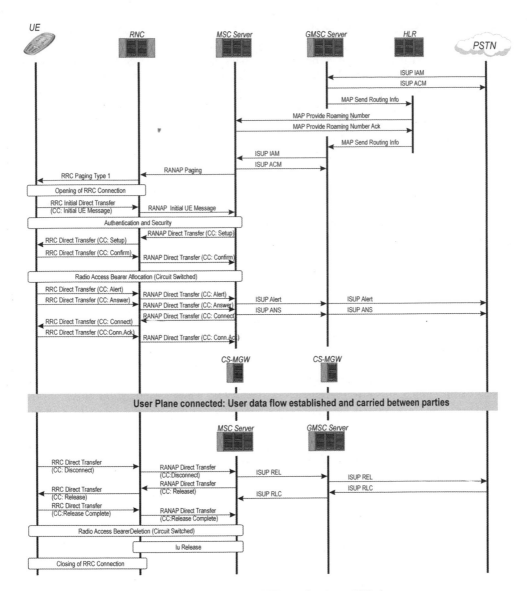

Figure 11.22 CS call—UE terminating—UE clears

establishment procedure. The actual response to paging is carried in a RANAP UE INITIAL message, on receipt of which the CN starts authentication and security activities for the call. If the security activities are successful, the CN initiates transaction setup by sending the UE a CC SETUP message about the incoming call and its nature.

If the UE accepts this call, the CN initiates RAB allocation. When the RAB is allocated, the UE alerts the subscriber and informs the network that the subscriber has been alerted by sending a CC ALERT message. This information is carried through the network(s) up to the calling subscriber exchange. When the subscriber answers, the

CC CONNECT procedure is triggered over the UTRAN and answering information is carried over the network(s) by sending an ISUP ANS (answering) message.

The CS call path is now in a position to pass through the CS Media Gateways (CS-MGWs) that were controlled by the MSC server and the GMSC server. Charging starts, related voice codecs start and the system is ready to transfer a user data flow since the user-plane connection is now open.

In this example we are assuming that the UE clears the connection as well.

A three-way clearing procedure is used by exchanging CC DISCONNECT, CC RELEASE and CC RELEASE COMPLETE messages over the UTRAN. At the ISUP end the same thing is taken care of by ISUP REL (release) and ISUP RLC (release complete) messages. These messages bring all charging to an end and clear the connection between nodes. After the CS path is cleared, the CN releases the RAB and RRC connection by invoking the Iu Release procedure.

11.5 Packet Data Example

Figure 11.23 shows a small sample of what happens during a packet session. We are again assuming that the UE is continuously on the move in the UTRAN area. Under these circumstances the user application in the UE wants to send data packets to the network (uplink). Since the UE is continuously moving it first has to carry out cell update. By doing this, the UE ensures that its location information in the SRNC is valid—it needs to be to the accuracy of a cell.

After cell update the UE is able to request PDP context activation through the current cell and an RAB can be allocated for packet data transfer. Figure 11.8 shows the complete message flow involved in this procedure. When the packet session is open, the UE transmits uplink data packets, which first arrive at the SRNC over the radio links forming a radio bearer for this session. Packet traffic is then tunnelled between the RNC and the GGSN by means of the GTP-U protocol; physically, this tunnel consists of two parts: the Iu bearer over the Iu interface and the PS domain backbone bearer between the SGSN and the GGSN. The tunnel ends at the GGSN, from where user data packets are further routed to an external packet data network, according to a method described in the connection-type parameter of the PDP context. This parameter may have values like the Point-to-Point Protocol (PPP) or X.25. When there is no longer anything to transfer, the PDP context is deactivated and RAB is released. The complete message flow involved in this deactivation is shown in Figure 11.11.

In the downlink direction the GGSN may later realise that packets are coming in from other networks. This triggers the SM PDU NOTIFICATION procedure, which leads to network-initiated PDP context activation. Upon receiving the SM PDU NOTIFICATION REQUEST message the SGSN sends paging to the RNCs controlling the RA where the UE carried out its latest RAU. The RNCs send the paging on towards the UE as RRC PAGING TYPE I. Since the UE is continuously moving it has to perform cell update again so that valid cell information is recorded at the RNC.

After cell update, PDP context is activated, as is RAB and the data packets start to flow from the GGSN to the UE. If the UE has any data packets to send, then they can be sent at the same time. Data packets transferred in the uplink direction are treated as

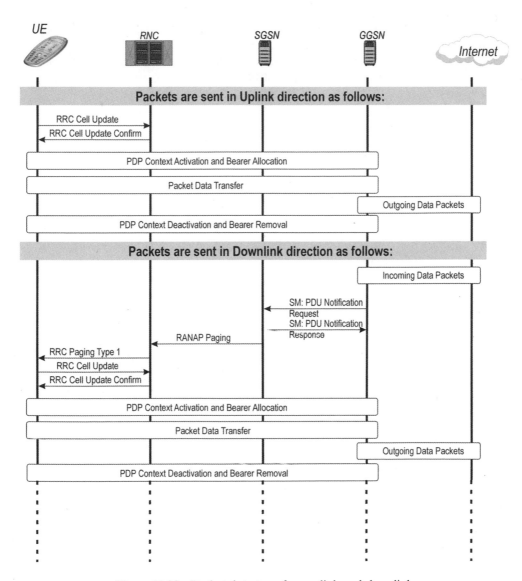

Figure 11.23   Packet data transfer—uplink and downlink

described earlier. When there are no data to be transferred the PDP context is once again deactivated and the RAB released.

11.6 IMS Examples

Up to now we have introduced the IMS architecture (Chapter 6) and presented some examples of the services IMS could offer. In this section we will introduce two simplified examples of IMS functionality. IMS as such is a relatively complex entity; for further details, we can thoroughly recommend *IMS—IP Multimedia Concepts and Services in the Mobile Domain* written by Poikselkä et al. (2004).

The two procedures we have selected as examples are, first, IMS registration and, second, a simplified version of an IMS multimedia session.

11.6.1 IMS Registration Example

The objective of IMS registration is to register and authorise an IMS user so that multimedia sessions can be established. Figure 11.24 shows the main topics involved in the IMS registration procedure.

In this example we assume that the IMS is implemented on top of the existing GPRS, where operators typically do not allow the use of a "wildcard Access Point Name (APN)" (i.e., the user is forced to establish GPRS connection through a GGSN located in the home network of the user).

Before a signalling PDP context can be established, the UE and the network need to establish an RRC connection in the same way as explained earlier in this chapter. According to the above-mentioned assumption, the SGSN resolves with help from the Domain Name Server (DNS) the GGSN address in the home network of the subscriber. If the subscriber is entitled to utilise IMS services, some parameters are required for IMS connection establishment. The very first is the Proxy CSCF (P-CSCF) address. Although this could be preconfigured and stored in the IMS Identity Module (ISIM), it would be a very limiting way to use the address space. Instead, the UE would be better off by using its ability to request the P-CSCF address during PDP context establishment. When requested, the GGSN returns the IPv6 address prefix of the P-CSCF to the UE during this PDP context activation request. Another alternative would be to use the dynamic address allocation scheme, where the Dynamic Host Configuration Protocol (DHCPv6) returns the P-CSCF address. In this case any further P-CSCF address-resolving can be done by means of the DNS.

When the UE knows the address of the P-CSCF, it sends a SIP REGISTER message to the P-CSCF. Construction of a SIP REGISTER message requires some parameters, which are available on ISIM within the UE. These parameters are the various identities used in context of the IMS:

- User private identity.
- User public identity (since there could be several of these, a default public identity— first on the list—is used for IMS registration).
- Home network domain.
- The IP address allocated for the UE in the visited network.

Public and private identities were introduced in Chapter 6 in Sections 6.2.1.1.7 and 6.2.1.1.8. Private identity will be discussed shortly when we look at security-related associations.

For security reasons the ideal situation would be for the IMS environment to use its own IMS-specific identities. The ISIM, however, is not necessarily always present in the UE. In such cases, identities should be derived by means of the standard USIM. The

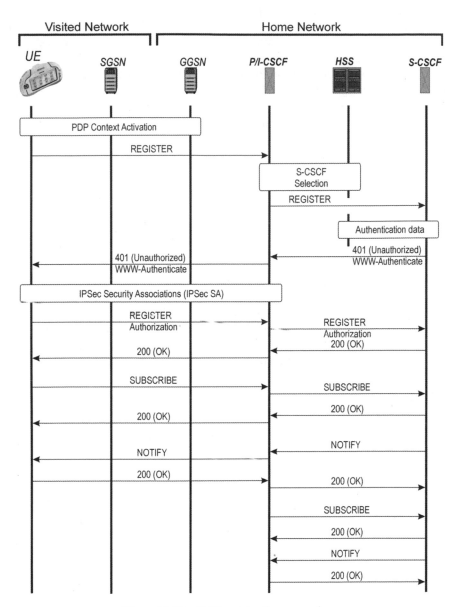

Figure 11.24 IMS registration procedure

results of this process should never be revealed outside the IMS environment for
security reasons. For example, were the private user identity constructed like this, it
would be very similar to the IMSI.

Before the P-CSCF is able to send the REGISTER message any farther on its way, it
must work out the route to the correct Serving CSCF (S-CSCF). For this purpose the
IMS contains an Interrogating CSCF (I-CSCF), which questions the HSS. In this
example (Figure 11.24) it is assumed that since the P-CSCF and S-CSCF are held by

the same IMS entity in the same physical network, I-CSCF functionality is integrated together with that of the P-CSCF. When the correct S-CSCF address is resolved, the REGISTER message is routed to the S-CSCF.

Based on the received user identity in the REGISTER message, the S-CSCF is able to obtain authentication data from the HSS. In the UMTS case, these data contain UMTS security definitions (as was described in Chapter 9). These security parameter sets are known as "authentication vectors" and each contains:

- A random challenge (RAND).
- Expected result (XRES).
- Network authentication token (AUTN).
- Integrity Key (IK).
- Ciphering Key (CK).

In this phase no precise information is available about the user: from the S-CSCF point of view all that has happened is that someone has made contact with a correct-looking address without any guarantees. In order to verify that the contactor is genuine (i.e., is who he or she claims to be) the S-CSCF sets a challenge. It sends a SIP 401 UNAUTHORIZED message back to the UE. The S-CSCF packs this message with authentication vector data (e.g., RAND and AUTN parameters). However, the IK and CK are placed in fields the P-CSCF reads and stores; these values will never reach the UE.

The UE computes the response to the P-CSCF's challenge and answers with a new SIP REGISTER message, which this time contains private user identity, RAND, AUTN and a computed response RES. At the same time the UE has calculated the IK. The P-CSCF and the UE are now in a position to verify the integrity of traffic between them.

This SIP REGISTER message reaches the S-CSCF, which checks and compares the XRES and the RES value sent by the UE. If they match, the user is authenticated. Authentication in the reverse direction was done when the UE received the challenge (e.g., that message contained an AUTN which enabled the UE to determine whether the network was real or fake). Successful authentication is finalised by a SIP 200 OK message, which the S-CSCF sends to the UE.

When security issues are handled a so-called "IMS subscription" is performed. Subscription is a procedure during which the lifetime of the registration is set. The default value for IMS registration lifetime is 600,000 seconds (about 7 days). In addition to this, the subscription procedure is responsible for registering public user identities other than the default (which is the only one registered for use so far).

Subscription is first done between the UE and the S-CSCF. This procedure makes the UE not only aware of all the public user identities this IMS user has but also the registration states of these identities.

The same information must also be present in the P-CSCF. Hence, subscription is also separately performed between the P-CSCF and the S-CSCF. When subscription is complete, the UE has a connection with the IMS and is able to establish multimedia calls and/or use other services accessible through the IMS.

11.6.2 IMS Session Example

In the second example we show how an IMS session is established (Figure 11.25), which network elements participate in user-plane communication (Figure 11.26) and how an IMS session is cleared (Figure 11.27).

We use the same assumption as in the previous example (i.e., UE *A* is on the move and the GPRS connection terminates at the home network GGSN). UE *B* is located in the home network. Both UE *A* and UE *B* have the same home network. They have both also carried out IMS registration and, thus, they are available to each other.

Figure 11.25 shows a signalling diagram of UE *A* contacting UE *B*. IMS Session establishment starts with a SIP INVITE request. This request's header contains two public identities: one identifying the caller party (*A* in this case) and the other identifying the called party (*B* in this case). This is received by the P-CSCF, which then routes it to the

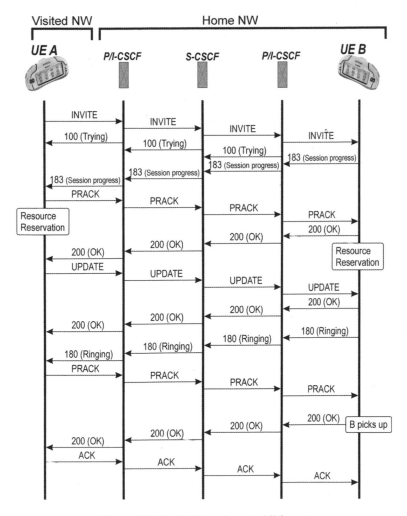

Figure 11.25   IMS session establishment

S-CSCF of UE *A*. In this phase the system only knows the final destination of the SIP INVITE request (i.e., UE *B*'s registered public identity). The I-CSCF functionality contacts the HSS in order to find out where this public identity was itself registered (i.e., under which S-CSCF it can be found). After finding this information, the INVITE request is then routed to UE *B*'s S-CSCF, which in turn routes it to UE *B* via UE *B*'s P-CSCF. For multimedia session establishment purposes, the INVITE request contains all the *desired* media and codec information. This is the so-called "first SDP offer" (SDP = Session Description Protocol).

During its journey the various CSCFs have added their address information to the INVITE request. When the INVITE request reaches UE *B*, UE *B* adds its own IP address to the list, indicating all the other elements involved, such as the addresses for any other messages within this dialogue. UE *B* also adds the port number it wants to use for this session. In this way all subsequent messages will use the established IP Security (IPSec) Security Association (SA).

When UE *A* sent the INVITE request it started a timer with a value of 2 seconds. During this time it awaits an answer from the network. The expected answer is SIP 183 SESSION PROGRESS. It is very probable that this answer will not be returned in the 2 seconds allocated; this is why the intermediate answer SIP 100 TRYING is used. The SIP 100 TRYING message indicates to UE *A* that there is no point in retransmitting the INVITE request because the P-CSCF will take care of all retransmissions, should they be required.

The SIP 183 SESSION PROGRESS message travels back towards UE *A*. The P-CSCF of UE *A* is the last point handling this message before it actually reaches UE *A*. In this phase the P-CSCF adds its port number to the message; by doing this the connection will utilise IPSec SA. When SIP 183 SESSION PROGRESS reaches UE *A*, it stores the IP address of UE *B* as well as the routing information that was present in SIP 183 SESSION PROGRESS. This message contains the response to the first SDP offer included in the original INVITE request; the response included in the SIP 183 SESSION PROGRESS message contains information about *supported* media and codecs; this is called a "first SDP answer".

Upon receiving the SIP 183 SESSUIB PROGRESS message, UE *A* answers with a PRACK (Provisional Response ACKnowledgement) message. This message is also called a "second SDP offer"; it contains information defining one codec per accepted media. UE *B* acknowledges the PRACK with SIP 200 OK, indicating that both parties have now agreed the codec and media types to be used. Likewise this is called a "second SDP answer".

When both parties know and have agreed the codecs and media types, they initiate resource reservation. Actually, resource reservation is the same as setting up a media PDP context, which is in turn no more than a collection of parameters describing the user plane for this multimedia session (e.g., QoS definitions).

So far, UE *B* has not indicated to its user that an incoming multimedia session is about to be established. This is as it should be, since there are still no guarantees about the session. These guarantees exist only after both parties have successfully carried out resource reservation. When UE *A* has done this, it sends a SIP UPDATE message towards UE *B*. This message is called a "third SDP offer" and indicates whether resource reservation was successful and whether all the requirements were met. UE *B* answers

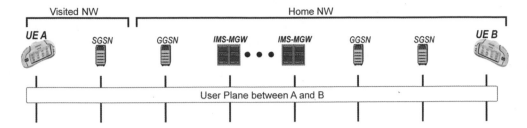

Figure 11.26   User plane of an IMS session

this message by sending SIP 200 OK if its resource reservation has also been successful; this message is called a "third SDP answer". After this message exchange everything is ready for user-plane routing.

UE *B* lets UE *A* know that it is going to inform its user about the incoming multi-media session and does so with a SIP 180 RINGING message. UE *A* acknowledges the state of the connection by sending a SIP PRACK. When the user of UE *B* picks up, UE *B* sends SIP 200 OK, indicating that the user plane can be opened. This is acknowledged by a SIP ACK message. The IMS session has now been established and UE *A* and UE *B* have a multimedia connection between each other with the desired and agreed media types and codecs. Figure 11.26 shows an example of how the user plane could go through various equipment.

In this example it is assumed that UE *B* clears the connection. Connection-clearing is started by sending a SIP BYE message. At the same time the SIP BYE is sent, UE *B* drops out of the media PDP context, thus causing the user plane to close. When the SIP BYE reaches UE *A*, it respectively drops out of the media PDP context as well. UE *A* responds to the SIP BYE by sending a SIP 200 OK. This tells the attached CSCFs to clear all information and address tables related to this multimedia session.

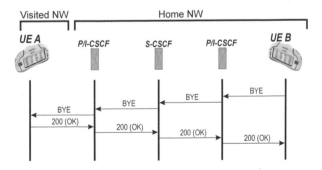

Figure 11.27   Clearing an IMS Session

List of Abbreviations

16QAM	Quadrature Amplitude Modulation
3GPP	Third Generation Partnership Project
3GPP2	Third Generation Partnership Project No. 2
8-PSK	Octagonal Phase Shift Keying
AAA	Authentication, Authorisation and Accounting
AAL	ATM Adaptation Layer
ABR	Available Bit Rate
AC	Admission Control
ACM	Address Complete Message
ACTS	Advanced Communications Technology and Services
ACTS FRAMES	ACTS Future Radio Wideband Multiple Access System
AGCH	Access Grant CHannel
AH	Authentication Header
AICH	Acquisition Indication CHannel
AK	An anonymity key
AKA	Authentication and Key Agreement
AM	Acknowledged Mode
AM-SAP	Acknowledged Mode SAP
AMC	Adaptive Modulation and Coding
AMF	Authentication Management Field
AMPS	American Mobile Phone System
AMR	Adaptive Multi Rate
AOA	Angle Of Arrival
AP	Access Point; Application Part
API	Application Programming Interface
APN	Access Point Name
ARIB	Association of Radio Industries and Business
AS	Application Server
ATD	Absolute Time Difference
ATM	Asynchronous Transfer Mode
AuC	Authentication Centre

UMTS Networks Second Edition   H. Kaaranen, A. Ahtiainen, L. Laitinen, S. Naghian and V. Niemi
© 2005 John Wiley & Sons, Ltd   ISBN: 0-470-01103-3

AUTN	An authentication token
AUTS	An authentication key in re-synchronisation
AV	Authentication Vector
BC	BroadCast
BCCH	Broadcast Control CHannel
BCFE	Broadcast Control Function Entity
BCH	Broadcast CHannel
BER	Bit Error Rate
BGCF	Breakout Gateway Control Function
BICC	Bearer Independent Call Control
BMC	Broadcast/Multicast Control
BML	Business Management Layer
BMP	BitMap Picture
BS	Base Station (or Node B)
BSC	Base Station Controller
BSS	Base Station Subsystem; Basic Service Set
BSSI	Base Station Subsystem Identifier
BSSMAP	Base Station Subsystem MAP
BTS	Base Station Transceiver Station
CA	Certification Authority
CA-ICH	CPCH Assignment Indication CHannel
CAMEL	Customised Application for Mobile Network Enhanced Logic
CAP	CAMEL Application Part
CAPEX	CAPital EXpenditure
CBR	Constant Bit Rate
CC	Country Code; Call Control
CCA	Clear Channel Assignment
CCCH	Common Control CHannel
CCF	Charging Collection Function
CCH	Communication CHannel
CCK	Complementary Code Keying
CCPCH	Common Control Physical CHannel
CD-ICH	Collision Detection Indicator CHannel
CDMA	Code Division Multiple Access
CDR	Charging Data Record
CF	Connection Frame
CFN	Connection Frame Number
CFP	Connection Free Period
CGF	Charging Gateway Function
CGI	Cell Global Identity
CGW	Charging GateWay
CI	Cell Identity
CK	Ciphering Key
CLP	Cell Loss Priority
CLPC	Closed Loop Power Control

CM	Code Management; Communication Management; Configuration Management
CN	Core Network
COMC	COMmunication Control
COPS	Common Open Policy Service
COUNT	A frame-dependent input to A5/3
COUNT-C	A frame-dependent input to f8
COUNT-I	A frame-dependent input to f9
CP	Contention Period
CPC	Centralised Power Control
CPCH	Common Packet CHannel
CPICH	Common PIlot CHannel
CPM	Continuous Power Mode
CQI	Channel Quality Indication
CRC	Cyclic Redundancy Check
CRNC	Controlling RNC
C-RNTI	Cell RNTI
CS	Circuit Switched; Coding Scheme; Convergence Sublayer
CS-MGW	Circuit Switched Media GateWay
CSCF	Call Session Control Function
CSE	CAMEL Service Environment
CSICH	CPCH Status Indicator CHannel
CSMA/CA	Carrier Sense Multiple Access with Collision Avoidance
CSPDN	Circuit Switched Public Data Network
CTCH	Common Traffic CHannel
CTS	Client To Server
CW	Call Waiting; Carrier Wave; Continuous Wave
CWTS	China Wireless Telecommunication Standard Group
DBPSK	Differential Binary Phase Shift Key
DCCH	Dedicated Control CHannel
DCF	Distributed Coordination Function
DCFE	Dedicated Control Function Entity
DCH	Dedicated CHannel
DCH-FP	DCH Frame Protocol
DES	Data Encryption Standard
DGPS	Differential GPS
DHCP	Dynamic Host Configuration Protocol
DiffServ	Differentiated Services
DIFS	DCF InterFrame Space
DL	DownLink
DNS	Domain Name Server
DoS	Denial of Service
DPCCH	Dedicated Physical Communication CHannel
DPCH	Dedicated Physical CHannel
DPDCH	Dedicated Physical Data CHannel
D-RNTI	Drift RNC Radio Network Temporary Identity

DQPSK	Differential QPSK
DRNC	Drifting RNC
DRX	Discontinuous Reception
DS-CDMA	Direct Sequence CDMA
DSCH	Downlink Shared CHannel
DSCH-RNTI	DSCH Radio Network Temporary Identity
DSCP	Differentiated Services Code Point
DSL	Digital Subscriber Line
DSSS	Direct Sequence Spread Spectrum
DS-WCDMA-FDD	DS Wideband CDMA Frequency Division Duplex
DS-WCDMA-TDD	DS Wideband CDMA Time Division Duplex
DTCH	Dedicated Traffic CHannel
DTX	Discontinuous Transmission
E-GPRS	Enhanced GPRS
E-OTD	Enhanced Observed Time Difference
E1	E1 system; European Digital Signal 1
E2E, e2e	End-to-End
ECT	Explicit Call Transfer
EDGE	Enhanced Data for GSM Evolution
EIR	Equipment Identity Register
EMC	ElectroMagnetic Compatibility
EML	Element Management Layer
ESP	Encapsulation Security Payload
ETSI	European TelecommunicationS Institute
FACCH	Fast Association Control CHannel
FACH	Forward Access CHannel
FBI	FeedBack Information
FCH	Frequency correction CHannel
FDD	Frequency Division Duplex
FDMA	Frequency Division Multiple Access
FER	Frame Error Rate
FH-CDMA	Frequency Hopping CDMA
FHSS	Frequency Hopping Spread Spectrum
FM	Fault Management
FPS	Fast Packet Scheduling
FPLMTS	Future Public Land Mobile Telephony System
FR	Frequency Reuse
FRAMES	ACTS Future Radio Wideband Multiple Access System
FRESH	A one-time random number chosen by the RNC
FW	FireWall
GERAN	GSM/EDGE Radio Access Network
GGSN	Gateway GPRS Support Node
GIF	Graphic Interchange Format (image file)
GMLC	Gateway Mobile Location Centre
GMM	GPRS Mobility Management
GMSC	Gateway MSC

GMSK	Gaussian Minimum Shift Keying
GPRS	General Packet Radio Service
GPS	Global Positioning System
GRX	GPRS Roaming Exchange
GSM	Global System for Mobile communication
GSMS	GPRS SMS
GT	Global Title
GTD	Geometric Time Difference
GTP	GPRS Tunnelling Protocol
HARQ	Hybrid Automatic Repeat reQuest
HCS	Hierarchical Cell Structure
HDLC	High-level Data Link Control
HEC	Header Error Control
HFN	HyperFrame Number
HLR	Home Location Register
HM-CDMA	Hybrid Modulation CDMA
HO	HandOver
HON	HandOver Number
HS-DPCCH	High Speed DPCCH
HS-DSCH	High Speed DSCH
HS-DSCH-FP	High Speed DSCH Frame Protocol
HS-PDSCH	High Speed PDSCH
HS-SCCH	High Speed SCCH
HSCSD	High Speed Circuit Switched Data
HSDPA	High Speed Downlink Packet Access
HSS	Home Subscriber Server
HSUPA	High Speed Uplink Packet Access
HTML	HyperText Markup Language
HTTP	HyperText Transfer Protocol
HW	HardWare
I-CSCF	Interrogating CSCF
I/O	Input/Output
IA	Interception Area
IAM	Initial Address Message
IBSS	Independent Basic Service Set
ICC	Integrated Circuit Card
IEEE	Institute of Electrical and Electronic Engineers
IETF	Internet Engineering Task Force
IFS	InterFrame Space
IK	Integrity Key
IKE	Internet Key Exchange
IM	IP Multimedia
IMA	Inverse Multiplexing for ATM
IMEI	International Mobile Equipment Identity
IMEISV	IMEI and Software Version
IMPI	IP Multimedia Private Identity

IMPU	IP Multimedia Public User identity
IMS	IP Multimedia System
IMS-MGW	IMS Media GateWay
IMS-SF	IMS Switching Function
IMSI	International Mobile Subscriber Identity
IMT	International Mobile Telephony
IN	Intelligent Network
IP	Internet Protocol
IPDL	Idle Period DownLink
IPSec	IP Security
IPSP	IP Signalling Point
IRC	Internet Relay Chat
ISCP	Interference Signal Code Power
ISDN	Integrated Services Digital Network
ISHO	InterSystem HandOver
ISIM	IMS Identity Module
ISM	Industrial, Scientific and Medical
ISO	International Organization for Standardization
ISUP	ISDN User Part
ITU	International Telecommunication Union
JPEG	Joint Photographic Experts Group (format for image compression)
KASUMI	An encryption operation
LA	Location Area
LAC	Location Area Code
LAI	Location Area Identity
LAN	Local Area Network
LCS	LoCation Service
LDP	Label Distribution Protocol
LIF-MLP	LIF Mobile Location Protocol
LMU	Location Measurement Unit
LOS	Line Of Sight
LSP	Label Switched Path
LSR	Label Switching Router
M3UA	MTP3 User Adaptation
MAC	Medium Access Control; Message Authentication Code
MAC-I	A message authentication code for integrity protection
MAC-S	A message authentication code for re-synchronisation
MAP	Mobile Application Part
MAPSec	MAP Security
MC-CDMA	MultiCarrier CDMA
MCC	Mobile Country Code
MCS	Modulation Coding Scheme
ME	Mobile Equipment
MEHO	Mobile Evaluated HandOver
MGCF	Media Gateway Control Function

MGCP	Media Gateway Control Protocol
MGW	Media GateWay
MIB	Master Information Block
MM	Mobility Management
MMS	Multimedia Messaging Service
MNC	Mobile Network Code
MO	Mobile Originated (Originating)
MOBC	MOBility Control
MOC	Mobile Originated (Originating) Call
MPDU	MAC Protocol Data Unit
MPLS	Multi Protocol Label Switching
MRC	Maximum Ratio Combining
MRFC	Multimedia Resource Function Controller
MRFP	Multimedia Resource Function Processor
MS	Mobile Station
MSC	Mobile service Switching Centre
MSIN	Mobile Subscriber Identification Number
MSISDN	Mobile Subscriber ISDN
MSN	Mobile Subscription Number
MSRN	Mobile Station Subscriber Roaming Number
MT	Mobile Terminated (Terminating)
MT	Mobile Termination
MTC	Mobile Terminated (Terminating) Call
MTP-3B	Message Transfer Part-3B (Broadcast)
MTP3	Message Transfer Part Layer 3
MTU	Maximum Transmission Unit
N-ISDN	Narrowband ISDN
NAS	Network Application Server; Non Access Stratum
NAV	Network Allocation Vector
NBAP	Node B Application Protocol
NDC	National Destination Code
NE	Network Element
NEHO	Network Evaluated HandOver
NEL	Network Element Layer
NLOS	Non Line Of Sight
NMC	Network Management Centre
NML	Network Management Layer
NMS	Network Management Subsystem
NMT	Nordic Mobile Telephone
NRT	Non Real Time
NSS	Network SubSystem
NT	Network Termination
O&M	Operation and Maintenance
OCS	Online Charging System
OFDMA	Orthogonal Frequency Division Multiple Access
OHG	Operator Harmonisation Unit

OLPC	Open Loop Power Control
OMA	Open Mobile Alliance
OMC	Operation and Management Centre
OPEX	Operation and Maintenance EXpenditure
OSA	Open Service Architecture
OSA/SCS	OSA/Service Capability Server
OSI	Open System Interconnection
OTD	Observed Time Difference
OTDOA	Observed Time Difference Of Arrival
PA	Paging Area
PACCH	Packet Access Control CHannel
PAGCH	Packet Access Grant CHannel
PAR	Peak Average Rate
PBCC	Packet Binary Convolution Code
PBS	Physical Bearer Service
PC	Power Control
PCCH	Paging Control CHannel
P-CCPCH	Primary CCPCH
PCF	Point Coordination Function
PCH	Paging CHannel
PCM	Pulse Code Modulation
PCPCH	Physical Communication Packet Channel
PCS	Personal Communication System
P-CSCF	Proxy CSCF
PD	Protocol Discriminator
PDC	Pacific Digital Communications
PDCP	Packet Data Convergence Protocol
PDF	Policy Decision Function
PDH	Plesiochronous Digital Hierarchy
PDP	Packet Data Protocol
PDSCH	Physical Downlink Shared CHannel
PDTCH	Packet Data Traffic CHannel
PDU	Packet Data Unit; Protocol Data Unit
PGP	Pretty Good Privacy
PICH	Paging Indicator CHannel
PIFS	Point coordination function IFS
PKI	Public Key Infrastructure
PLMN	Public Land Mobile Network
PLMN-ID	PLMN Identity
PM	Performance Management
PMM	Packet MM
PN	Pseudo Noise
PNCE	Paging and Notification Control Entity
PNG	Portable Network Graphics
POC	PSTN Originated (Originating) Call
PoC	Push to talk over Cellular

PPCH	Packet Paging CHannel
PPP	Point-to-Point Protocol
PRACH	Packet Random Access CHannel
PRACK	Provisional Response ACKnowledgement
PRN	Pseudo-Random Numerical
PS	Packet Switched; Packet Scheduler
P-SCH	Primary SCH
PSM	Power Save Mode
PSPDN	Packet Switched Public Data Network
PSTN	Public Switched Telephone Network
PT	Payload Type
PTC	PSTN Terminated (Terminating) Call
P-TMSI	Packet TMSI
QoS	Quality of Service
QPSK	Quadrature Phase Shift Keying
RA	Routing Area; Receiver Address
RAB	Radio Access Bearer
RAC	Routing Area Code
RACE	Research in Advanced Communications in Europe
RACH	Random Access CHannel
RADIUS	Remote Authentication Dial In User Service
RAI	Routing Area Identity
RAN	Radio Access Network
RANAP	RAN Application Part
RAND	A random (authentication) challenge
RAP	Random Access Procedure
RAU	Routing Area Update
RES	An authentication response
RFC	Request For Comment
RFE	Routing Function Entity
RL	Radio Link
RLC	Radio Link Control
RLC-SQN	RLC SeQuence Number
RNBP	Reference Node Based Positioning
RNC	Radio Network Controller
RNS	Radio Network Subsystem
RNSAP	Radio Network Subsystem Application Part
RNTI	Radio Network Temporary Identity
ROHC	RObust Header Compression
RR	Radio Resource
RRC	Radio Resource Control
RRM	Radio Resource Management
RSVP	ReSerVation Protocol
RT	Radio Termination; Real Time
RTD	Relative Time Difference
RTP	Real Time Protocol

RTS	Request To Send
RTT	Round Trip Time
Rx	Receiving
S-CCPCH	Secondary Communication Control Physical CHannel
S-MIME	Secured Multipurpose Internal Mail Extension
S-RNTI	SRNC Radio Network Temporary Identity
S-SCH	Secondary SCH
SA	Security Association
SAC	Service Area Code
SACCH	Slow Associated Control CHannel
SAI	Service Area Identity
SAP	Service Access Point
SAR	Segmentation And Re-assembly
SAW	Stop And Wait
SB	Scheduling Block
SBLP	Service Based Local Policy
SCCP	Signalling Connection Control Part
SCH	Synchronisation CHannel
SCCH	Synchronisation Control CHannel
SCI	Subscriber Controlled Input
SCR	System Chip Rate
SCS	Service Capability Server
SCTP	Stream Control Transport Protocol
SDCCH	Stand-alone Dedicated Control CHannel
SDH	Synchronous Digital Hierarchy
SDP	Session Description Protocol
SDU	Signalling Data Unit
SEG	SEcurity Gateway
SF	Spreading Factor
SFN	System Frame Number
SGSN	Serving GPRS Support Node
SGW	Signalling GateWay
SHA-1	Secure Hash Algorithm #1
SIB	System Information Block
SIFS	Short IFS
SIGTRAN	Special Interest Group TRAN
SIM	Subscriber Identity Module
SIP	Session Initiation Protocol
SLA	Service Level Agreement
SLF	Subscription Locator Function
SML	Service Management Layer
SMLC	Serving Mobile Location Centre
SMS	Short Message Service
SMSC	Short Message Service Centre
SN	Serving Network
SN	Subscriber Number

SNR	Serial NumbeR
SPC	Signalling Point Code
SQN	SeQuence Number
SRB	Signalling Radio Bearer
SRNC	Serving RNC
SS	Supplementary Service
SS7	Signalling System #7
SSDT	Site Selective DiversiTy
SSL	Secure Socket Layer
STM	Synchronous Transport Module
SVN	Software VersioN
SW	SoftWare
TA	Timing Advance; Transmitter Address
TAC	Type Allocation Code
TCAP	Transaction Capabilities Application Part
TCH	Traffic CHannel
TCP	Transmission Control Protocol
TDD	Time Division Duplex
TDM	Time Division Multiplexing
TDMA	Time Division Multiple Access
TDOA	Time Difference Of Arrival
TE	Terminal Eequipment
TEID	Tunnel Endpoint IDentifier
TFCI	Transport Format Combination Indicator
TFRC	Transport Format and Resource Combination
TFRI	Transport Format and Resource Indicator
TH-CDMA	Time Hopping CDMA
THIG	Topology Hiding Inter-network Gateway
TI	Standardisation Committee of TI Communications
TI	Transaction Identifier
TLS	Transport Layer Security
TMN	Telecommunications Management Network
TM-SAP	Transparent Mode SAP
TMSI	Temporary Mobile Subscriber Identity
TOA	Time Of Arrival
TPC	Transmission Power Command; Transmit Power Control
Tr	Transparent mode
TRAU	Transcoding and Rate Adaptation Unit
TRX	Transmitter–Receiver
TTA	Telecommunications Technology Association
TTC	Telecommunications Technology Committee
TTI	Transmission Time Interval
TTP	Traffic Termination Point
Tx	Transmitting
U-RNTI	UTRAN Radio Network Temporary Identity
UA	User Agent

UBR	Unspecified Bit Rate
UDP	User Datagram Protocol
UE	User Equipment
UICC	Universal Integrated Circuit Card
UL	UpLink
UM	Unacknowledged Mode
UM-SAP	Unacknowledged Mode SAP
UMSC	UMTS Mobile service Switching Centre
UMTS	Universal Mobile Telecommunication System
UNI	User Network Interface
UNII	Unlicensed National Information Infrastructure
URA	UTRAN Registration Area
USAT	UMTS SIM Application Toolkit
USIM	Universal Subscriber Identity Module
UTRA	Universal Terrestrial Radio Access
UTRAN	UMTS Terrestrial Access Network
UWB	Ultra WideBand
VAS	Value Added Service
VBR	Variable Bit Rate
VC	Virtual Channel
VCI	Virtual Channel Identifier
VLR	Visitor Location Register
VMS	Voice Mail System
VMSC	Visited MSC
VoIP	Voice over IP
VP	Virtual Path
VPI	Virtual Path Identifier
VPN	Virtual Public Network
VSF	Variable Switching Function
WAP	Wireless Application Protocol
WCDMA	Wideband CDMA
WEP	Wired Equivalent Privacy
Wi-Fi	Wireless Fidelity Alliance
WLAN	Wireless LAN
WML	Wireless Markup Language
WRC	World Radiocommunication Conference
WTLS	Wireless Transport Layer Security
XHTML	Extended HTML
XMAC	An expected MAC value
XRES	An expected RES value

Bibliography

This bibliography includes the technical specification and other literature that the interested reader may find useful when more information is needed about:

- Radio communications, especially CDMA systems.
- 3GPP technical specifications.
- ATM networks and protocols.
- IP networks and protocols.
- Mobile positioning.
- Security algorithms and protocols.

The material listed is a good starting point for further reading.

1 Technical Specifications

In our modern world, publicly available technical specifications can be easily found on the World Wide Web. The following URLs will gain access to:

- 3GPP technical specifications: *http://www.3gpp.org*
  The 3GPP specifications referred to in this book are mostly from Release 5. The references given in number format, such as TS 12.345, can be used to browse the list of specifications on the 3GPP website.
- Internet Engineering Task Force (IETF) specifications: *http:// www.ietf.org*
  The IETF specifications referred to in this book are indicated with their RFC (Request For Comments) numbers and can be found on the IETF website using those numbers.
- IEEE specifications: *http://standards.ieee.org*
  The IEEE specification series, such as 802.11, can be found on the IEEE website.

UMTS Networks Second Edition   H. Kaaranen, A. Ahtiainen, L. Laitinen, S. Naghian and V. Niemi
© 2005 John Wiley & Sons, Ltd   ISBN: 0-470-01103-3

2. Other Literature

Ahonen T and Barrett J (eds) (2002). *Services for UMTS* (373 pp.). Chichester, UK: John Wiley & Sons.

Black U (1998). *ATM*, Vol. II: *Signaling in Broadband Networks* (224 pp.). Englewood Cliffs, NJ: Prentice Hal.

Black U (1999). *ATM*, Vol. I: *Foundation for Broadband Networks* (450 pp.). Englewood Cliffs, NJ: Prentice Hall.

Comer DE (2000). *Internetworking with TCP/IP*, Vol. I: *Principles, Protocols and Architecture* (4th edn, 755 pp.). Englewood Cliffs, NJ: Prentice Hall.

Doraswamy N and Harkins D (1999). *IPSec: The New Security Standard for the Internet, Intranets, and Virtual Private Networks* (216 pp.). Englewood Cliffs, NJ: Prentice Hall.

Ericsson Telecom AB (1997). *Understanding Telecommunications* (I-2, Vol. 1 = 493 pp., Vol. 2 = 677 pp.). Studentlitteratur AB.

Glisic S and Vucetic B (1997). *Spread Spectrum CDMA Systems for Wireless Communications* (383 pp.). Boston: A-H Publishers.

Halonen T, Romero J and Melero J (2002). *GSM, GPRS and EDGE Performance* (585 pp.). Chichester, UK: John Wiley & Sons.

Händel R, Huber MN and Schröder S (1994). *ATM Networks, Concepts, Protocols, Applications* (2nd edn, 287 pp.). Reading, MA: Addison-Wesley.

Heine G (1998). *GSM Networks: Protocols, Terminology and Implementation* (416 pp.). Boston, MA: Artech House.

Hillebrand F (2002). *GMS and UMTS: The Creation of Global Mobile Communication* (580 pp.). Chichester, UK: John Wiley & Sons.

Holma H and Toskala A (2004). *WCDMA for UMTS* (3rd edn, 450 pp.). Chichester, UK: John Wiley & Sons.

Laiho J, Wacker A and Novosad T (2002). *Radio Network Planning and Optimization for UMTS* (484 pp.). Chichester, UK: John Wiley & Sons.

Kilkki K (2001). *Differentiated Quality of Service for the Internet* (384 pp.). Basingstoke, UK: Macmillan.

Lee WCY (1995). *Mobile Cellular Telecommunications: Analog and Digital Systems* (2nd edn, 664 pp.). New York: McGraw-Hill.

Manterfield R (1999). *Telecommunications Signalling* (435 pp.). The Institution of Electrical Engineers.

Menezes AJ, van Oorschot PC and Vanstone S (1996). *Handbook of Applied Cryptography* (780 pp.). Boca Raton, FL: CRC Press.

Mouly M and Pautet M-B (1992). *The GSM System for Mobile Communications* (701 pp.). M. Mouly and MB Pautet.

Niemi V and Nyberg K (2003). *UMTS Security* (273 pp.). Chichester, UK: John Wiley & Sons.

Ojanperä T and Prasad R (1998). *Wideband CDMA for Third Generation Mobile Communications* (439 pp.). Boston: Artech House.

Patil, B et al. (2003). *IP in Wireless Networks* (372 pp.), Upper Saddle River, NJ: Prentice Hall.

Poikselkä M, Mayer G, Khartabil H and Niemi A (2004). *IMS IP Multimedia Subsystems Concepts and Services* (440 pp.). Chichester, UK: John Wiley & Sons.

Rantalainen T, Spirito MA and Ruutu V (2000). Evolution of location services in GSM and UMTS networks. *Proceedings of the Third International Symposium on Wireless Personal Multimedia Communications (WPMC 2000), November, Bangkok* (pp. 1027–1032).

Redl SH, Weber MK and Malcolm WH (1995). *An Introduction to GSM* (379 pp.). Boston: Artech House.

Spirito MA (2000). Mobile stations location estimation in current and future TDMA mobile communication systems. Ph.D. thesis, Politecnico di Torino, Facoltà di Ingegneria.

Stallings W (2000). *Data and Computer Communications* (810 pp.). Englewood Cliffs, NJ: Prentice Hall.

Viterbi AJ (1995). *CDMA Principles of Spread Spectrum Communications* (254 pp.). Reading, MA: Addison-Wesley.

Index

DATE DUE